Wooden Boats to Build and Use

Wooden Boats to Build and Use

by John Gardner

Mystic Seaport Museum
Mystic, Connecticut
1996

Manufactured in the United States of America

ISBN 0-913372-78-1

Designed by Sharon Brown
Software consultation by Dan Connors
Typed in Garamond Book
Printed on Finch Opaque

With the following exceptions all drawings in the book are by John Gardner:
Walter M. Wales (20, 21), George Anderson Chase (28, 29, 30, 32, 33, 34, 36, 38)
and E. I. Schock, Mystic Seaport (*Cuspidor*, 59).

Wooden Boats To Build And Use *is dedicated to Sharon Brown—designer and editor of the book, supervisor of Mystic Seaport's Boathouse for the past eight years, research assistant for small-craft studies, and John Gardner's devoted associate for the last six years of his life.*

John Gardner
1905-1995

"... the way to preserve small craft is not to embalm them for static exhibits or to tuck them away in mothballs, but to get their reproductions out on the water, use them, wear them out and replace them anew.

"Treated in that way, small craft are immortal or as near immortal as anything can be... Historic small craft are for the young and old and the in-between. They are to use and enjoy, and to pass on for future generations to use and enjoy, *ad infinitum.* Preservation through use, in the long run, that is the only way."

Mary Anne Stets photograph, M.S.M. 90-3-103

CONTENTS

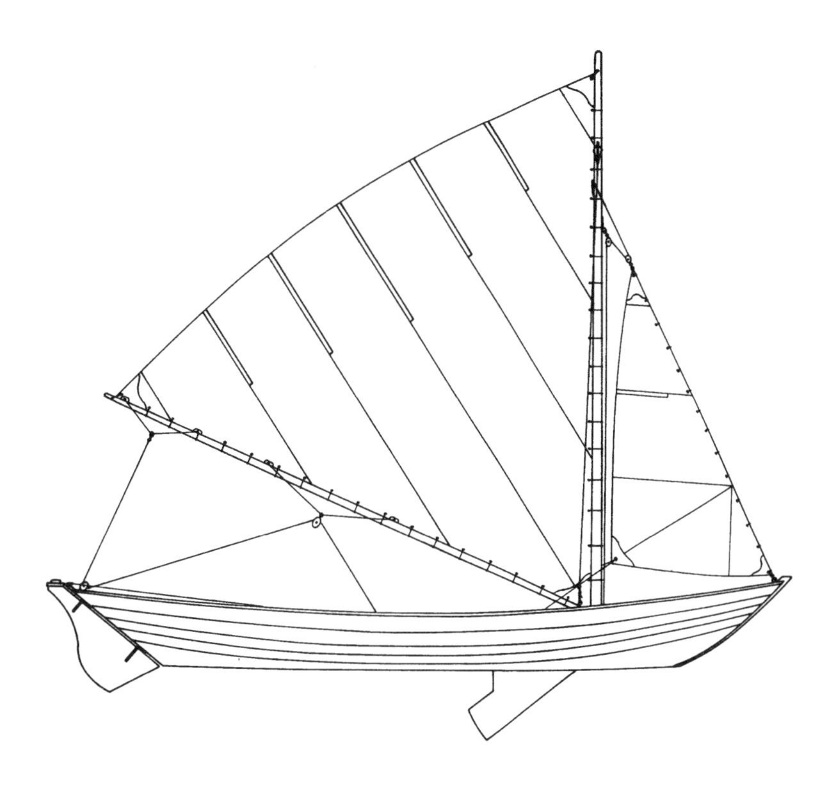

CHAPTER ONE

THE FUTURE OF WOODEN BOATS

The Museum Small Craft Association was conceived and brought forth at Mystic Seaport 20 years ago.

An article entitled, "Maritime Museums: Obligation and Objectives," published later the same year in Volume 1, No. 6 of *WoodenBoat*, reported and commented on that first meeting. The reasons for holding the meeting, set forth in the opening paragraphs of this report, are as valid and as applicable today as they were then, in spite of the changes that have taken place, and the progress that has been made since.

"The future of wooden boats, if they are to have a future," the final paragraph concludes, "lies to a considerable extent with marine museums. History has placed in their hands a unique opportunity and a unique obligation. The boats in their collections constitute a treasury of small craft knowledge and design, which could not now be found anywhere else. And because these museums attract millions of visitors, they have an unparalleled opportunity to make this historic resource available to the widest reaches of the public for their appreciation and use. Indeed, in view of the museum's acknowledged educational function, any other course would be a violation of their public trust."

This was the first time, I believe, that museum professionals had come together solely for the purpose of considering what could be done with small craft and what the role and function of small craft in an active museum program could and should be. The passage of time has brought changes, but there have been no radical departures, so far as I am aware, from the basic conclusions that came out of that first meeting, which acknowledge that small craft have a unique and vital contribution to make to the active programs of the living museum.

A team of competitors from East Harlem Maritime School pull together at the oars of a Mystic Seaport-built whaleboat during the November 1993 Youth Rowing Race. Rowing teams came from Connecticut, Rhode Island, Massachusetts and New York (Sharon Brown photograph).

And what is unique and vital about small craft? Basically it is this. It is only by means of small craft, defined as ranging from the small dinghy up to 40 feet and even somewhat more, that many people today, especially young people, have any chance of getting on the water in an active way, have any chance of experiencing by themselves and on their own something of the pleasures and thrills, as well as the hardships, dangers and responsibilities of going to sea. The youngster in his small boat, caught out in a sudden squall, learns to respond with a resourcefulness and self-control that will stand him in good stead later in life, whether at sea or ashore. Without that direct personal experience that comes from using boats on the water—yes in caring for them, repairing and even building them—youngsters will have missed something important in growing up. Nor can they be expected in later life, if they have been so deprived, to have any deep interest in the sea and things maritime. Were that to happen on a broad scale, as is now threatening to happen, maritime history, despite our best efforts otherwise, would be in danger of losing social relevance, to become the retreat of a few socially detached antiquarians.

Maritime museums are not numerous, and for the greater part quite recent in inception. The oldest in this country by far is the Peabody Essex Museum in Salem, Massachusetts, founded in 1799 as the East India Marine Society. The individual most responsible was the Reverend William Bentley, minister of Salem's East Church, an immensely able and learned man of wide-ranging scientific interests and progressive views. He encouraged his parishioners, many of whom were engaged as seamen in the newly opened trade with the far east, to be observant, and to bring back curios, oddities and specimens of the strange and different things they saw. This they proceeded to do in increasing quantities, and it was to house and exhibit such items for the information and instruction of the public, as well as for their entertainment, that these various collections were brought together in the East India Marine Hall on Essex Street. By 1821 the Society's collections had grown to number an imposing 2,269 items.

In the days before the invention of photography and such later developments as moving pictures, television and theme parks, such exhibits had an impact they do not have today. They opened up fascinating new worlds to the viewers, they transported their viewers to faraway places that otherwise they would have had no conception of. The educational effect was profound, and while no formal statement of educational intent may have been made, it was implicit and operative, nonetheless, to a high degree.

The effects to be obtained now from exhibits intended merely for walk-by viewing, no matter how cleverly and lavishly laid out, are bound to be much weaker, and even feeble in comparison with the response achieved by the outside competition museums must contend with. And this is most especially true for small craft exhibits where viewers have had little or no prior experience with small craft or the opportunity to use them.

For maritime museums to attempt to compete with commercialized mass

Eagerly rising to the challenge in preparation for Mystic Seaport's third annual Youth Rowing Race, two youngsters practice their rowing skills against a stiff head wind in one of the* L. A. Dunton *dories built by Will Ansel (Sharon Brown photograph).

entertainment would be a fatal mistake, a trap to be avoided. This is well stated in a recent issue of the *New York Review* in a commentary on *Jurassic Park* by Stephen Jay Gould, Harvard Professor, MacArthur fellow and author of numerous prize-winning books.

"I happen to love theme parks," he writes, "so I do not speak from a rarefied academic post in a dusty museum office. But theme parks are in many ways the antithesis of museums. If each institution respects the other's essence and place, this opposition poses no problems. But theme parks belong to the realm of commerce, and museums to the world of education—and the first is so much bigger than the second. Commerce will swallow museums, if educators try to copy the norms of business for immediate financial reward." Nothing could be more true or to the point, yet unfortunately there are indications that what Gould warns against is beginning to happen.

Leaving aside the assemblage of models that comprise the Watercraft Collection at the U. S. National Museum in the Smithsonian Institution as hardly within the scope of our considerations here, 130 years were to elapse before a second maritime museum was to appear in this country, followed by a third in little more than a year, namely the Marine Historical Association at Mystic, Connecticut, in 1929 and the Mariner's Museum at Newport News in Virginia in 1930.

These are and were history museums from the beginning, as the Peabody Museum at Salem was not in the beginning, 200 years ago when the glories of America's maritime achievements lay ahead in the future, and had not yet faded into the past.

As Carl Cutler, maritime historian, explains in the "Statement of Plan and Purposes of the Marine Historical Association," which he issued as secretary of the organization at the time of its formation, the museum originated as a "by product," as he put it, of efforts by himself and others to salvage historical material related to New England's seafaring past that were fast disappearing, being discarded, thrown out and sold as junk—customhouse records by the ton, manuscripts, log books, photographs and wagon loads of builder's half-models, sawed up and burned as firewood.

"From the acquisition of a collection to the problem of proper housing for it, is a short, imperative step," Cutler wrote. "And from problems of housing to the advisability of developing other ways and means of putting this and similar future acquired property to work in an effective way, is a question which has already suggested itself. *It is one thing to have a fine collection and exhibit it to the public in an interesting manner. It is wholly another matter to make that collection the nucleus of a vital, growing force which shall be the rallying point of an organization capable of playing a worthy part in a living America with a future to face*" [emphasis mine].

What we have here is a powerful declaration of purpose. Clearly the intent of the museum's founding fathers was an active and influential involvement on the part of the museum in the ongoing social process.

The Mariner's Museum at Newport News, Virginia had quite a different origin. Archer M. Huntington, who founded it, envisioned the museum as a nonprofit educational institution intended to help make America sea-minded. The inscription on the museum's great bronze doors reads: "This museum is devoted to the culture of the sea and its tributaries, its conquest by man, and its influence on civilization."

As the son of Collis P. Huntington, this country's greatest railroad builder, Archer M. Huntington had unlimited financial means at his disposal. This enabled him when he set up his small craft collection, the first of any size in this country, to assemble representative small craft from all over the world, one of the most unusual and striking being a 51-foot example of a Portuguese fishing type, a survival from ancient times propelled solely by oars. This collection was first displayed for a short time in the open, but was soon moved under cover to protect it from the weather.

The depression years of the 1930s and the war years of the 1940s were almost entirely blank for maritime museums, except for the opening of a small marine museum in Searsport, Maine, in 1936.

This was also a thin time for small craft, although interest was building up beneath the surface, so to speak, as evidenced by the resounding popular response to L. Francis Herreshoff's contributions to *Rudder* in the late 1940s that resurrected and revitalized for a short time that all-but-defunct boating magazine.

Then in 1952 the Adirondack Museum was instituted and funded by Harold Hochschild. Hochschild had spent much of his childhood and youth in the Adirondacks which he had come to know well and to love. To serve as director for the new museum Hochschild picked Dr. Robert

Barry Thomas, Chris Rawlings and Kevin Leonard launch* Fly*, a 10-foot Chaisson dory tender built by 14-year old Forest Lowry in Mystic Seaport's Boat Shop for the livery. This model is excellent for youngsters to row or scull, and easily adapted for sail (Sharon Brown photograph).

Bruce Inverarity who had studied West Coast Indian canoes extensively. From this background he was able to appreciate the singular qualities of the Adirondack guide-boat and the contribution that remarkable craft had made to the culture and development of the Adirondack region. In view of this, Inverarity chose the guide-boat as the focal center around which to organize the new museum.

The previous year had seen the publication by W. W. Norton of Howard Chapelle's *American Small Sailing Craft*, a revised and extended version of material first treated by him in articles in *Yachting* in 1932. The book was an instant success, and its influence on the boating public was immediate and far-reaching. Also, I might add, in that same year, 1951, my first article on dories appeared in the old *Maine Coast Fisherman*.

Small craft at the Mariner's Museum during the 1950s was quiescent. There is no evidence of activity at Searsport. The

Peabody in Salem (and I was spending a lot of time there as a volunteer in the 1950s and 1960s), to the best of my knowledge, had only one full-size small boat, a Chaisson-built sailing dory donated by the Phillips family, stored away out of sight in the basement. But at Mystic Seaport during that decade an active small craft program under the direction of James Kleinschmidt, Marine Maintenance Superintendent, had made remarkable progress.

The rescue of endangered examples of historic New England working craft had begun in earnest. Efforts were made to search out such boats and secure them for collections. An example is the Kingston lobsterboat, one of an all-but-vanished type, found by Kleinschmidt in 1956, rotting in a field in Scituate, Massachusetts. This find, with a photograph, was reported in the *Maine Coast Fisherman*. Kleinschmidt brought the boat back to Mystic, where it was restored and added to the permanent collection, and is now an exhibit.

Small craft activity at Mystic Seaport during the 1950s was extensive and productive. An account of how things stood at the end of the decade is to be found in *Small Craft At Mystic Seaport*, issued in December 1959, the first catalog of museum small craft to be compiled and published. This small 82-page paperback book included 54 of the 61 boats classified as small craft then at Mystic with photographs, descriptive text and line drawings of nine of these boats, mostly the work of James Kleinschmidt.

The 1960s was a decade of explosive growth for marine museums. It saw the inception of no fewer than nine new maritime museums that lost little time in starting their own small craft collections. Seven of them were later to participate in compiling the *Union List of Museum Watercraft* sponsored by the Museum Small Craft Association.

According to my count, more maritime museums were established in that ten years than came into being in the two and a half decades that followed. I confess I cannot satisfactorily account for this sudden burst of museum proliferation, although I suspect that the surge of tourism that peaked in the early 1970s before the gas crunch hit, had something to do with it. Prior to that setback, endless crowds of ordinary Americans with the yen and the means took to their cars and the open road each summer to see America for the first time, and loved it. Maritime museums drew them in by the thousands.

In the summer of 1969, my first year at Mystic Seaport, I was invited to be one of the judges at the Antique Boat Show at Clayton, New York, along with Howard Chapelle and representatives from *Rudder*, *Yachting*, one of the motor boating publications and others.

One summer a few years previous, a married couple had restored an old, broken-down power launch and were so pleased with the results that they celebrated by inviting friends who came in a variety of powered craft. The gathering was repeated the following year, and by 1969 had become an annual one-day event involving the whole town of Clayton, New York.

In 1969 so many interesting non-powered small craft put in an appearance in addition to the larger power boats, that we suggested the show henceforth be made a two-day affair with one of the days devoted solely to antique small craft, particularly St. Lawrence skiffs. This was done.

It was from the Clayton show that we got the idea of holding a boat show of our own at Mystic Seaport, only it was to be limited to traditional non-powered small craft. In order to arouse interest in recreational rowing, then at a very low ebb, we sent out announcements for a rowing workshop. The response far exceeded our expectations. Many came, and had the time of their lives on the water trying out each other's boats in spite of the pouring rain. Although the call had gone out for rowing boats, so many of the boats that came were also rigged for sail that sailboats as well as rowing craft were invited the following year, and the name was changed to Small Craft Workshop. Powerboats were excluded, however, and have been to this day with a few special exceptions.

This was the first small craft boat show, the granddaddy, if you will, of all the various and sundry small craft shows and meets that now take place each summer across the length and breadth of this country. To attempt to list them all would be more than we could possibly undertake here.

Toward the end of 1966, Edmund E. Lynch, with a masters degree in Museum Administration from Cooperstown and practical experience as curator at the Adirondack Museum, took over the curator's job at Mystic Seaport. Bursting with ideas, energy and drive, Ed Lynch hit Mystic with the force of a seismic shock. The reforms he instituted and the programs he put in place are with us yet. One of his first accomplishments was a complete restoration of the sandbagger, *Annie*, Mystic's premier small craft attraction, that sun, wind, water and neglect had reduced for a second time to a rotting hulk.

Discussing *Annie*'s restoration in *The Log of Mystic Seaport* for December 1969, Lynch made some observations foreshadowing developments soon to take place that would greatly enlarge museum involvement with small craft.

"Although we attempt to exhibit small craft now (only about one-third of the total we own)," Lynch wrote, "there is a further dimension that I have felt is more important than their display. If the technology of boat building is lost, we will begin to possess a collection of curios and oddities... As a viable museum we want to encourage small boat building in the twentieth century. We would like to give the home craftsman a practical knowledge of the past and have our collection of small boats used more actively by researchers, scholars, students and builders. Everyone fears that a museum has a tendency to become a mausoleum. We have the capacity of instilling life into our small craft collection. We ask only the opportunity to begin."

And begin Mystic Seaport did very shortly in the winter of 1970-71 by initiating an evening class in traditional boat construction for the home, or non-professional, builder. It was the first organized instruction in recreational boat building to be offered in this country, the very first. I would like to think of it as the seed from which have sprung all the various and sundry nonprofessional boat building classes and schools, whether directly sponsored by museums or not, now scattered from Maine to the state of Washington, from upstate New York to the Carolinas, Louisiana and Texas. More likely it was only by chance, so to speak, that Mystic happened to be first. A widely pervasive need for such instruction had been building up for some time and otherwise would undoubtedly have broken through somewhere else very shortly.

For a variety of reasons not wholly clear or easy to explain, more and more of those nurtured on the writings of Howard Chapelle, L. Francis Herreshoff and others were turning to the classic small craft of

Despite rain, 200 people attended the first Rowing Workshop at Mystic Seaport June 6-7, 1970. Forty-five boats were in the water including the 10-foot pram (center) built by W. Southworth of Egypt, Massachusetts, from Herreshoff lines and redesigned construction details published by John Gardner in the National Fisherman in 1968 (Lester D. Olin photograph, M.S.M.).

A distinguished panel including Mystic Seaport Museum's Curator, Edmund Lynch (standing) and Atwood Manley, author of Rushton And His Times In American Canoeing (seated) addressed an appreciative audience at the first Rowing Workshop which considered the topic, Rowing for Pleasure and Health, and featured an exhibit of the 14-foot pulling boat Lawton designed and built by John Gardner. The success of this workshop led to a subsequent gathering at Mystic on October 24, 1970 of 35 individuals interested in pulling boat design, and to the event becoming an annual meet (Lester D. Olin photograph, M.S.M.).

yesteryear, and their construction, partly as a means of self-expression, partly in order to get on the water and partly as a reaction against ill-designed, over-powered, over-priced fiberglass. The unlikely but instant and spectacular success of a new publication started on a shoestring, and devoted to classic small craft, namely *WoodenBoat*, is an indication of the way the wind was blowing in the early 1970s.

What I have attempted so far to show is what a recent phenomenon the maritime history museum actually is, and to sketch against that background the even more recent emergence of small craft studies and activity as a serious part of the museum program. We have now arrived in our considerations at the 1970s, the decade in which the action really begins to take off. To lay it all out here in full detail would take us into next week. We must be content with calling attention to a few salient points. We might begin by mentioning that at the first World Congress of Maritime Museums in London in 1972 small craft got equal billing with yachts and larger vessels for the first time ever.

With the arrival of the 1970s, a substantial small craft literature began to build up with solidly researched studies of lasting value, including Peter J. Guthorn's *Sea Bright Skiff*, Pete Culler's *Skiffs And Schooners*, Atwood Manley's *Rushton And His Times In American Canoeing* and Kenneth Durant's *The Adirondack Guide Boat*. Roger Taylor's International Marine Publishing Company came into the picture early on. We have already mentioned *WoodenBoat*. The *National Fisherman* kept up a steady output of small craft articles and designs. The Traditional Small Craft Association was started, and although the Catboat Association goes back to 1963, its activities and membership were greatly expanded in the 1970s. Lance Lee's Apprenticeshop, a ground-breaking experiment in hands-on small craft education, was initiated and carried on at Maine Maritime Museum during this period. And in Olympia, Washington, in 1976, a small craft conference at Evergreen State College, where the announced subject was "Wooden Boats and the New Craftsman," opened up and set in motion organized

With his assistant Syl Costelloe holding the backing iron, John Gardner demonstrated rivet fastening of the steam-bent white oak frames in the 13'5" Chamberlain dory skiff for the first Recreational Boat Building Class in March 1971. This is the same boat known as Harry Williams which has been available for rent at Mystic Seaport's Boathouse since the summer of 1988 (Lester D. Olin photograph, M.S.M.).

small craft activity in the Pacific Northwest. The following year saw the formation of the Center For Wooden Boats in Seattle. This is hardly a complete summary of all that took place during the 1970s, but it is enough to give the picture.

The course was set. In the years that have followed, small craft activity has deviated very little from patterns laid down in the 1970s except for one promising and exciting new development. At the same time the number of maritime museums seems to have stabilized. The only other maritime museum to be founded in the 1970s besides the Center For Wooden Boats was the North Carolina Maritime Museum at Beaufort, unless the Hudson River Maritime Museum at Kingston, organized in 1980, is also included. And since then only two new museums with a maritime orientation have appeared, both highly specialized, neither of which has done very much with small craft or seems likely to. Perhaps a point of saturation has been reached in the maritime field, at least for the immediate future. Museums are very much dependent on outside forces, and considering the present depressed state of the economy, retrenchment rather than expansion seems to be the order of the day.

Small craft in maritime museums have come a long way in a short time. From relative obscurity that has verged at times on neglect, small craft have come to occupy an important place in the museum program. They have come, or are fast coming, to stand on an equal footing with the larger vessels, no longer to be treated like poor relations and to be fed by crumbs from the museum table.

Once considered merely as objects to be exhibited passively for walk-by viewing—and it is not my intention to denigrate exhibits, which have a worthy and time-honored function to perform—small craft are now coming to be treated in the museum as boats to be exhibited and used as boats on the water, in their natural element.

Looking back, I believe we can take satisfaction in what has been accomplished. Looking ahead, I see many opportunities and much to do. We have gone about the task of building up our boat collections with enthusiasm and diligence and have assembled imposing numbers of specimen craft, as documented in the *Union List* compiled by the Museum Small Craft Association. It may just be that the time has arrived to slack off a bit in collecting. It could be that in some instances we have more boats than we need or can use to advantage. It is possible there may be duplicates or inferior specimens that could be culled to advantage. The housing and maintenance of huge collections of full-size boats can be expensive, absorbing an undue share of limited financial resources better applied elsewhere in the small craft program. I submit this is something in immediate need of thorough ongoing consideration and analysis, and for me at least it looms as a painful task I have no desire to undertake.

We have already touched upon the weakened state of the economy. Who can deny that the standard of living for middle America has not declined appreciably in the last few decades until now with two wage earners working, many families are hard pressed to make both ends meet. One consequence is that, more and more, middle class Americans are being excluded from recreational boating. The only way that many of us can get on the water these days is in small, simple, inexpensive boats, such boats as maritime museums have in their collections of traditional small craft.

In view of the maritime museum's acknowledged educational function, this would seem to present maritime museums with an educational opportunity they can hardly avoid. Maritime museums are in an ideal situation to put visitors into suitable small boats, to let them try out these boats on the water, to use them, to get to know and compare them, preliminary to choosing boats for themselves. Boathandling and elementary seamanship are taught in the process. For those who decide they want to build a reproduction of one of these heritage craft for themselves, there are plans and instruction in the museum's small boat shop close by.

Two museums are already so engaged with successful programs in place, and there may be others I am not aware of. The Center For Wooden Boats in its ideal situation on

The Woods Hole spritsail boat Sandy Ford *was built in the Small Craft Laboratory in 1972 by John Gardner and Syl Costelloe from lines taken from the donated Crosby-built vessel* Explorer *(ex. T. C.). Guy Hermann is at the helm; his son Jammer and Walter Ansel and his son Douglas round out the crew during the summer racing series sponsored by The Boathouse at Mystic Seaport (Sharon Brown photograph).*

The 20-foot Crosby Cat reproduction* Breck Marshall, *built in Mystic Seaport's Boat Shop by Barry Thomas, Bret Laurent and Clark Poston, is the flagship of The Boathouse livery program. Shown here under reef with Capt. John Phelan at the helm, this working catboat has carried more than 34,000 passengers on half-hour excursions on the Mystic River since her launching in 1987 (Sharon Brown photograph).

Lake Union in the center of Seattle was built up around such a program based on experience derived from a boat rental facility operated on Lake Union by Dick and Colleen Wagner prior to the inception of the Center. The program the Center has developed is state-of-the-art, and a model to aim for.

At Mystic Seaport we have offered instruction in wooden boat building for the past 25 years, but it wasn't until the summer of 1988 that we opened the boat livery at The Boathouse, putting visitors on the river in reproductions of classic wooden boats that they could row and sail themselves–dories, whitehalls, peapods, sharpie skiffs, Beetle cats and others–20 different boats in all. The Boathouse was enthusiastically received from the start, to become, early on, one of the museum's most popular attractions. In the eight summers the livery has been in operation, 18,073 visitors all told have been put on the water in small boats, and *Breck Marshall*, the Seaport-built 20-foot Crosby Cat reproduction, has carried in excess of 34,000 paying passengers on half-hour trips on the river.

Visitors have come from as far away as Canada, Australia and Europe to try out livery boats. Children, especially, have been thrilled by the boats, frequently to experience for the first time what it is like to be on the water in a small boat, and to have a try at the oars, an educational experience that will be with them all their lives.

Children naturally and instinctively take to boats, are fascinated and enthralled by them. They delight in building them and working on them with their own hands. They should be given that opportunity. And only if our children are introduced to boats at an early age and grow up using them will what we are doing today have any relevance for the future. Otherwise our efforts to collect and preserve heritage boats will have been for nothing.

As museum professionals, preservation is our business, our forte, our thing, so to speak, so we had better understand that the

Youth eagerly rise to the challenge of small boat construction (Gregory A. Kriss photograph).

way to preserve small craft is not to embalm them for static exhibits or to tuck them away in mothballs, but to get their reproductions out on the water, use them, wear them out and replace them anew.

Treated in that way, small craft are immortal or as near immortal as anything can be, while, unfortunately, large vessels are not, having been condemned to oblivion in the not-far-distant future by the harsh realities of modern economies.

Just because small craft are small does not make them any less a joy. Pete Culler, who from early manhood sailed, built and designed large vessels, turned for his yachting pleasure in later years to a Swampscott dory which he rigged to sail. And his friend Waldo Howland, who had owned and sailed his share of large yachts in his day, got his fun on the water as he grew older in *Dixie Belle*, a trim 17-foot, flat-bottom sprit-rigged Chesapeake skiff that Pete Culler designed and built for him.

Historic small craft are for the young and old and the in-between. They are to use and enjoy, and to pass on for future generations to use and enjoy, *ad infinitum.* Preservation through use, in the long run, that is the only way. *(Edited text of a talk presented by John Gardner at the Annual Meeting of the Museum Small Craft Association, held at Mystic Seaport Museum, Mystic, Connecticut, in October, 1993.)*

Young Riverfront Rangers built flatiron skiffs by the Old State House in Hartford, Connecticut, in the summer of 1988 (Gregory A. Kriss photograph).

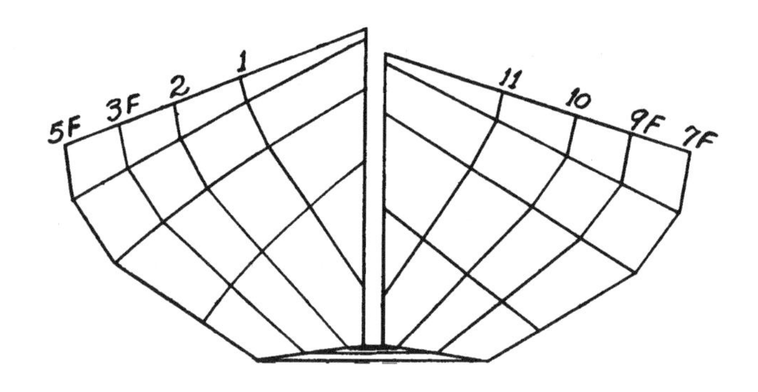

Chapter Two

Marblehead Gunning Dory

Not every man's boat to be sure, yet for what it was conceived and built for, the Marblehead gunning dory is unexcelled. For rough-water ability, easy performance under oars, respectable speed and capacity, capped by its handsome appearance, the gunning dory is one of a kind. In the estimation of those who own them, these boats come close to being the ultimate embodiment of that perfection of form and function attained by traditional small craft in the final years of the nineteenth century. It is the ultimate dory, yet, prior to 1960, hardly more than a handful, so to speak, of gunning dories had ever been built, and the type was all but unknown beyond the limits of the town from which it takes its name. It was my write-up in the February issue of that short-lived publication, *Outdoor Maine*, that first turned attention to the Marblehead gunning dory.

Back in 1942 I had gone to work on the second shift at the James E. Graves Beacon yard down at Barnegat in Marblehead, planking McInnis-designed 38-foot picket boats for the Coast Guard for the war effort. There was a U-boat scare at the time, and a crying need for boats to patrol the coast.

The Beacon Street yard was situated on a small cove where lobstermen moored their boats. Along the shore was a row of shacks where they stored their gear. Hauled up on the beach lay an assortment of punts and flatiron skiffs, and among them such a boat as I had never seen before. Battered and abused though she was, she had retained her shape, and the sheer beauty of her lines all but took my breath away.

The boat, they told me, was Will Chamberlain's old gunning dory, his own personal boat, the boat he had built years ago with Percy Bamford's assistance, for the two to go gunning in for sea ducks, amidst

Captain Gerald Smith at work on the 17-foot gunning dory,* Republican, *in his Pond Street shop in Marblehead in 1960. The bottom plank of native white pine is 19 1/2" wide at its maximum width and braced from the floor and ceiling to add rocker. The four sawn frames, held in place with cross spalls, are of natural-crook elm, and limbers are cut where the garboards will land (photograph courtesy of Captain Gerald Smith's family).

the ledges and islands of Marblehead's outer harbor. Charlie Briggs had found the boat stored away somewhere and abandoned, tucked under a building, perhaps, or in some old shed, had hauled her out, put her overboard and was using her to go lobstering in part time. He had nailed a couple of braces across her gunwales, forward, to stiffen her up when he piled her full of waterlogged pots. Considering her light construction, it is a wonder she held together, but she had. She was still intact, and had not lost an iota of her original handsome shape, testimony to the boat building genius of the man who had put her together.

In those days I had fallen under the spell of Howard Chapelle, whose pioneer accomplishments in rescuing our small craft heritage from oblivion were attracting wide attention in the pages of *Yachting* and elsewhere, and I had begun to emulate his efforts in measuring and taking off the lines of heritage small craft whenever the opportunity presented. Here lay such a chance as I had dreamed about. Charlie Briggs said he didn't mind, go ahead and measure her to your heart's desire. I did not let the grass grow green under my feet. One Saturday was spent and most of another, as I recall, with my templates and my tape, as well as whatever odd moments I could steal before and after work, coming back again and again to check this and that until her lines were drawn and faired to my satisfaction. I was not publishing then, so my drawing and notes were filed away where they remained buried, as it were, for 17 years. Not until Owen Smith, after he was eased out of the *Maine Coast Fisherman* in 1959, when he was putting *Outdoor Maine* together in a hurry and was sending out urgent appeals for copy, did they again see the light of day. Wracking my brain to come up with something appropriate in the way of a story, I remembered the gunning dory. It was a natural, no less. Besides, a report had come to me that Captain Gerald Smith over in Marblehead had a gunning dory framed out and set up for planking in his little shop behind his home on Pond Street. Sure enough, a trip to Marblehead confirmed it.

Jerry was a good friend of mine from way back. We had worked together on the same bench planking Coast Guard picket boats at Graves at the beginning of World War II, before Jerry left Graves to go lobstering. Yachting skipper, boat builder, lobster fisherman, Jerry had gone gunning since boyhood and had already used up two gunning dories which he had built himself. One of these had been smashed on the rocks and broken up in a storm, the other is in Mystic Seaport's small craft collection. This was to be his third, built on a set of Chamberlain molds, the real thing. How he had come by a set of these molds he didn't say. Will Chamberlain didn't lend his molds around, and when he was growing blind in his old age, as I was told by Frank Caswell who used to come to the Chamberlain shop as a boy to sweep the floor and clean up, Will became bitter and misanthropic and broke up most of his molds to keep them out of the hands of imitators and latter-day

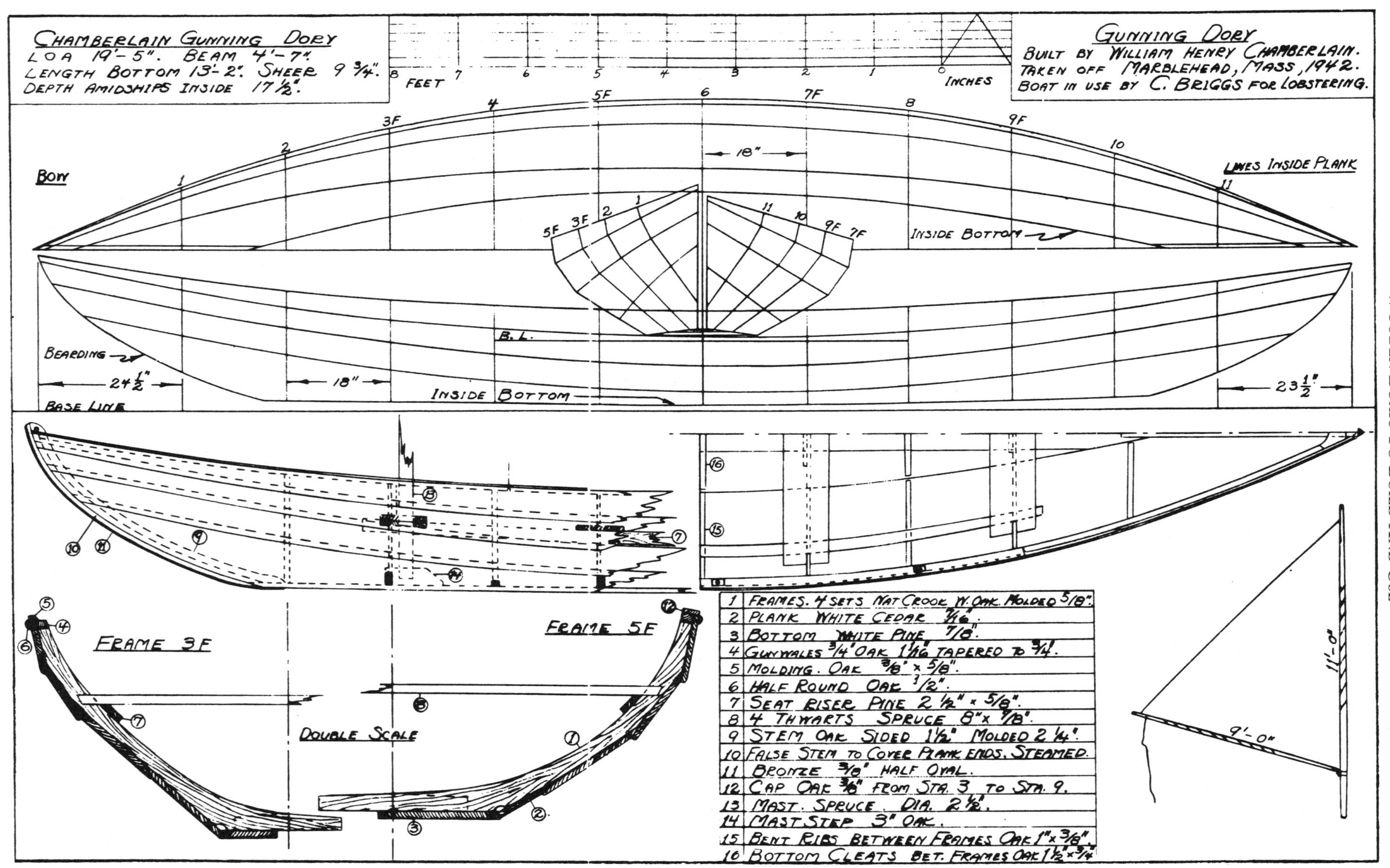
CHAMBERLAIN GUNNING DORY
LOA 19'-5". BEAM 4'-7".
LENGTH BOTTOM 13'-2". SHEER 9 3/4".
DEPTH AMIDSHIPS INSIDE 17 1/2".
GUNNING DORY
BUILT BY WILLIAM HENRY CHAMBERLAIN.
TAKEN OFF MARBLEHEAD, MASS, 1942.
BOAT IN USE BY C. BRIGGS FOR LOBSTERING.
FEET
INCHES
BOW
LINES INSIDE PLANK
INSIDE BOTTOM
B.L.
BEARDING
BASE LINE
FRAME 3F
FRAME 5F
DOUBLE SCALE
11'-0"
9'-0"
1 FRAMES. 4 SETS NAT CROOK W. OAK MOLDED 5/8".
2 PLANK WHITE CEDAR 7/16".
3 BOTTOM WHITE PINE 7/8".
4 GUNWALES 3/4" OAK 1 1/16" TAPERED TO 3/4".
5 MOLDING. OAK 3/8" x 5/8".
6 HALF ROUND OAK 1/2".
7 SEAT RISER PINE 2 1/2" x 5/8".
8 4 THWARTS SPRUCE 8" x 7/8".
9 STEM OAK SIDED 1 1/2" MOLDED 2 1/4".
10 FALSE STEM TO COVER PLANK ENDS. STEAMED.
11 BRONZE 3/8" HALF OVAL.
12 CAP OAK 3/8" FROM STA. 3 TO STA. 9.
13 MAST. SPRUCE. DIA. 2 1/2".
14 MAST STEP 3" OAK.
15 BENT RIBS BETWEEN FRAMES OAK 1" x 3/8".
16 BOTTOM CLEATS BET. FRAMES OAK 1 1/2 x 3/4

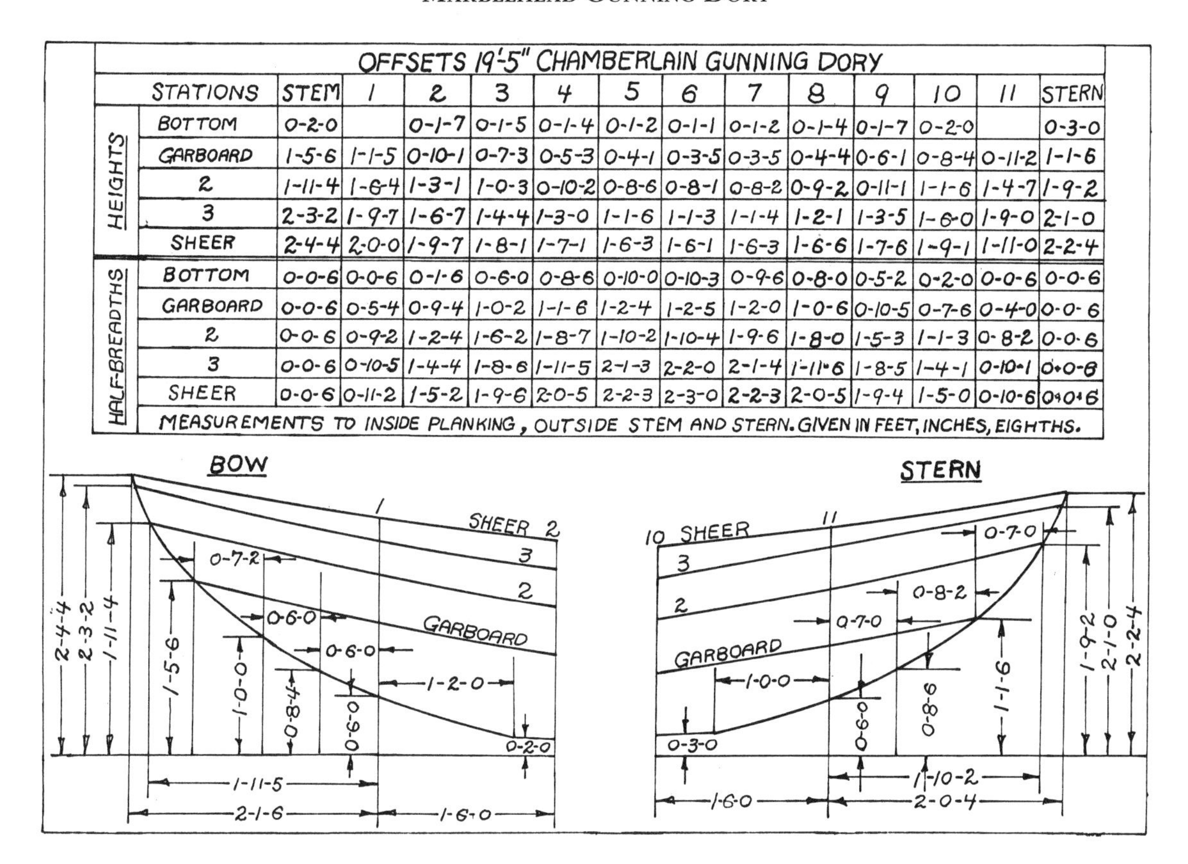

OFFSETS 19'-5" CHAMBERLAIN GUNNING DORY

	STATIONS	STEM	1	2	3	4	5	6	7	8	9	10	11	STERN
HEIGHTS	BOTTOM	0-2-0		0-1-7	0-1-5	0-1-4	0-1-2	0-1-1	0-1-2	0-1-4	0-1-7	0-2-0		0-3-0
	GARBOARD	1-5-6	1-1-5	0-10-1	0-7-3	0-5-3	0-4-1	0-3-5	0-3-5	0-4-4	0-6-1	0-8-4	0-11-2	1-1-6
	2	1-11-4	1-6-4	1-3-1	1-0-3	0-10-2	0-8-6	0-8-1	0-8-2	0-9-2	0-11-1	1-1-6	1-4-7	1-9-2
	3	2-3-2	1-9-7	1-6-7	1-4-4	1-3-0	1-1-6	1-1-3	1-1-4	1-2-1	1-3-5	1-6-0	1-9-0	2-1-0
	SHEER	2-4-4	2-0-0	1-9-7	1-8-1	1-7-1	1-6-3	1-6-1	1-6-3	1-6-6	1-7-6	1-9-1	1-11-0	2-2-4
HALF-BREADTHS	BOTTOM	0-0-6	0-0-6	0-1-6	0-6-0	0-8-6	0-10-0	0-10-3	0-9-6	0-8-0	0-5-2	0-2-0	0-0-6	0-0-6
	GARBOARD	0-0-6	0-5-4	0-9-4	1-0-2	1-1-6	1-2-4	1-2-5	1-2-0	1-0-6	0-10-5	0-7-6	0-4-0	0-0-6
	2	0-0-6	0-9-2	1-2-4	1-6-2	1-8-7	1-10-2	1-10-4	1-9-6	1-8-0	1-5-3	1-1-3	0-8-2	0-0-6
	3	0-0-6	0-10-5	1-4-4	1-8-6	1-11-5	2-1-3	2-2-0	2-1-4	1-11-6	1-8-5	1-4-1	0-10-1	0-0-6
	SHEER	0-0-6	0-11-2	1-5-2	1-9-6	2-0-5	2-2-3	2-3-0	2-2-3	2-0-5	1-9-4	1-5-0	0-10-6	0-0-6

MEASUREMENTS TO INSIDE PLANKING, OUTSIDE STEM AND STERN. GIVEN IN FEET, INCHES, EIGHTHS.

builders. But Jerry's set were the authentic thing all right, there could be no doubt about it.

For all his deference to Will Chamberlain, Jerry, like any true-born Yankee and hard-nosed Marbleheader, had his own ideas, grounded in experience. For one thing, Jerry had decided on several inches more bottom rocker or fore-and-aft camber than Chamberlain, who made his dories closer to straight on the bottom. It was Jerry's opinion that a few inches of rocker made a better boat to row and an easier one for getting off the beach with a load. It must be said, however, that the straighter a dory is on the bottom, the shallower the displacement will be, and the higher it will run up and ground out on a shelving beach like Fisherman's Beach in Swampscott where George Chaisson built his big dories dead straight on the bottom for the local fishermen.

Jerry had shortened his new boat to 16 feet, and had filled out its after lower quarters somewhat to give it more bearing on the water aft. The full-size laydown of the lines on the shop floor showed an especially handsome up-sweep to the sheer, characteristic of Chamberlain's boats, but something that other builders sometimes failed to achieve. In fact the second gunning dory built by Jerry, the *Annie*, now in Mystic Seaport's small craft collection, is somewhat flatter on top than she might be, and so were the couple of "Sharpshooters" reputedly built on Chamberlain molds by Jerry's Norman Street neighbor, Albert Cloutman, ex-yard boss at the Graves yard. Cloutman made other changes, too. His Sharpshooter was a foot shorter on the bottom than the Chamberlain model.

For Captain Jerry, an avid duck hunter and gunning enthusiast from way back, and now an "old timer," as he once somewhat ruefully admitted, building this new boat for his own use was a labor of love. Enjoying every minute of it, he was in no hurry. As for the materials that went into her, they were the best, selected with the greatest care. Some of the lumber had been put by and saved for years just for this purpose. Frames were long-seasoned natural-crook elm with twisted and interlocking grain that makes just about the toughest frames going, as tough but lighter than the apple limb crooks also prized by the old-time builders for their frames. One single board of native white pine, $19^{1}/_{2}$ inches at its point of greatest

Annie, *16' 1" x 54 1/2", was built on Chamberlain's molds in 1940 by Captain Gerald Smith and donated to Mystic Seaport in 1969 by Richard Parker of Marblehead. Smaller, with less sheer and rocker than* Republican, Annie *was successfully raced by Parker's son in dory competitions prior to her retirement (Maynard E. Bray photograph, M.S.M.).*

width amidships, made the bottom. For his planking Jerry had settled on light, tough northern white cedar, or *arbor vitae*. Although white pine is the standard material for planking dories, this special dory rated cedar.

Gunning in Marblehead parlance signifies duck hunting in the outer harbors and adjacent waters among the ledges and small islands, and is a very old sport indeed, extending back to earliest times. William Wood in *New Englands Prospect*, a true, lively, experimental description of that part of America commonly called "New England," printed in London in 1634, mentions that the settlers in the general area where Marblehead and adjoining towns lie today "go afowling two leagues to sea," adding that "such as would eate Fowle, musta not forget their six-foote Gunnes, their good powder and shot, of all sorts."

If the gunners of 300 years ago needed great "six-foote Gunnes" to fill their cooking pots with tasty duck, the latter-day gunners of Jerry's day had their ten-gauge shotguns that belched clouds of black powder when they went off with a recoil that all but kicked the dories backward and sent careless gunners overboard.

The waters around the scattered islands lying off the outer harbor—Tinkers, Ram, the Dry Breakers, Coney and Inner and Outer Pigs—are the natural habitat of numerous duck species that since earliest days have afforded game in fall and early winter for town gunners. At the beginning of the season, "shadows" are put out for coots, and flight shots are taken from the water. But for the rest of the year decoys are set off the rocks and the gunners shoot from cover on the islands. Dories must be hauled up on shore among the rocks, which is why they should be light in weight. Black ducks, sheldrake, buffleheads and mallards are among the species taken. Whistlers come in December.

Years ago when fishermen went gunning in the fall after lobstering all summer, they would naturally have gone in the dories they had been using, but these had been soaking up water all summer, making them heavy to drag over the rocks to keep above the tides. Besides, some would be leaking and in need of repair. So more often than not gunning enthusiasts would plan to buy a new dory in the fall to have it as light as possible during the gunning season. In the early days, according to what

Jerry told me, before Chamberlain, Graves and others started building dories in town, most Marbleheaders got their dories in Swampscott from Warren Small or E. Gerry Emmons.

Gunning was not the only form of diversion for which Marbleheaders employed these dories before automobiles and powerboats appeared on the scene. Weekend excursions in dories as well as rowing trips lasting for one or more weeks were not uncommon in the summertime. Down around Cape Ann to Straitsmouth. Across Massachusetts Bay to Plymouth. And one famous trip under oars went as far as Portland, Maine, and back, with the dory mates cooking and sleeping out on convenient beaches. Once a favorite gunning excursion for two or three dories was to go to Camp Coot on the Annisquam River, a good row home of some 15 or 16 miles. Weekend gunning trips were not infrequent in the fall, with dory mates hustling their oars to get back in time to go to work in the shoe factory on Monday morning. By the time the last century had come to a close, merchant ships manned by Marblehead seamen no longer plied the seven seas. Some displaced sailors turned to lobstering or yachting, but for others the shoe factory offered the only employment. Yet they did not lose their affinity for saltwater and spent as much time on it as they could.

In all likelihood Will Chamberlain built his first perfected gunning dory before 1900. Born in 1864, he was an ardent gunner himself and quite likely had hunted from boyhood. His big retriever curled up on a pile of cedar shavings at the head of his work bench or sitting in the stern sheets of his sailing dory at the start of a race was a sight still remembered in Marblehead when I first went there back in the 1940s.

Both a sport and a perfectionist, Will Chamberlain built boats as much for the joy of it as the profit. Designer and builder of the 21-foot Beachcomber sailing dory for the Beachcomber Club, he raced one himself and for a time was winning all the races until the Brown brothers, Bill and Sam, barely into their teens (Sam later to become the naval architect who designed winning yachts for John Alden), discovered a trick centerboard in Will's Beachcomber that got him disqualified for a time, although it was not long before all was forgiven. Boat building for Will came close to being a religion. It has been told how once when a dory had been finished and was waiting delivery, Will, on noticing some inferior workmanship one of his workmen had

Jerry Smith stands at the oars in* Republican *in 1964 off the shore at Marblehead and works three strings of duck decoys–scoters and sheldrakes. A gunning enthusiast, he continued to hunt ducks with this graceful dory until he reached 85 years of age. In this photograph the dory had only three thwarts, the aft one being removed, two sets of oars, an anchor and gear stowed in the bow (photograph courtesy of Captain Gerald Smith's family).

slipped by him, hove the new boat out of the top floor of the shop down onto the rocks below where it smashed and broke up. Little wonder he and his business were nearly broke when blindness overtook him in his old age.

One of the readers who happened to pick up a copy of the newly published *Outdoor Maine* back in February, 1960, was James S. Rockefeller, Jr. Jim Rockefeller, who had once whimsically referred to himself as "just one of the pebbles," had taken up year-round residence in the scenic upland on the side of Mt. Blue just outside the Penobscot Bay village of Camden, Maine. Jim was an inveterate duck hunter who found his sport in the waters of lower Penobscot Bay among the offshore sunken ledges around Vinalhaven where a nasty chop can kick up in a hurry without much prior warning in the winter time. Jim needed a reliable boat that could take it, yet one that was easy to handle, and big enough to carry two, or possibly even three, men comfortably with their guns, decoys, a dog perhaps, plus the usual assorted gear. In addition it would be nice, he thought, if the boat might double as a summer rowboat for his wife and two children, particularly if it was light enough to haul up on the stern of his 38-foot lobsterboat with the aid of rollers. He had not considered the possibility of sailing the boat, but when that was suggested, he accepted the idea with enthusiasm.

Nearing completion,* Republican *rests right side up fully planked with five planks copper-rivet fastened, breasthooks, rubrails, gunwale caps, floors, cleats, thwart risers, frames and four thwarts in place. Five bent frames fall at the ends and between the sawn frames on each side, adding extra support for the thin cedar planking (photograph courtesy of Captain Gerald Smith's family).

Would I work over the design, he asked, to give him the features he wanted, cutting down the length to 16 feet? Yes, I would work over the design, making the desired changes, I told him, all except cutting down the length to 16 feet. That would spoil the boat, and I wouldn't think of it. Considerable argument back and forth ensued, but I stuck to my guns, and in the end, won my point.

Changes in the lines were actually slight. I gave the modified version a little more beam in proportion to the length and widened the bottom a trifle aft. In lowering the deadrise angle a very moderate amount, a somewhat harder knuckle at the juncture of the second and third planks was attained with the object of getting stiffer bearing for sailing. In fact all divergencies from Chamberlain's original lines, except for shortening the length by a foot, were made primarily to stiffen the boat for sailing, yet without impairing her easy rowing qualities to any significant degree.

Built during the winter of 1962-63, with Jim's assistance, by Veli Holmstrom of Vinalhaven, a pilot model was constructed which exceeded expectations, particularly in her sailing ability, in spite of a meager spread of canvas at first, and a relatively clumsy steering oar. Its extremely light construction proved adequately strong, consisting of planking of quarter inch marine plywood butted on seam battens, epoxy-glued and covered on the outside with a layer of Dynel fabric set in epoxy. Indeed, Rockefeller was so gratified with the performance of his new dory that when he set up a boat shop a few years later, to build peapods, Friendship sloops, power launches and auxiliaries, Marblehead gunning dories were included. In a small prospectus issued at the time celebrating the Marblehead gunning dory as "The Queen of All Double-Enders," Jim was not stinting in his praise.

"Gardner fulfilled his promise, and the dream of recreating Chamberlain's classic boat, using modern materials and construction, became a reality. Tested over a period of three years under the toughest conditions on the outer ledges of Penobscot Bay, this pilot model of the 'Chamberlain Queen' surpassed all expectations. Moreover, its unique characteristics made it an ideal family boat. Loaded with adults and children, dogs and gear, it still slipped easily through the water, and fitted with a centerboard and sail yet another dimension was added. A child could sit on the gunwale without catastrophe, and its ability to glide more than a boat length with one easy pull of the oars endeared it to those who enjoy rowing. Not least, it was practically indestructible, requiring almost no maintenance, dragged easily, and could be beached anywhere.

"Today, Will Chamberlain's classic boat is being recreated in Maine by Maine craftsmen—an honest and lineal descendant of the 'loveliest double-ender ever built.'"

Shortly after launching in 1960,* Republican *undergoes sea trials on Redd's Pond with Jerry Smith and five youngsters on board. The profile view clearly shows the upsweeping curve of the sheer, the long, narrow run aft and the bow fuller, with sharp entry and less overhang than the stern. Years before, Jerry Smith and Lincoln Hocks, bending on the oars of an old Banks dory, towed the barn pictured across the pond to the spot where it still sits today, the home of Redd's Pond Boatworks (photograph courtesy of Captain Gerald Smith's family).

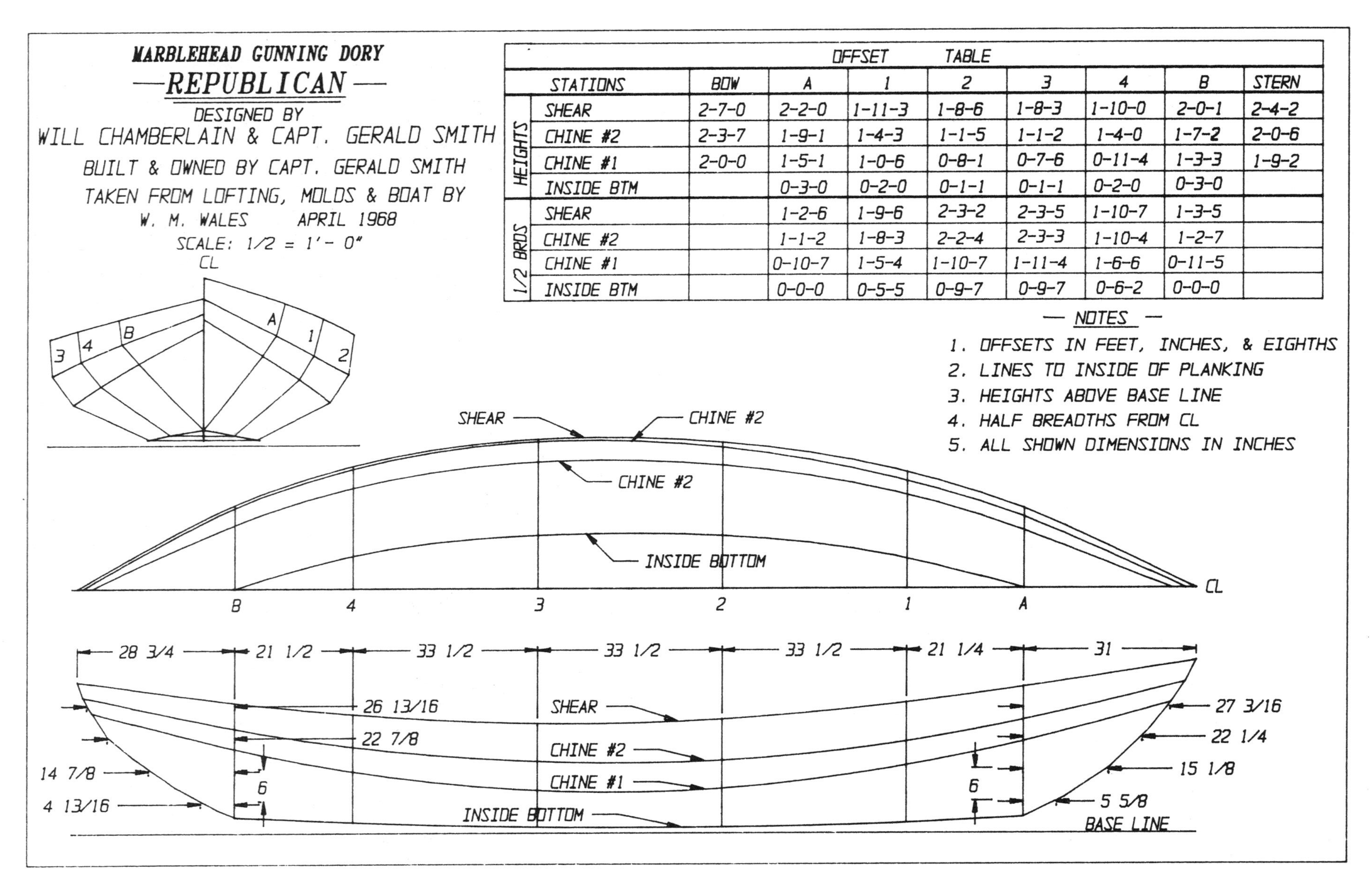

OFFSET TABLE

	STATIONS	BOW	A	1	2	3	4	B	STERN
HEIGHTS	SHEAR	2-7-0	2-2-0	1-11-3	1-8-6	1-8-3	1-10-0	2-0-1	2-4-2
	CHINE #2	2-3-7	1-9-1	1-4-3	1-1-5	1-1-2	1-4-0	1-7-2	2-0-6
	CHINE #1	2-0-0	1-5-1	1-0-6	0-8-1	0-7-6	0-11-4	1-3-3	1-9-2
	INSIDE BTM		0-3-0	0-2-0	0-1-1	0-1-1	0-2-0	0-3-0	
1/2 BRDS	SHEAR		1-2-6	1-9-6	2-3-2	2-3-5	1-10-7	1-3-5	
	CHINE #2		1-1-2	1-8-3	2-2-4	2-3-3	1-10-4	1-2-7	
	CHINE #1		0-10-7	1-5-4	1-10-7	1-11-4	1-6-6	0-11-5	
	INSIDE BTM		0-0-0	0-5-5	0-9-7	0-9-7	0-6-2	0-0-0	

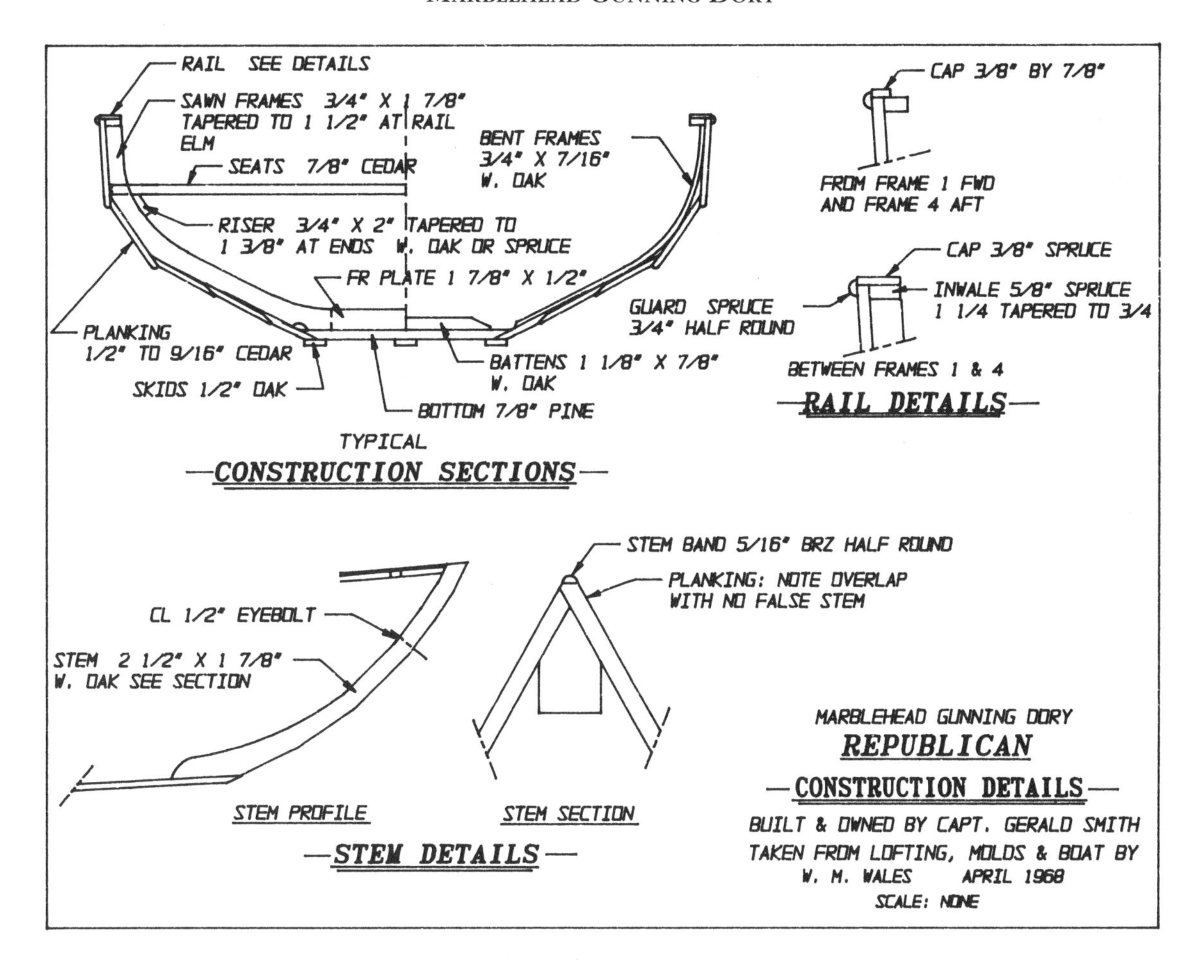

Unfortunately, as it turned out, demand did not prove great enough to support production of these dories as a profitable commercial enterprise, but word got around, in part from reports in the *National Fisherman* and elsewhere, and these dories have continued to be turned out in moderate numbers by home builders and independent craftsmen from the East to the West Coast, and from Alaska to as far away as New Zealand. As many as 70 sets of construction drawings were supplied to prospective builders until the plans with building instructions were included in *Building Classic Small Craft, Vol. 1*, in 1976, and later reprinted in *The Dory Book*.

Now a well-established recreational small craft type, very few of the scores of these boats scattered throughout the country, so far as I have been able to discover, were built or are being used for gunning, although they still retain the old name. Perhaps the last of the true Marblehead gunning dories to have been built and used exclusively for gunning was the boat Captain Jerry Smith had set up for planking in his Pond Street shop when I called in to see it, as previously related, back in early January 1960. When I revisited the shop that fall, the *Republican*, as she was shortly to be named, was planked, had been taken off the molds and was sitting upright on horses, a beautifully proportioned craft with a handsome sweep of sheer. Why Jerry decided to call her the *Republican* I do not know. I was never aware that Jerry had strong political leanings one way or another, although normal New England Yankees of Jerry's and my generation were born Republicans and usually died Republicans as a matter of course.

If anyone was qualified to follow in the footsteps of Will Chamberlain in the construction of Marblehead gunning dories, it was the late Captain Gerald Smith who passed away in 1994 at the age of 88. Jerry was exposed to the art of boat building from infancy, as you might say, acquiring its mysteries under the exacting tutelage of his father, Captain Charlton L. Smith, whom he revered and sought to emulate throughout his life—as well he might, for Captain

Walter Wales' double-ended gunning dory,* Independent, *underway on the Mystic River during the Rowing Workshop in June 1970 (Lester D. Olin photograph, M.S.M.).

Charlton Smith was a perfectionist, and, in other ways, a most remarkable man.

Born in Chelsea, Massachusetts, shortly after the Civil War, he went to sea as a young man as ship's carpenter in some of the last of the old square-riggers where, among other things, he acquired a fluent German as well as some facility in a couple of other European languages. Coming to Marblehead early in the century as skipper on some of the large sailing yachts of the time, he subsequently went to work as loftsman for Starling Burgess when Burgess and Packard were in the yacht building business in Marblehead. Burgess is reputed to have said that Charlton Smith was the only one at the yard he trusted to lay down his designs. It was then, no doubt, that Smith picked up a working knowledge of naval architecture that he was later to use to advantage in his own shop, known locally as the "Home of the Brutal Beast," a popular Starling Burgess one-design boat for youthful sailors, and particularly in Smith's rebuilding and restoration of the keel-cat *Nixie* designed by Edward Burgess. Apart from boat building, Charlton Smith was a frequent contributor of feature articles on yachting to the old *Boston Evening Transcript* as well as to various boating magazines. He published two books of juvenile adventure, *Bob Haskins in Southern Seas* and *Gus Harvey of Cape Ann*, and was a friend and correspondent of W. P. Stephens. Last but not least, he was accepted in Marblehead as a Marbleheader, for the Marbleheaders were a clannish lot, and if your grandfather, at the very least, had not been born a Marbleheader, you did not rate membership in the club.

So it should be plain that Jerry had much to live up to, and that a lot of knowhow, pride and tradition went into the fabric of the *Republican* along with the cedar, oak and pine, when Jerry put it together.

I was mistaken in my judgment that shortening the length of the Marblehead gunning dory to 16 or 17 feet would spoil it. The *Republican* has proved me wrong. For 20 years it saw its share of rough water and served Jerry well until his wife's anxieties persuaded him to give up gunning at the age of 85. Still sound and seaworthy, the boat is now at his son's place in New Hampshire, awaiting repainting in preparation for many long years of use yet to come.

As previously stated, Jerry built the boat from molds without recourse to lines or drawings, nor did he keep any systematic record of the dimensions with the needs of subsequent builders in mind. Fortunately a young student of naval architecture who lived nearby, whom Jerry had more or less taken under his wing, followed construction of the *Republican*

Independent, ***adapted for sail, has a spritsail main, wire shrouds and headstay, jib, daggerboard and steering oar (Lester D. Olin photograph, M.S.M.).***

which he built a fine gunning dory closely replicating the *Republican* for himself. That was the naval architect Walter M. Wales, known in those days to his friends and acquaintances as "Wally." Wally has said that when it came to naming his dory, he had an impulse to call her the *Democrat*, but he couldn't quite bring himself to do it. He called her the *Independent* instead.

Wales' lines plan with dimensions and offsets was too large, as originally laid out, to reduce for inclusion in this book and still remain legible. When this was pointed out to Wales, he graciously volunteered to redraw and rearrange the plan as it now appears and he also added a construction drawing, including details of the stem, rail and the hull in cross section. Additional details may safely be taken from those given on the accompanying drawing of Chamberlain's 19-footer, although it should be recalled that the Chamberlain boat was very lightly built. The *Republican*'s scantlings could well have been somewhat heavier—for example, sawn frames sided 3/4" instead of 5/8".

In working out scantling dimensions for the construction of the *Republican*, the builder would be well advised to consult various dory plans as shown in *The Dory Book*.

very closely, eventually producing, some years later, an accurate takeoff of lines and a table of offsets for the *Republican*, from

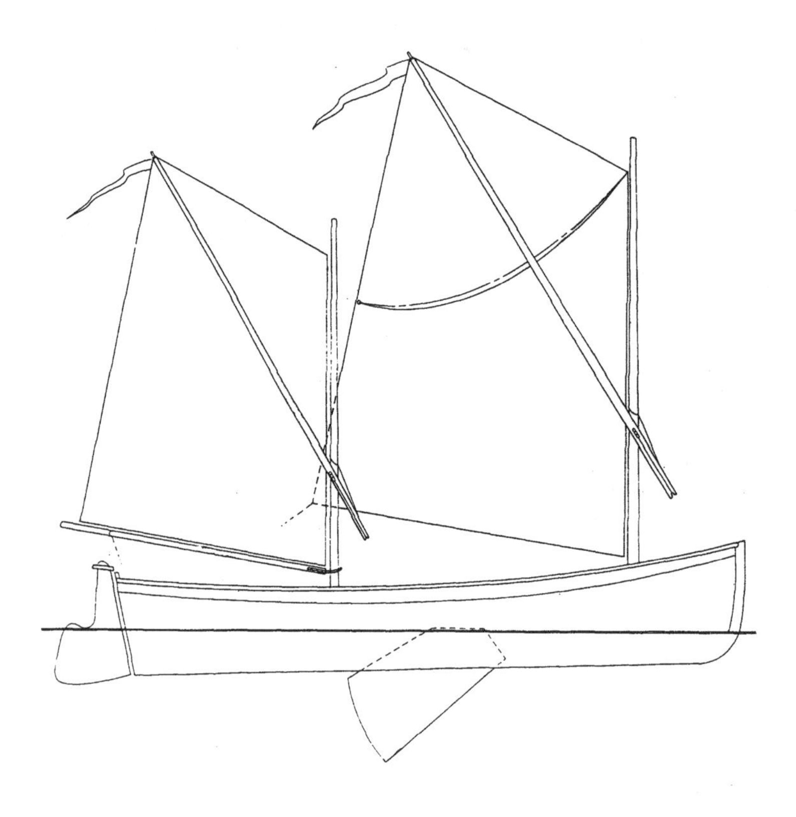

Chapter Three

Moosabec Reach Boat

In the spring of 1994 Mystic Seaport Museum added *Temporary* to its small craft collection, a fine example of what has been identified, whether correctly or not, as a Moosabec Reach Boat. Reach boat was one of the names formerly applied in Maine to fishermen's small, open work-boats, ranging in length from 10 to 20 feet and generally fitted with oars and spritsails, once employed in large numbers in the inshore fisheries of eastern Maine from the approaches to Penobscot Bay to Eastport.

Reach boats figured prominently in the commercial lobster fishery, highly successful from the beginning, that Goode in his *Fisheries and Fishery Industries of the United States* dates from 1841 when Captain Elisha Oakes of Vinalhaven undertook to supply lobsters to the Boston market from Harpswell in the well smack *Swampscott.*

By 1848 Captain Oakes was carrying 40,000 lobsters to Boston, part of which he bought in the Penobscot Bay region, and by 1850 he had left Harpswell where the lobsters had got too small and was buying entirely from the Muscle Ridges in Penobscot Bay where in 1850 lobsters were so plentiful in the shoal waters of the Ridges that four men could fully supply Captain Oakes' smack *Josephine.* Trips to Boston and back with the smack required about 10 days, and during that time each of the four men in Penobscot Bay would take between 1,200 and 1,500 lobsters. Marketable lobsters averaged about 3 lbs., with those under $10^{1}/_{2}$ inches being rejected. Prices were 3 cents each in March, $2^{1}/_{2}$ cents in April and 2 cents in May and June.

The rapid development of the lobster industry from about the middle of the nineteenth century on is typical of a like expansion of all manner of commercial enterprises along the Maine coast during this period. Fisheries, canneries, the sawing and shipping of lumber, shipyards, plaster mills and lime kilns, granite quarries and others as well, all depended to some extent on waterborne transportation. It was a time of great commercial activity. All along Maine's more than a thousand miles of indented and convoluted coastline there was increasing need for boats, with many of them small enough for one man to handle alone. The engine had not yet replaced oars. Boats that rowed well and handled easily were at a premium. This imposed limits on size and rewarded improvements in hull design and refinements in construction.

Later on, when motors became widely available, it was possible and not unusual to compensate for poor design and heavy, awkward construction by increasing the horsepower. So long as fuel was cheap and plentiful, it didn't matter what kind of a tub it was so long as it had a big enough engine in it. But when all a man had was his arms and his back to move ship, he was bound to apply skill and judgment to the attainment of hull shapes which moved easily and sweetly, either with oars or under sail.

During the second half of the nineteenth century, a number of related small craft, workboat types of superior performance, were developed, or at least improved and perfected, along the Maine coast. These were generally built in the location where they were used, from lumber grown locally and, to a considerable extent, by the men who used them. Those under 20 feet in length included dories, wherries, peapods and the so-called Reach boat.

The latter, according to Mitman in his *Catalogue of the Watercraft Collection in the U. S. National Museum*, took its name from Moosabec Reach near Jonesport, and was extensively employed in the general fisheries of eastern Maine. Although the average length was about 15 feet, they sometimes ran to more than 20 feet.

The model listed and described by Mitman was for a wooden, clinker-built, open keel boat, sharp at both ends with curved stem and straight stern post. Cat rigged, with a single spritsail, the model had two thwarts and carried one pair of oars. Dimensions were listed as: length 15', beam 4'9" and mast above the thwart 11$^{1}/_{2}$'.

Goode's description of the Reach boat, published in 1887, reads somewhat differently. Extensively used in the coastal fisheries of Maine as well as to some extent in lobstering, Reach boats, according to Goode, "range in length from 10 to 18 feet. They are sharp at the bow, round bilged, keeled, clinker or lap-strake and have a square, heart, or V-shaped stern, with two or three thwarts, according to their size; they are as a rule entirely open fore and aft, rarely having washboards. They are well adapted for rowing and sailing and all but the smallest usually carry one or more sprit sails."

Most of what H. I. Chapelle has to say about Reach boats in his *American Small Sailing Craft* is obviously based upon Goode and Mitman. Yet neither provides authority for his statement (p. 138) that "the Reach boat did not come into use in Maine until about 1870, when it began to be built in North Haven, Maine." This was a double-ender, according to Chapelle, but not, he goes on to say, a smaller counterpart of the whaleboat, as one historian has claimed.

How this purported North Haven origin can be made to jibe with Chapelle's statement (p. 202-203) that the Reach boat used in the 1870s and 1880s near Jonesport, Maine, was 14 to 18 feet long and similar to the "Cape Roseway" wherry mentioned by Goode is hard to see. Chapelle continues by describing this model as "much like a drag boat (Long Island Sound drag boat) but without a centerboard, and with a scantling keel, to which a deep false keel was added to enable it to work to windward. The Reach boat carried one or two small sprit sails. It took its name from Moosabec Reach near Jonesport. The name 'Reach boat' was also applied to some of the double-ended peapods used in this area. Boats very similar to the square-sterned Reach boat are to be seen in old photographs of Portland harbor."

Chapelle's attribution of similarity between the Reach boat and the "Cape Roseway" wherry mentioned by Goode requires some examination. Goode's description (Sect. V, Vol. 2, p. 672) is as follows: "The Cape Roseway wherry employed in the lobster and inshore fisheries of Penobscot Bay, Maine, especially

Salmon wherry at Lincolnville Beach, Maine, in 1940, owned by Lawrence Carver, who ran a lobster business on the waterfront. The photographer's father, Walter Leavitt, stands at the stern which is notched for sculling (John F. Leavitt photograph).

in the vicinity of Castine, is a lap-strake boat with sharp bow, round bilge, narrow flat bottom, and very narrow, heart-shaped stern. It ranges in length from 12' to 18', is entirely open, and is seldom provided with sails."

One important difference between the Reach boat as described by Chapelle and Goode's wherry is in the keels. The Reach boat, according to Chapelle, has a keel, or that is to say what he calls a "scantling keel," while Goode's wherry has a "narrow, flat bottom." Also the wherry's "very narrow heart-shaped stern," as described by Goode, is not the same as the wider sterns of Chapelle's "square sterned Reach boats." Furthermore, it should be mentioned here, to set the record straight, that there is no "Cape Roseway" in the Eastern Bay of the Penobscot, although there is a Cape Rosier.

Referring to Goode's "Cape Roseway" wherry, Chapelle states: "This type is now extinct, and no model has been found." The fact is, however, that directly across from Cape Rosier in Lincolnville several wherries formerly used there in netting salmon have survived, and that wherries were still being built there in the mid-1970s. Furthermore, at Ash Point, not so far down the Bay on the western shore near Rockland, a number of wherries were still in use in the early years of the present century, and detailed descriptions of these boats with line sketches are to be found in an article by Alfred A. Brooks in the October 1942 issue of the *American Neptune*. The Ash Point wherry belonging to Paris Ratcliff fits Goode's description closely. While Goode stated that the "Cape Roseway" wherries used particularly in the vicinity of Castine were "seldom provided with sails," both the Lincolnville and the Ash Point wherries were frequently sailed, particularly the latter.

Brooks has this to say: "The wherries were a common type there which I have not seen very much elsewhere. They had a flat bottom board the same as the double-enders, and the lower half of the hull was somewhat similar. The stern had a straight, sloping stern post and a fairly wide transom about half the length of the stern post in height. The hull was a little deeper at the stern than at the bow. There were two general types, the rowing wherry primarily for rowing, although often fitted with a centerboard and rigged with a sprit sail, and the sailing wherry, a little larger and intended for sailing and always fitted with a centerboard and rigged with a boom and gaff mainsail, bowsprit and jib."

One other observation by Brooks is relevant here and should be noted. "The double-enders, wherries, and sloops were either carvel built which they called 'smooth skin,' or clinker built which they called 'lap-strake.'" None of the Lincolnville salmon wherries, so far as we know, were carvel built, yet it appears that boats could be planked "smooth skin" or "set works" at Ash Point, and still be considered wherries. Two things that a wherry had to have apparently to be considered a wherry: a bottom board instead of a keel and a transom stern instead of being double-ended.

Our only sources for the Reach boat, *i.e.*, Goode and Mitman, for Chapelle clearly derives his information from them, identify this boat as clinker planked, with a keel instead of a bottom board. Mitman seems to have generalized from a single instance, and from a model at that whose authenticity might be open to question.

Goode also might well have been generalizing from a sample too limited to be representative. His description of his "Cape Roseway" wherry as having a "very narrow heart-shaped stern" suggests this might have been the case, for we know from the Lincolnville wherries and those at Ash Point that these boats were built with sterns which ranged in width all the way from very narrow to quite wide.

Furthermore, small boats built along the Maine coast were built until quite recently without plans. Quite often they were built by the men who used them, and naturally and inevitably they reflected the needs and preferences of their owners. Although the overall characteristics of tried and accepted models were generally adhered to, minor deviations and variations from the norm were bound to creep in. And not only did such variations occur from builder to builder, but also from locality to locality.

According to whatever limited information we have, differences between wherries and Reach boats were not always as clear-cut and unmistakable as some have supposed. I am not at all sure that it would

Shown here is a Penobscot Bay salmon wherry rigged for sail with rudder, centerboard, two mast positions and single spritsail as drawn by John F. Leavitt from a photograph taken at Lincolnville, Maine, in 1940 (painting courtesy of John Gardner).

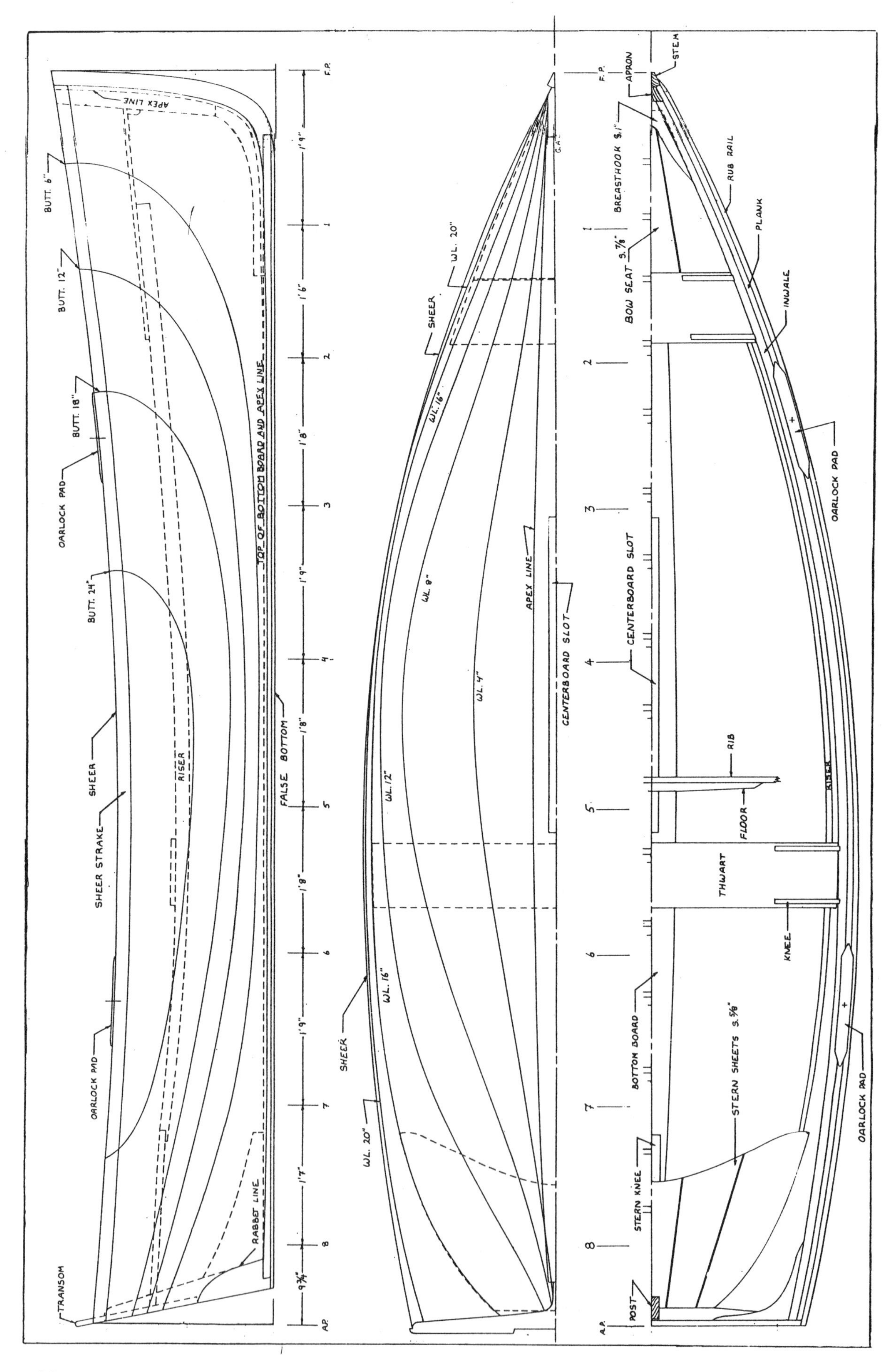
F.P.
APEX LINE
BUTT. 6"
BUTT. 12"
BUTT. 18"
OARLOCK PAD
BUTT. 24"
SHEER
SHEER STRAKE
RISER
TOP OF BOTTOM BOARD AND APEX LINE
FALSE BOTTOM
OARLOCK PAD
RABBET LINE
TRANSOM
A.P.
1'9"
1'6"
1'8"
1'9"
1'8"
1'8"
1'9"
1'7"
9 3/4"
1
2
3
4
5
6
7
8
WL. 20"
SHEER
WL. 16"
WL. 8"
APEX LINE
WL. 4"
CENTERBOARD SLOT
WL. 12"
SHEER
WL. 16"
WL. 20"
STEM
APRON
BREASTHOOK S. 1"
RUB RAIL
PLANK
BOW SEAT S. 7/8"
INWALE
OARLOCK PAD
CENTERBOARD SLOT
RIB
FLOOR
RISER
THWART
KNEE
BOTTOM BOARD
STERN SHEETS S. 5/8"
OARLOCK PAD
STERN KNEE
POST

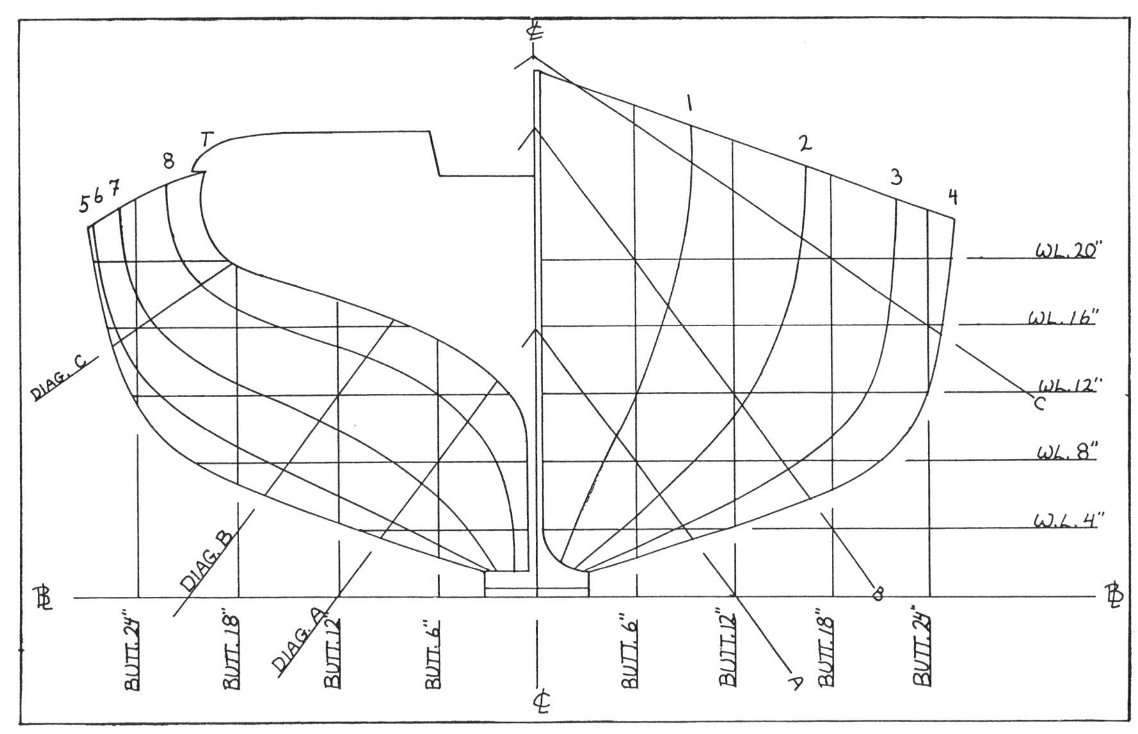

be possible to draw any precise, hard and fast line between these two types. What passed for a Reach boat in some localities might well have been known as a wherry in others, and vice versa.

The boat we are about to consider has some of the characteristics of both types. I might have been satisfied to call it a carvel-planked wherry and let it go at that, had not a man who had fished in a similar boat in Pleasant Bay, Maine, as a boy in the 1920s, identified it as a Reach boat when he saw it under repair at Mystic Seaport early in the spring of 1974. Lawrence Crowley is a life-long resident of Addison, Maine, situated at the head of Pleasant Bay. Addison is the next town to Jonesport, only a few miles away.

The origin of this boat is not known exactly. It was already old when it was purchased in Castine about 1950 by Nathaniel French when he had the Alamoosook Island Camp for boys at Bucksport. About this time, the tender of the camp yacht, the auxiliary schooner *Alamar*, was smashed, being replaced by the boat under consideration which received the name of *Temporary*, which it still retains.

Although he cannot say for sure, French believes he first saw *Temporary* in the 1920s at Castine where the boat was used in and about the harbor by a very old man who rowed in a semi-upright position braced against a board laid across the gunwales. *Temporary*'s ample freeboard, her stability and her age all tend to confirm French's recollection. But in any case, it should be noted that French did buy the boat at Castine, which is where Goode wrote that boats like the "Cape Roseway" wherry were much used.

I first learned about *Temporary* in the winter of 1973 when George Anderson Chase, or "Andy," who owned her then, wrote me about her, enclosing a sketch of her lines which he had made. What struck me immediately was the definite wherry shape and the narrow bottom board instead of a keel. Besides, Andy lived at South Brookfield on the easterly end of Penobscot Bay, not far from Cape Rosier. It seemed to me there was a chance that *Temporary* might be related to Goode's "Cape Roseway" wherry, which Chapelle believed to be extinct.

Correspondence ensued, and the upshot was that arrangements were made for Andy, who was a senior at the Putney School in Putney, Vermont, to bring *Temporary* to Mystic Seaport for an interim six-week work-study period between school semesters which took place during March and part of April in 1974.

During his stay at Mystic, Andy measured the boat, taking off her lines and recording these and her construction details in the drawings reproduced here. In addition, he carried out an extensive restoration and repairs, making her sound and tight once again, as well as preparing for the reinstallation of a centerboard, which she once had.

When Andy acquired *Temporary* several years before, her bottom was in bad shape, and she leaked about her garboards. He was advised to cover her with fiberglass, which he did, and which he subsequently regretted. The first step in the restoration undertaken at Mystic was the removal of this worse-than-useless disfigurement. The glass fabric, which was coming loose in spots, stripped off quite easily, but some of the resin did not loosen as easily, requiring repeated coats of strong paint remover and much scraping and sanding—a messy business, but a transformation and a rebirth when accomplished.

With the weight of fiberglass removed, *Temporary* recovered her old lightness and much of her former good looks. Her garboard seams were so badly enlarged that it was impossible to caulk them, and instead of removing the cedar garboards which were otherwise sound, we elected to reinforce them and make them tight with phenolic microballoons and polypropylene tape set with epoxy adhesive. This produced a tight, strong job which was hardly noticeable after painting, and the added weight was negligible. Microballoons and epoxy are light enough to float, and polypropylene fabric is one-third the weight of glass.

The condition of *Temporary* indicated that she was old. If she is the boat French saw at Castine in the 1920s, she would have been at least 50 years old, but I should judge she is considerably older than that. She could easily date back a couple of decades into the nineteenth century.

Her materials and certain features of her construction indicate that she was built down east, quite likely in Washington County, where old-growth yellow birch was commonly used instead of oak. Her transom, stern framing and inner and outer bottom boards were made of birch.

OFFSETS IN FEET-INCHES-16THS. TO INSIDE OF PLANKING

	STATION	TRANS	8	7	6	5	4	3	2	1	APEX
HALF-BREADTHS FROM CENTER LINE	SHEER	1-7-12	1-10-0	2-0-14	2-2-8	2-2-13	2-1-10	1-10-3	1-4-8	0-9-9	0-0-6
	WL 20"	1-6-6	1-9-2	2-0-8	2-2-4	2-2-8	2-1-8	1-10-0	1-4-0	0-8-13	0-0-6
	WL 16"	0-7-10	1-3-9	1-10-9	2-1-14	2-1-12	2-0-15	1-9-9	1-3-0	0-7-10	0-0-6
	WL 12"	0-1-9	0-5-15	1-4-9	1-10-7	2-0-3	2-0-0	1-8-4	1-1-0	0-6-1	0-0-6
	WL 8"	0-0-8	0-2-7	0-8-10	1-4-2	1-8-9	1-9-1	1-4-2	0-9-7	0-4-3	0-0-6
	WL 4"	0-0-8	0-1-5	0-4-3	0-7-12	0-10-12	0-11-5	0-8-7	0-5-1	0-2-7	0-0-6
	APEX LINE	0-0-7	0-1-5	0-2-8	0-2-14	0-3-2	0-3-3	0-2-14	0-2-5	0-1-8	—
HEIGTHS FROM BASE LINE	SHEER	2-1-3	2-0-7	1-11-1	1-10-2	1-9-15	1-10-7	1-11-9	2-1-8	2-3-13	2-7-1
	BUTT. 24"	—	—	1-5-15	1-1-8	0-11-7	1-0-0	—	—	—	—
	BUTT. 18"	1-7-10	1-5-1	1-0-11	0-6-9	0-8-15	0-6-6	0-9-7	—	—	—
	BUTT. 12"	1-5-8	1-2-12	0-9-15	0-6-0	0-4-5	0-4-3	0-5-11	0-10-13	—	—
	BUTT. 6"	1-3-4	1-0-0	0-5-13	0-3-1	0-2-5	0-2-5	0-2-14	0-4-12	1-0-1	—
	APEX LINE	0-1-8	0-1-8	0-1-8	0-1-8	0-1-8	0-1-8	0-1-8	0-1-8	0-1-9	—
DIAG.S	DIAG. A	1-9-15	2-0-3	2-3-12	2-6-1	2-6-10	2-5-13	2-2-3	1-7-8	0-11-9	—
	DIAG. B	1-2-7	1-5-2	1-9-13	2-1-7	2-3-7	2-3-11	2-1-9	1-8-12	1-1-6	0-0-10
	DIAG. C	0-3-15	0-6-15	0-11-9	1-2-6	1-3-8	1-3-13	1-2-14	1-0-7	0-8-2	0-0-10

BOTTOM OF KEEL IS HORIZONTAL BASE LINE. FACE OF STEM AT SHEER IS AT F.P. AND 31" ABOVE BASE LINE. AFT FACE OF TRANSOM AT SHEER IS AT A.P. DISTANCE BETWEEN A.P. AND F.P. IS 14'-3"

G.A. CHASE, MARCH 24, 1974

When relieved of the fiberglass covering,* Temporary, *despite her 14' 3" length overall and ample freeboard, proved to be surprisingly agile and easy to row due to her double-ended waterline profile and light construction.* Temporary *is shown here on the Mystic River shortly after her spring overhaul in 1974 with Andy Chase at the oars (photograph courtesy of John Gardner).

Her original floors, very lightly and daintily made, as well as her breast-hook and quarter knees, were apple, sawn out of limb crooks, but so badly discolored by age and hard use that we were not sure at first. A sample sent to the Forest Products Laboratory at Madison, Wisconsin, confirmed our identification. The planking was tough, knotty, down-east cedar which had taken some abuse over the years but in 1974 was essentially as sound as the day it was cut.

Temporary's original framing was quite light, and quite likely was pre-bent and beveled at the bench before setting up for planking, as was commonly done down east years ago. Some time after, an additional set of slightly larger steamed timbers were bent in, probably as the boat began to age. *Temporary* was built with a centerboard and trunk, which was later removed and plugged as the drawing shows. This may have been done when she was retimbered, presumably at that time being generally overhauled and reinforced late in life to make her serviceable as a rowboat.

Under some recent construction at the turn of the forefoot, close to the stem, the old mast step was found, suggesting that she was cat rigged with a single spritsail, as were the Ash Point rowing and sailing wherries and the Reach boat model described by Mitman. The opening out of the top of the stern transom, which we thought at first might have been recently removed to facilitate hanging an outboard motor, could just have well have been cut out originally to allow for the swing of a tiller. A rusty rudder gudgeon still remained in the boat's stern post.

What were apparently the original inwales were still in the boat, and going by the holes left by old fastenings, it appeared that there was only one rowing station located forward of the centerboard trunk. Evidently the boat was built to sail, and undoubtedly she sailed well with her easy, essentially double-ended, underwater lines, her ample freeboard, solid bearing, powerful after quarters and original light construction.

Regardless of the outcome of possible arguments as to whether *Temporary* is a wherry or a Reach boat, there can be no doubt of her exceptional qualities, both for sailing and rowing. She is of particular historic interest, and her lines and details now recorded in Andy's drawings make a significant addition to our increasing store of small craft knowledge.

After we got her tight and primed with a coat of red lead, and just prior to Andy's loading her on a trailer for the trip back to Putney, where he was getting ready to convert to sail, we launched her in the Mystic River for a trial spin under oars. In spite of her wide transom, she is a double-ender on the water with any reasonable load, and moves with ease. She trimmed beautifully, as the photograph shows.

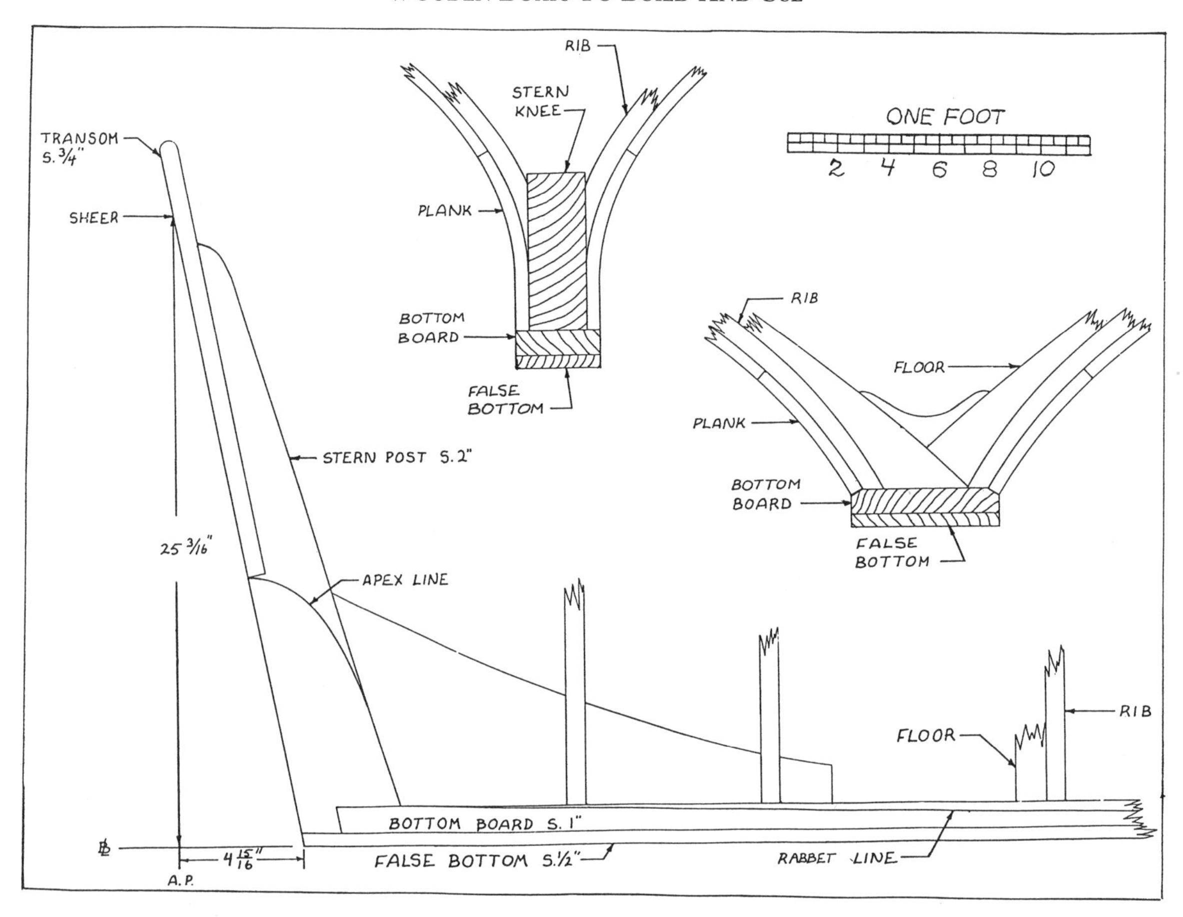

She was remarkably stable without being stiff. Possibly her considerable freeboard, an advantage in carrying sail, would not be an advantage at times when rowing in strong winds. After all, one can't have everything. In any case, I was both surprised and pleased when I first rowed her. Her easy performance surpassed my most sanguine expectations.

After Andy's return to Putney at the end of April, he lost no time in drawing up the lines reproduced here, which he completed in time for them to appear with my article on Reach boats and wherries in the July, 1974, issue of the *National Fisherman.*

Following Andy's graduation at Putney he enrolled in the Maine Maritime Academy at Castine, which left little time to devote to *Temporary.* However, in a letter to me dated September 9, 1976, he began, "Needless to say, I'm writing to ask a few more questions about rigging *Temporary.* I'm still plodding along, working an hour or two every month.

"I built the centerboard and trunk in her last winter, and it came out quite nicely. I cannot detect any leaks as yet. I used native cedar (very native) for the trunk with white oak frames, and pine bed logs. As you suggested, I went out of the way to make it solid and well braced. The board is red oak with steel drifts."

Temporary had been built with a centerboard, but sometime in the past the centerboard trunk had been removed and the slot through the plank keel plugged. Andy removed this plug and installed his replacement trunk in the old slot.

Also in this letter Andy enclosed some preliminary drawings of the rig he had decided on, of the rudder and tiller he was considering and of his replacement centerboard trunk and centerboard. As he planned on sitting on the after rowing thwart when sailing, he figured he'd need a reasonably long tiller.

One of his questions was, "How much support must I give the unstayed masts? They will pass through the thwarts. Is that sufficient? The sticks I'm using are quite husky."

Evidently the support supplied by the thwarts alone proved to be adequate, but it

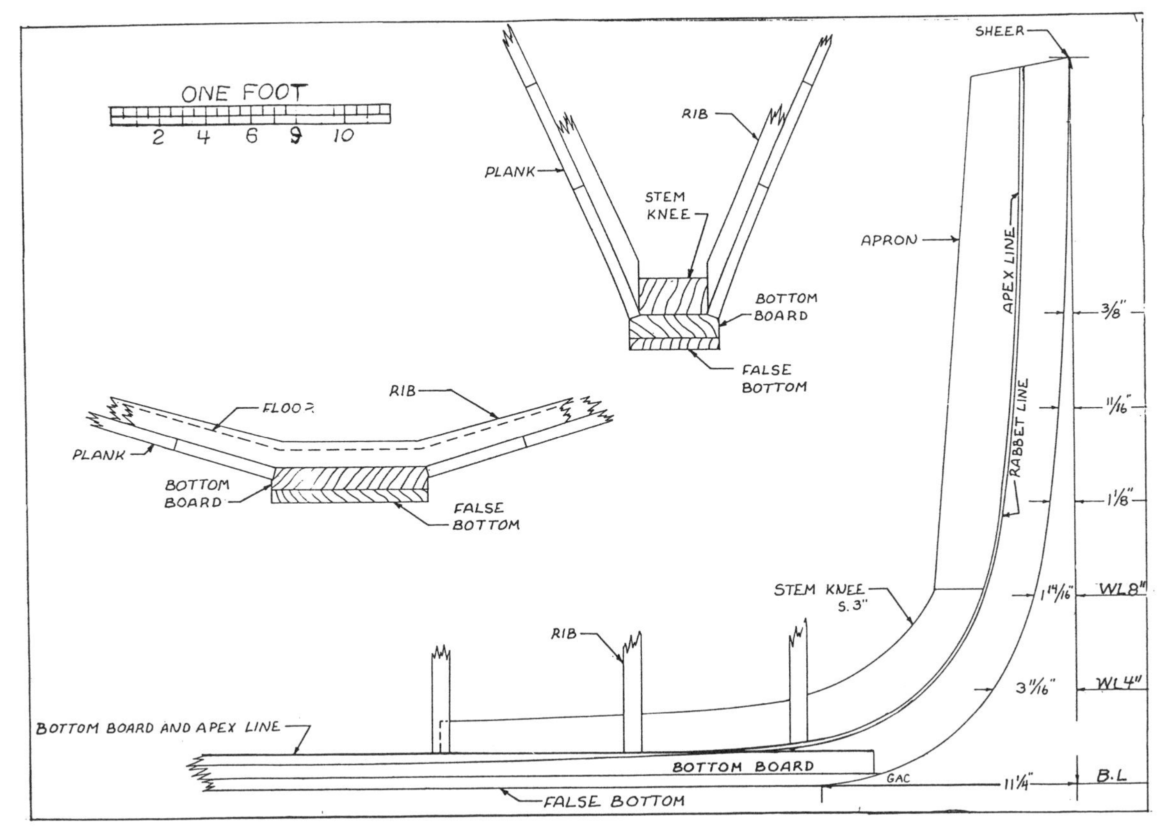

should be noted that the thwarts are well braced to the sides of the boat at either end by two stout knees.

In commenting on the rudder design, Andy wrote, "I think it looks quite nice. It would be pieces of 4" x 3/4" red oak drifted together. I may have a nice piece of locust for the tiller.

"As for the sail plan, I went by the 100 sq. ft. figure you suggested. Other than that, I just tried to keep the center a little aft of the center of the centerboard, and drew what looked pretty—any suggestions there? I've got all the spars made—they came out well—odds and ends of sticks I found around."

It was not until more than a year later that I heard from Andy again in the letter that follows, dated December 5, 1977.

"At long last I've managed to get *Temporary* sailing. The enclosed photos show her on trials in a good breeze.

"For all the worrying I've done in putting the rig together, she sails wonderfully. My chief concern was that she would be over-canvassed, but she carries all 96 sq. ft. with ease. The rig is a handful, admittedly, but the fun is easily proportional to the work. I just recently realized that by stepping the mainsail in the second thwart she should balance almost perfectly as a sloop for times when a simpler rig is desirable. Thus with two sails, three mast steps, and one set of reef points the possibilities are almost limitless.

"Reefed down she's nothing but fun, and ready for a good deal of wind. I guess it was blowing 15-plus that day, and she could have still carried the full rig if I'd been up to it myself. Though I took great pains to brace the rig so as to distribute the strains evenly, I still don't feel like pushing her any.

"The sails were made for me by Grant Gamble of Camden (he used to work in the rigging shop at the Seaport). I have nothing but praise for his work and price. They are of elegant Egyptian cotton, with nicely worked grommets and cringles all around. My brother just gave me a box of tanbark he brought back from Ireland that I'll treat them with when I get a chance. Besides a personal preference for the color of tanbarked sails, I think something is necessary to preserve them, as they spend a good deal of time furled on the spars at her mooring here in Castine.

"Perhaps the most valuable item in the rig is the brail I installed for the main (or fore, as you prefer). This leads from a cleat

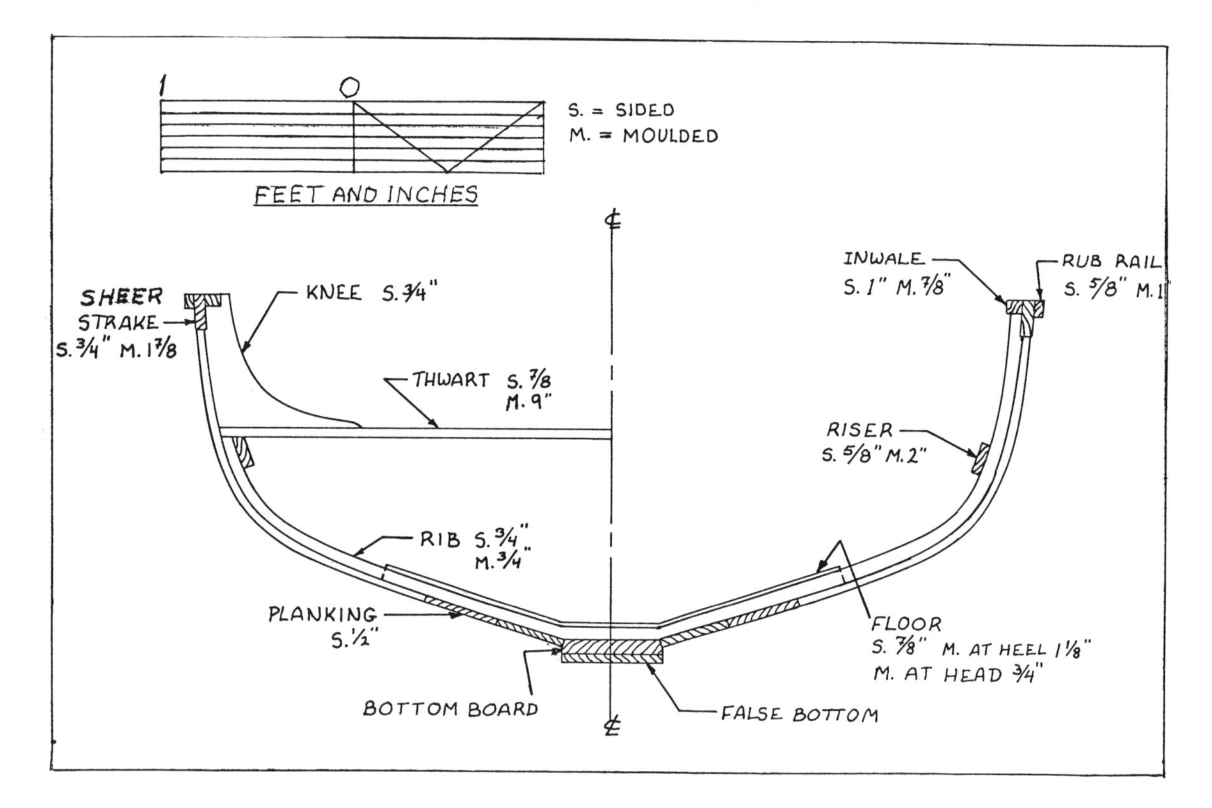

on the mast, up through the throat grommet, down through a small grommet about half way down the leach, and back to the throat grommet where it is made fast. This allows me to reach forward, grab it, and in one yank brail the main right up to the mast. With the mizzen made fast close in she becomes instantly hove to (or rather, in irons), with the forward rowing station clear for rowing if I wish. This was especially handy for me as a sort of panic-string on that first windy day...

"Well, I'm sure you can tell I'm thrilled by her performance. She is both fast and reasonably weatherly for a sprit rig. Although her initial stability is a bit shaky, once she leans over on to one of her bilges she is steady as can be. Even with the rig in (sails furled to the mast), I can stand on her rail without taking in any water.

"It's been a long time coming, but I guess the longer you wait for something the better it is to finally get it."

A month later Andy wrote again. His letter, dated January 9, 1978 follows.

"Here's the sail plan. I was just getting started on it when I got your letter.

"I haven't had a chance to try out the sloop rig yet, but I think it should work out alright. I've found the reefed-down rig (reefed main, scandalized mizzen) to balance just about right in use, while the full rig balances just a little aft. This shows in the drawings where the reefed rig's center of area comes right at the center of the centerboard area, while both the sloop rig and full rig come just a little aft of this. However, you might notice the new shape of the rudder, which I have had to enlarge, as the one I made was too small to be effective. This larger rudder may help out some to bring the center of effective lateral plane aft a little. We'll see...

"I'm not sure I like the shape of the new rudder quite so much, and I would have used a narrower, deeper rudder except that I didn't want it protruding below the hull, as I often land her (intentionally!) on shore.

"Since I sent you the pictures, I've had my two-week Christmas vacation, and with my brother's boat shop temporarily vacant I put her in for some more work. I replaced both sheer strakes, gunwales and inwales, quarter knees, breasthook, and half the transom. This gave her back a good deal of strength, as they were all cracked and chewed up a bit. It also gave me the chance to varnish her rails—my first concession to the 'yacht' in her. The sheer planks are local cedar, rails red oak, breasthook and quarter knees natural crook locust, and the new transom piece locust as well.

Temporary*'s forward thwart is heavily braced to support the unstayed mainmast. The thwart, 8 7/8" wide and resting on the thwart risers, is supported on each end by two stout thwart knees, sided 3/4", and athwartships by two lodging knees or mast partners, each butting on the thwart riser. The 2 7/8"-diameter mast hole is directly over the step, which is flush with the aft edge of the stem knee. The step is mortised so that the mast does not rotate (Sharon Brown photograph).*

"I'm sending you the original drawing of the sail plan for you to copy."

For the next 13 years Andy sailed *Temporary* when he could find the time and opportunity in his very full life at sea as an officer in the Merchant Marine and as a teacher at the Maine Maritime Academy at Castine. However by 1991 *Temporary* was showing her age and was leaking badly. After various attempts to bring her back, including what Andy calls "the roofing tar approach," all to no avail, *Temporary* was retired permanently, finally coming three years later to snug harbor at Mystic Seaport.

Summing it all up in the final paragraph of the letter which accompanied *Temporary* to Mystic, Andy wrote:

"She's an incredible model. If I ever get it together, I intend to build one from her plans. If anyone down there wants a superb boat, they should build it. She carries a tremendous load, rows and sails like a dream, is very fast, and is practically unsinkable. To flood her in the spring I had to stand on the gunwale and jump up and down to get the rail under. And she's twice the weight she was new. The only hitch will be planking her stern."

Not the easiest boat to build, to be sure, not one for the inexperienced beginner to tackle off hand, yet well within the capability of more than a few who would find *Temporary* a challenge and a delight to put together. If boats like *Temporary* were frequently built in the old days from rough sawn native lumber with simple hand tools by the fishermen who used them, surely we have not become so effete that there are not craftsmen among us today equal to the task.

And where today would you get a boat like *Temporary* with such unrivaled qualities and character unless you had her custom built at great expense or built her yourself?

And whether she's a Reach boat or a wherry what difference does it make when we have the boat herself to marvel at and reproduce?

When Andy replaced the centerboard and centerboard trunk in the winter of 1975-76, using the original slot through the bottom board or plank keel, he must have added the thwart at the forward end of the trunk, although he makes no mention of having done this in his letters. His drawing of lines and construction details made in 1974 does not show this thwart, but it is clear that this thwart had to have been in place when he first tried out the boat fully rigged for sail, for, as he explained, it reassured him to know that in a pinch, if the new rig proved more than he could handle in a blow, he could always brail in the main close to the mast, and with the mizzen in irons, row from the intermediate or second thwart. Also as he pointed out in this same letter, if a simpler rig is desired, the boat under its main alone stepped through the second thwart will balance nicely as a sloop.

It should be said here that building a boat like *Temporary* is no undertaking for an inexperienced beginner. Lining out, shaping and fitting *Temporary*'s carvel planking call for special knowledge and skills that generally take time and boat building experience to acquire, although it should also be said that once in a while the exceptional amateur does turn up with the natural ability to produce an acceptable job in spite of his lack of prior experience. So much in the way of due warning.

For the prospective builder who is still undeterred, there follows a listing of essential measurements and dimensions that do not appear in the plans reproduced here or are too small in reduction to be easily discernible.

Construction Details

First, the mast positions. The mainmast centers 2'8" from the forward perpendicular; the mizzenmast centers 9'4" from the forward perpendicular; the intermediate mast position centers 5'2" from the forward perpendicular.

The unstayed masts step through the thwarts. The hole for the mainmast through the forward thwart is 2⁷/₈" in diameter. The thwart is 8⁷/₈" wide, and its after edge is 3'0" from the forward perpendicular.

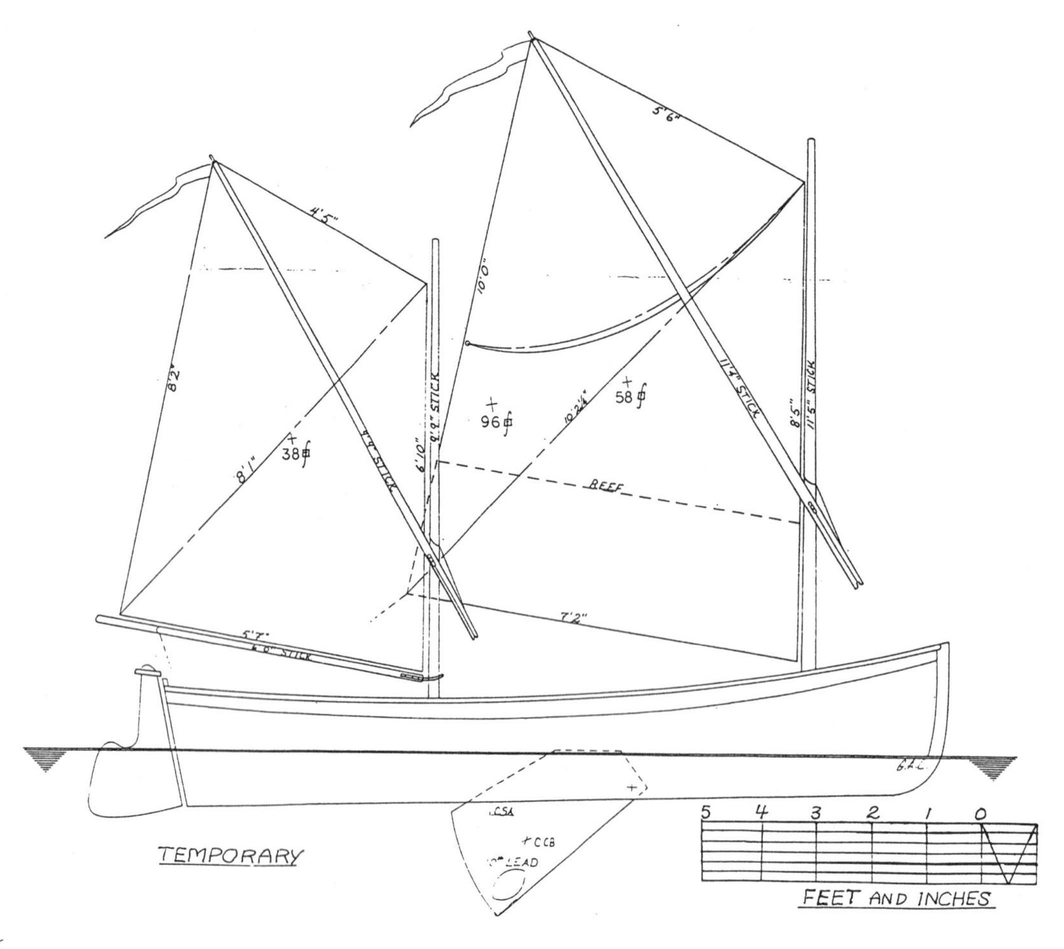

Temporary ***has two rowing stations and three thwarts all supported by natural-crook knees. The mainmast can be stepped in the bow thwart or at the forward edge of the middle thwart, and the mizzenmast steps through the aft thwart (Sharon Brown photograph).***

The intermediate thwart is 8 1/2" in width and its after edge is 5'10 1/2" from the forward perpendicular. This locates the mast in the forward edge of the thwart, exposing the forward half of the mast which is held in place by a circular iron band, attached to the thwart, through which the mast passes. There seems to be no good reason for duplicating this arrangement in the replication of this boat. Instead widen the thwart to 10" and move it forward 2" to bring the mast hole within the thwart.

The after thwart is 8" wide and the hole for the mizzenmast is 2 1/2" in diameter. The after edge of this thwart is 9'7" from the forward perpendicular.

The after end of the centerboard trunk is 9'2" from the forward perpendicular, and it is 3'10" in length.

The dimensions of the mainsail are as follows. The diagonal distance from the clew to the throat is 10'2 1/4". Length at the head 5'6"; leach 10'0"; hoist 8'5"; foot 7'2". Mast diameter at the bury through the thwart is 2 3/4" tapering to approximately half that at the head. Length of sprit is 11'4".

Dimensions of the mizzen: diagonal, clew to throat, 8'1"; head 4'5"; leach, 8'2"; hoist 6'10"; foot 5'7". Boom 8'10". Mizzenmast, diameter at the bury through the thwart 2 7/16"; 9'9" long, tapering to slightly more than half that diameter at the head. Length of sprit is 9'4".

Overall length of the boat 14'3". From forward perpendicular to Station 1, 1'9"; from Station 1 to Station 2, 1'6"; from Station 2 to Station 3, 1'8"; from Station 3 to Station 4, 1'9"; from Station 4 to Station 5, 1'8"; from Station 5 to Station 6, 1'8"; from Station 6 to Station 7, 1'9"; from Station 7 to Station 8, 1'7"; and from Station 8 to the after perpendicular, 10".

Offsets measured to the inside of the planking.

Planking, cedar, sided 1/2" except the sheer blank which is sided 3/4".

Thwart knees on either side of the thwarts, sided 3/4". Inwales sided 7/8" and molded 1". Ribs sided 3/4", molded 3/4". Thwart risers sided 5/8", molded 2". Floors

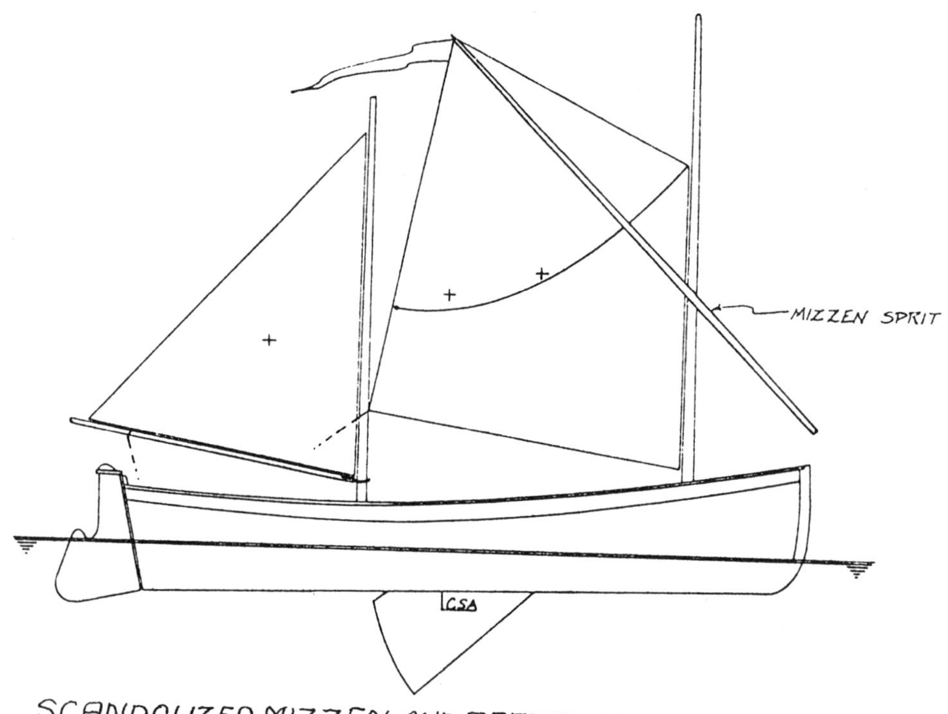
MIZZEN SPRIT
CSA
SCANDALIZED MIZZEN AND REEFED MAIN

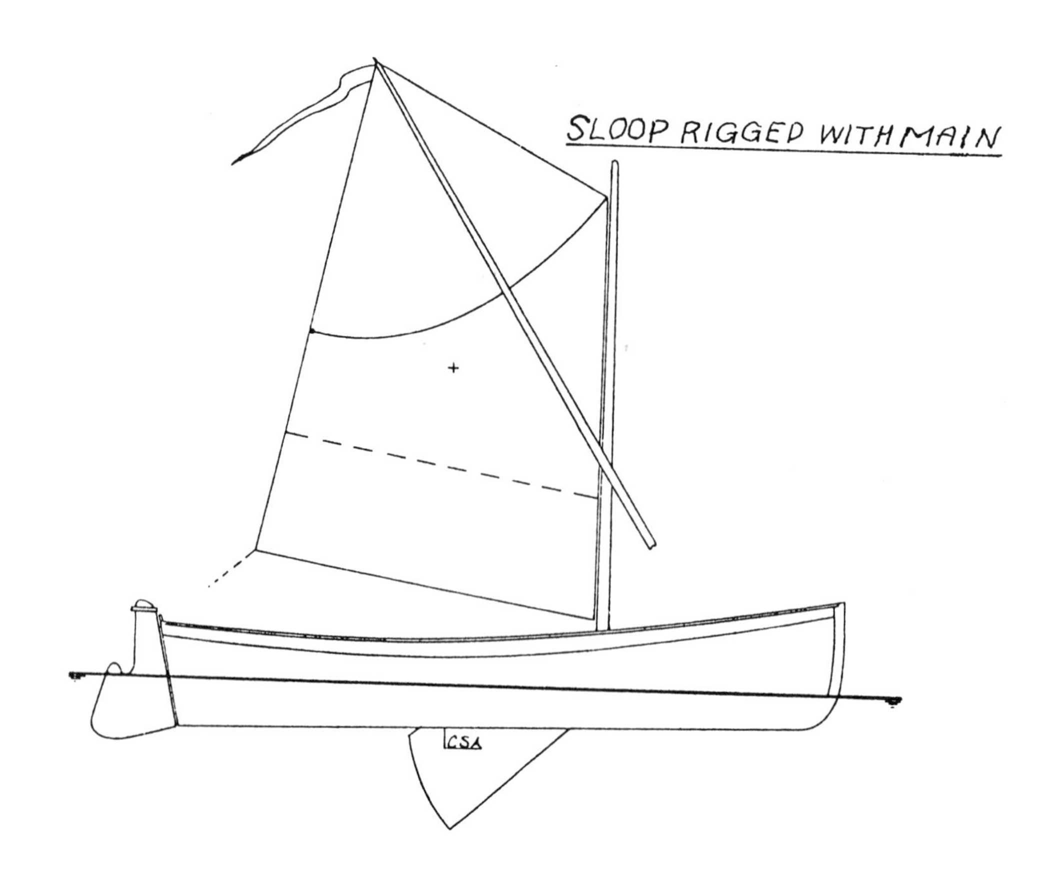
SLOOP RIGGED WITH MAIN
CSA

amidships sided 7/8", molded $1^1/_8$" in the middle and 3/4" at the ends. Bottom board or plank keel sided 1", molded widths given in the offset table. False or sacrificial bottom board sided 1/2".

Stem sided 2". Apron sided 2". Breasthook sided 1". Stem knee sided 3". Transom sided 3/4". Stern post sided 2". Stern knee sided $2^1/_2$".

The drawings reproduced here were sent to me by Andy in 1974 and 1976, as previously mentioned, except they have been cut up and rearranged to fit the pages of this book. Otherwise they have not been changed. Had they been reduced as drawn their detail would have reduced too small to be legible, and they would have been of little use to the reader.

It hardly needs to be said that the foregoing is by no means complete, yet it does provide enough information to build from. None of the traditional nineteenth-century small craft that were developed along the Maine coast, the wherries, Reach boats, peapods and others were ever exactly duplicated. There was plenty of variation from builder to builder and from boat to boat. The fishermen and country boat builders who built them did not have lines and offsets to work from. Sometimes they had molds borrowed from other builders or handed down from those who had gone before. But mostly they built from "eye," from visual images stored in memory and acquired over a lifetime of using and observing the boats upon which their livelihood and lives depended. When they looked at boats they saw much the casual observer missed entirely. There is so much that can't be put into words or included in drawings. For those who would replicate these classics there is nothing like being able to see and study examples of the actual boats themselves. This is, or should be, a principal motivation for preserving them in museum collections, just as *Temporary* is now able to be seen and studied in Mystic Seaport's small craft collection.

Temporary, *rebuilt for sail in 1975-76, swings at her mooring in Castine, Maine. Here the sprit-rigged mainsail is stepped in the middle thwart, the so-called sloop position, and the Egyptian cotton sail is brailed and furled on the mast (photograph courtesy of Andy Chase).*

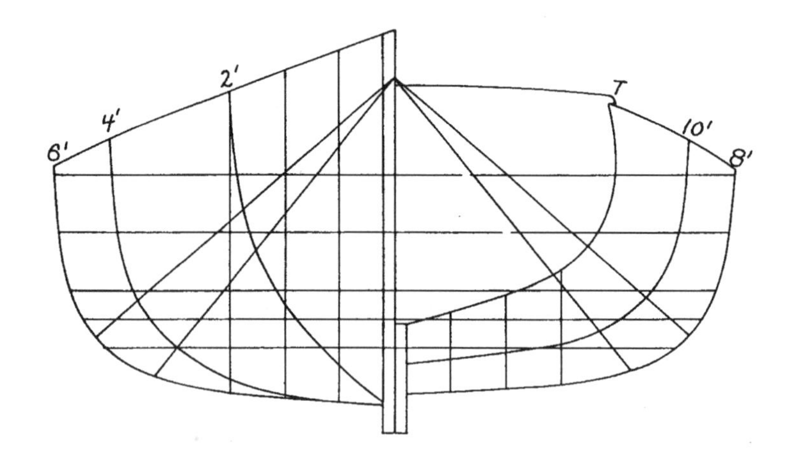

Chapter Four

Dion Tender

The boat from which the accompanying lines and construction details were taken was Fred Dion's favorite tender. He considered it the best model for heavy-duty, general-purpose service he had ever seen.

That is high praise, indeed, for Fred had seen and rowed a lot of yacht tenders in his more than 50 years at the helm of the Frederick J. Dion Yacht Yard in Salem, Massachusetts. Built up from scratch with Fred's own hands, so to speak, the yard became one of the finest on the Northeast coast.

One incident illustrating the superior seakeeping qualities of this tender was related by Rich St. Pierre, Fred's nephew. During the height of the 1938 hurricane that devastated the East Coast, Fred and Rich, who was then in his late teens, were out in this tender checking mooring lines and assessing damage along the waterfront.

Gusts of over 100 m.p.h. made it impossible for one man to row unassisted, for the wind tended to lift the oars out of the locks and blow them straight up in the air. Rich, facing his uncle, had to grasp the oars close to the blades to hold them down, pulling on them at the same time. Only then were the two of them able to keep the oars in the water and make some forward progress. Meanwhile, the boat rode the waves like a duck, with no solid water coming aboard.

The builder of this boat is not known. Sometime in the early 1930s, I was told, Fred brought several of these boats down from Maine. He disposed of all except this one as tenders or general-purpose rowing craft for fishing and similar use. They make excellent boats for this, due to their stiffness, stability and carrying capacity, all of which are exceptional for boats in this size and range. And, of course, they also row well and are steady and dry.

This classic 12-foot yacht tender built by master boat builder Arthur M. Spurling of Islesford, Little Cranberry Isle, Maine, is very similar to the Dion tender in principal dimensions, although it is a little narrower and not as flat in the floors (Robert C. Sutter photograph).

Judging from the Spurling rowboat and a similar 12-foot skiff whose lines were published in the September 1972 issue of *National Fisherman* and reprinted in my *Building Classic Small Craft, Vol. 1,* the Dion tender could have been built by Arthur M. Spurling of Islesford, Little Cranberry Isle, Maine. Spurling, born in 1878, is still remembered for his superlative rowboats, which he continued to build until he was 70 or thereabouts. Some of them are still in use, I am told, and are highly prized by their owners.

The Dion boat and the Spurling boat in my book are virtually identical in their principal dimensions. Fred's tender, however, differs in being a bit wider and flatter in its floors. It is also slightly harder and quicker in the turn at the bilge, as well as a shade sharper in its entrance. The Dion boat, in consequence, would probably be a little stiffer and steadier, besides being able to carry somewhat more. At the same time, it should lose nothing in ease of rowing and handling.

Another Maine-built boat with some similarity to this tender is the North Haven peapod described and discussed at length in the 1986 January, February and March issues of the *National Fisherman* and in my book *Classic Small Craft You Can Build.* This resemblance is not so unusual, since Islesford, where the Spurling boats were built, and North Haven are only about 35 or 40 miles apart as the seabirds fly.

Local boating conditions are nearly identical in both places. Leaving aside the fact that the peapod is a double-ender, the two models have much the same midsectional shape, which accounts in large measure for their stiffness, steadiness and load-carrying capacity.

But even more notable is a similarity in construction. In neither boat is the keel rabbeted to take the garboards, which fit snugly against it (with a slightly beveled seam for caulking) and lie flush inside with the top of the keel. Likewise, neither of these boats has any cross floors. Instead, bent frames passing across—and fastened to—the top of the keel hold the two halves of the boat together.

In the peapod, short intermediate cross timbers were located between the main, full length timbers, both to reinforce the boat and to provide something to step on in place of inside bottom boards. There are no like intermediates in the Dion tender, but in the drawing I show optional ones. These short cross timbers **(4)** are easily installed, and I recommend them.

Inspection of the boat after long years of use shows that reinforcement here is needed and that short pieces similar to butt blocks had been added at some time between the frames, on top of the keel. Having gone into the boat with little difficulty, such intermediates will be out of sight under the floorboards and will not add enough extra weight to matter.

As carvel-planked boats go, the Dion tender is not a difficult one to plank. The shape of the hull, with its fine entrance, is

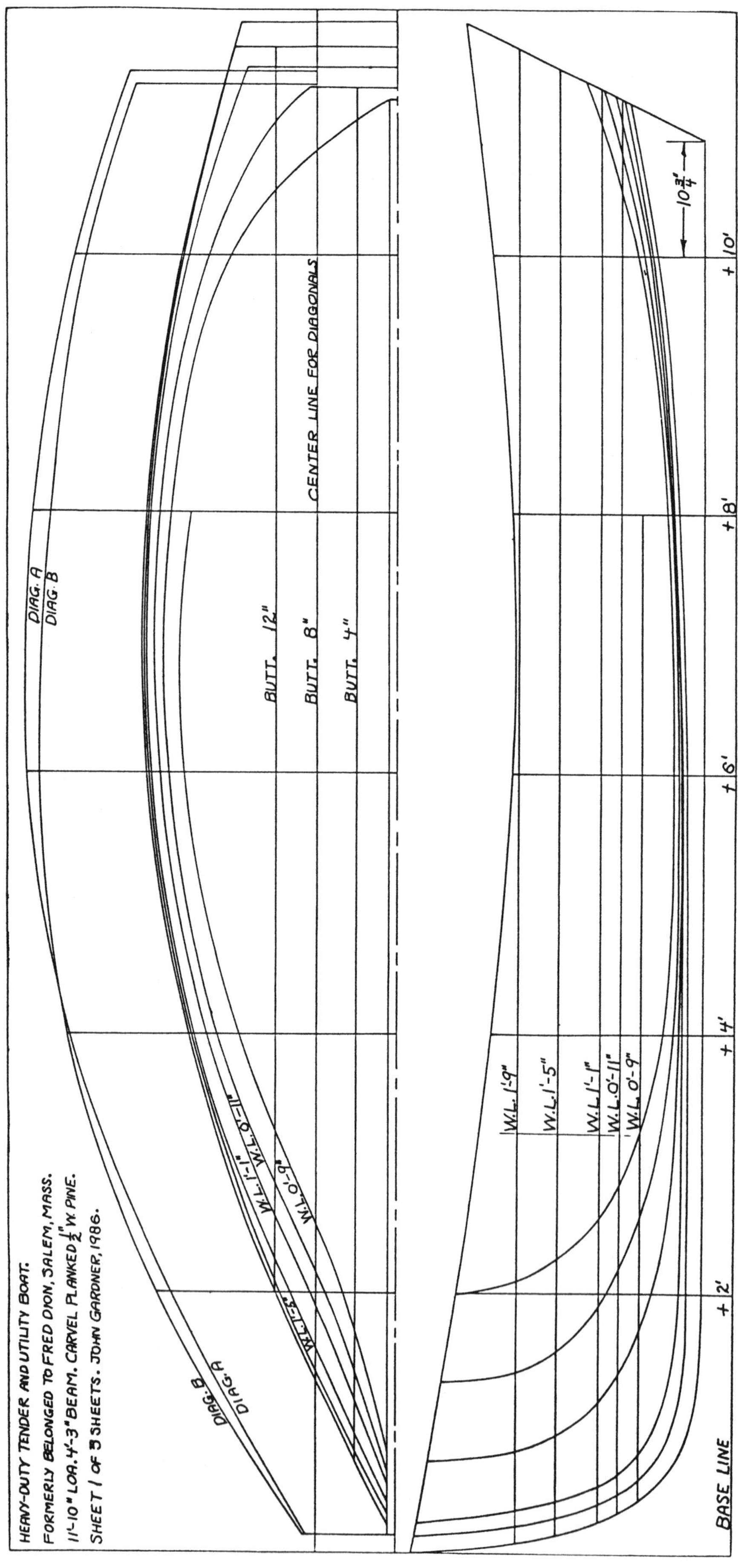
HEAVY-DUTY TENDER AND UTILITY BOAT.
FORMERLY BELONGED TO FRED DION, SALEM, MASS.
11'-10" LOA. 4'-3" BEAM. CARVEL PLANKED 1/2" W. PINE.
SHEET 1 OF 3 SHEETS. JOHN GARDNER, 1986.
DIAG. A
DIAG. B
BUTT. 12"
BUTT. 8"
BUTT. 4"
CENTER LINE FOR DIAGONALS
W.L. 1'-5"
W.L. 1'-1"
W.L. 0'-11"
W.L. 0'-9"
W.L. 1'-9"
W.L. 1'-5"
W.L. 1'-1"
W.L. 0'-11"
W.L. 0'-9"
BASE LINE
2'
4'
6'
8'
10'
10 3/4"

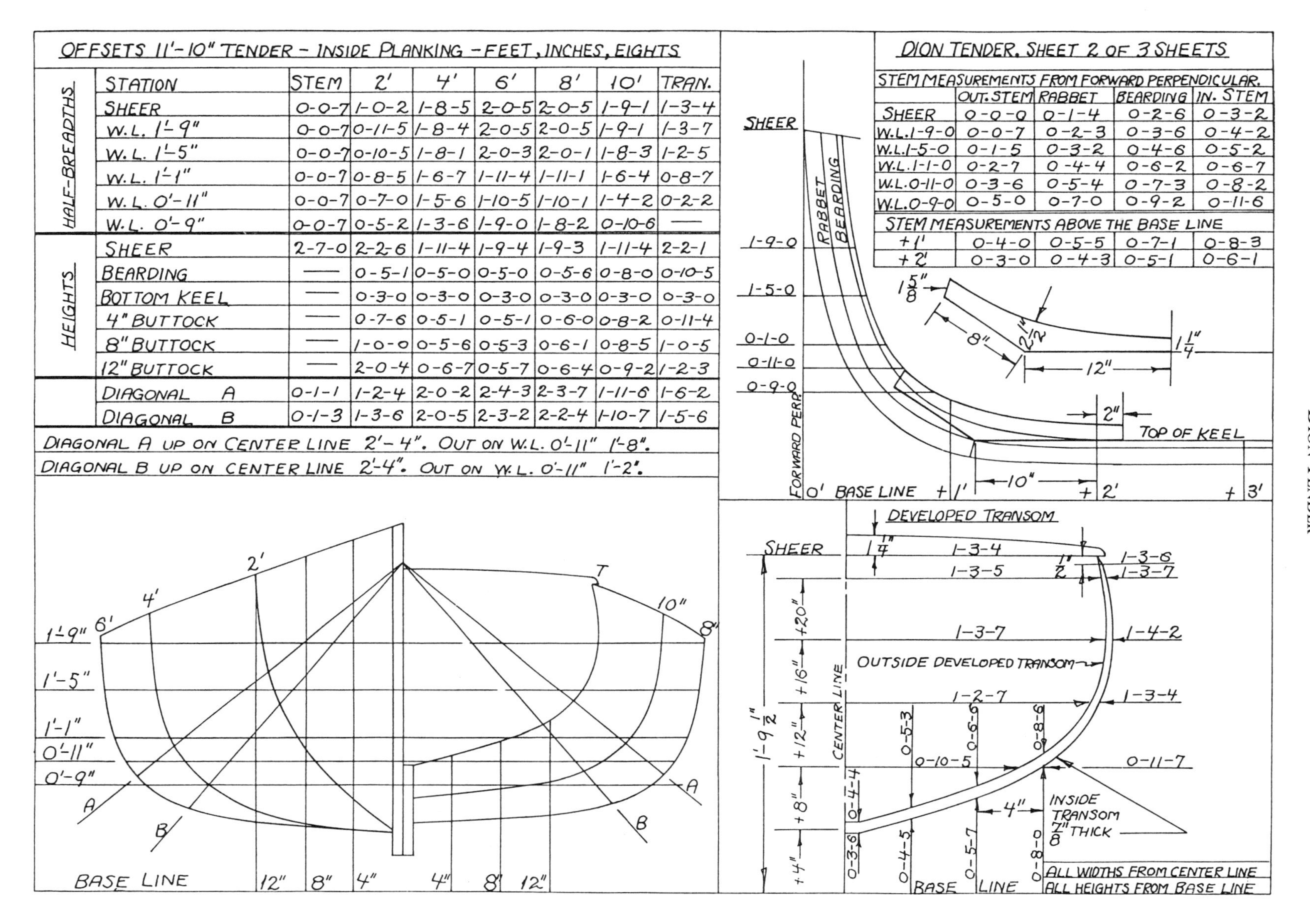

OFFSETS 11'-10" TENDER - INSIDE PLANKING - FEET, INCHES, EIGHTS

	STATION	STEM	2'	4'	6'	8'	10'	TRAN.
HALF-BREADTHS	SHEER	0-0-7	1-0-2	1-8-5	2-0-5	2-0-5	1-9-1	1-3-4
	W.L. 1'-9"	0-0-7	0-11-5	1-8-4	2-0-5	2-0-5	1-9-1	1-3-7
	W.L. 1'-5"	0-0-7	0-10-5	1-8-1	2-0-3	2-0-1	1-8-3	1-2-5
	W.L. 1'-1"	0-0-7	0-8-5	1-6-7	1-11-4	1-11-1	1-6-4	0-8-7
	W.L. 0'-11"	0-0-7	0-7-0	1-5-6	1-10-5	1-10-1	1-4-2	0-2-2
	W.L. 0'-9"	0-0-7	0-5-2	1-3-6	1-9-0	1-8-2	0-10-6	—
HEIGHTS	SHEER	2-7-0	2-2-6	1-11-4	1-9-4	1-9-3	1-11-4	2-2-1
	BEARDING	—	0-5-1	0-5-0	0-5-0	0-5-6	0-8-0	0-10-5
	BOTTOM KEEL	—	0-3-0	0-3-0	0-3-0	0-3-0	0-3-0	0-3-0
	4" BUTTOCK	—	0-7-6	0-5-1	0-5-1	0-6-0	0-8-2	0-11-4
	8" BUTTOCK	—	1-0-0	0-5-6	0-5-3	0-6-1	0-8-5	1-0-5
	12" BUTTOCK	—	2-0-4	0-6-7	0-5-7	0-6-4	0-9-2	1-2-3
	DIAGONAL A	0-1-1	1-2-4	2-0-2	2-4-3	2-3-7	1-11-6	1-6-2
	DIAGONAL B	0-1-3	1-3-6	2-0-5	2-3-2	2-2-4	1-10-7	1-5-6

DIAGONAL A UP ON CENTER LINE 2'-4". OUT ON W.L. 0'-11" 1'-8".
DIAGONAL B UP ON CENTER LINE 2'-4". OUT ON W.L. 0'-11" 1'-2".

DION TENDER. SHEET 2 OF 3 SHEETS

STEM MEASUREMENTS FROM FORWARD PERPENDICULAR.	OUT. STEM	RABBET	BEARDING	IN. STEM
SHEER	0-0-0	0-1-4	0-2-6	0-3-2
W.L. 1-9-0	0-0-7	0-2-3	0-3-6	0-4-2
W.L. 1-5-0	0-1-5	0-3-2	0-4-6	0-5-2
W.L. 1-1-0	0-2-7	0-4-4	0-6-2	0-6-7
W.L. 0-11-0	0-3-6	0-5-4	0-7-3	0-8-2
W.L. 0-9-0	0-5-0	0-7-0	0-9-2	0-11-6
STEM MEASUREMENTS ABOVE THE BASE LINE				
+1'	0-4-0	0-5-5	0-7-1	0-8-3
+2'	0-3-0	0-4-3	0-5-1	0-6-1

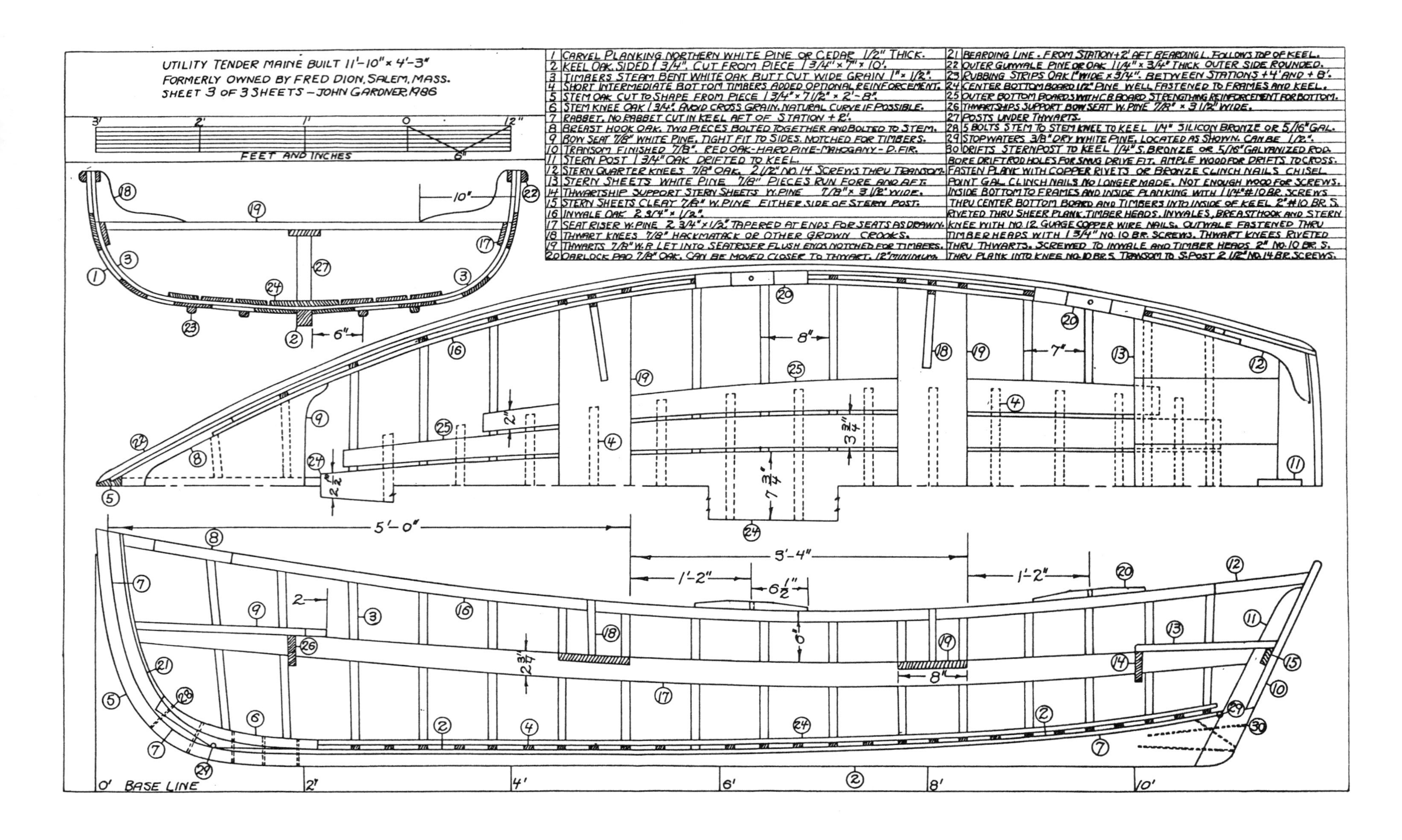
UTILITY TENDER MAINE BUILT 11'-10" × 4'-3"
FORMERLY OWNED BY FRED DION, SALEM, MASS.
SHEET 3 OF 3 SHEETS – JOHN GARDNER, 1986
FEET AND INCHES
1 CARVEL PLANKING NORTHERN WHITE PINE OR CEDAR 1/2" THICK.
2 KEEL OAK, SIDED 1 3/4". CUT FROM PIECE 1 3/4" × 7" × 10'.
3 TIMBERS STEAM BENT WHITE OAK BUTT CUT WIDE GRAIN 1" × 1/2".
4 SHORT INTERMEDIATE BOTTOM TIMBERS ADDED OPTIONAL REINFORCEMENT.
5 STEM OAK CUT TO SHAPE FROM PIECE 1 3/4" × 7 1/2" × 2'-8".
6 STEM KNEE OAK 1 3/4". AVOID CROSS GRAIN. NATURAL CURVE IF POSSIBLE.
7 RABBET. NO RABBET CUT IN KEEL AFT OF STATION + 2'.
8 BREAST HOOK OAK. TWO PIECES BOLTED TOGETHER AND BOLTED TO STEM.
9 BOW SEAT 7/8" WHITE PINE. TIGHT FIT TO SIDES. NOTCHED FOR TIMBERS.
10 TRANSOM FINISHED 7/8". RED OAK-HARD PINE-MAHOGANY-D.FIR.
11 STERN POST 1 3/4" OAK DRIFTED TO KEEL.
12 STERN QUARTER KNEES 7/8" OAK. 2 1/2" NO.14 SCREWS THRU TRANSOM.
13 STERN SHEETS WHITE PINE 7/8" PIECES RUN FORE AND AFT.
14 THWARTSHIP SUPPORT STERN SHEETS W.PINE 7/8" × 3 1/2" WIDE.
15 STERN SHEETS CLEAT 7/8" W.PINE EITHER SIDE OF STERN POST.
16 INWALE OAK 2 3/4" × 1/2".
17 SEAT RISER W.PINE 2 3/4" × 1/2". TAPERED AT ENDS FOR SEATS AS DRAWN.
18 THWART KNEES 7/8" HACKMATACK OR OTHER GROWN CROOKS.
19 THWARTS 7/8" W.P. LET INTO SEATRISER FLUSH ENDS NOTCHED FOR TIMBERS.
20 OARLOCK PAD 7/8" OAK. CAN BE MOVED CLOSER TO THWART. 12" MINIMUM.
21 BEARDING LINE. FROM STATION + 2' AFT BEARDING L. FOLLOWS TOP OF KEEL.
22 OUTER GUNWALE PINE OR OAK 1 1/4" × 3/4" THICK OUTER SIDE ROUNDED.
23 RUBBING STRIPS OAK 1" WIDE × 3/4". BETWEEN STATIONS + 4' AND + 8'.
24 CENTER BOTTOM BOARD 1/2" PINE WELL FASTENED TO FRAMES AND KEEL.
25 OUTER BOTTOM BOARDS WITH C.B. BOARD STRENGTHING REINFORCEMENT FOR BOTTOM.
26 THWARTSHIPS SUPPORT BOW SEAT W.PINE 7/8" × 3 1/2" WIDE.
27 POSTS UNDER THWARTS.
28 5 BOLTS STEM TO STEM KNEE TO KEEL 1/4" SILICON BRONZE OR 5/16" GAL.
29 STOPWATERS 3/8" DRY WHITE PINE. LOCATED AS SHOWN. CAN BE 1/2".
30 DRIFTS STERNPOST TO KEEL 1/4" S.BRONZE OR 5/16" GALVANIZED ROD.
BORE DRIFT ROD HOLES FOR SNUG DRIVE FIT. AMPLE WOOD FOR DRIFTS TO CROSS.
FASTEN PLANK WITH COPPER RIVETS OR BRONZE CLINCH NAILS CHISEL
POINT GAL. CLINCH NAILS NO LONGER MADE. NOT ENOUGH WOOD FOR SCREWS.
INSIDE BOTTOM TO FRAMES AND INSIDE PLANKING WITH 1 1/4" #10 BR. SCREWS
THRU CENTER BOTTOM BOARD AND TIMBERS INTO INSIDE OF KEEL 2" #10 BR. S.
RIVETED THRU SHEER PLANK, TIMBER HEADS, INWALES, BREASTHOOK AND STERN
KNEE WITH NO.12 GUAGE COPPER WIRE NAILS. OUTWALE FASTENED THRU
TIMBER HEADS WITH 1 3/4" NO.10 BR. SCREWS. THWART KNEES RIVETED
THRU THWARTS. SCREWED TO INWALE AND TIMBER HEADS 2" NO.10 BR. S.
THRU PLANK INTO KNEE NO.10 BR.S. TRANSOM TO S.POST 2 1/2" NO.14 BR. SCREWS.
5'-0"
3'-4"
1'-2"
6 1/2"
8"
7"
10"
6"
0' BASE LINE
2'
4'
6'
8'
10'

such that hard twists and bends in the planking are avoided. It is therefore unnecessary, I would judge, to steam any of the planks, which finish 1/2" thick. Northern white cedar or select northern white pine (without sapwood) are recommended. Other timber of more or less equivalent characteristics may be substituted when cedar and pine are unavailable.

Several of the planks will require considerable "backing," or hollowing, on the inside to make them fit the curvature of the frames at the turn of the bilge. In laying out the shape and width of the eight planks, those that come at the turn of the bilge should be made narrower than the rest.

It is obvious that the wider these planks are, the more wood will have to be planed away in backing them out. Increased width, then, means more labor and thicker boards to start with. Yet even when these planks are kept quite narrow, as much as 1/4" to 3/8" of extra thickness is needed because of the hard turn in the bilge. It is a common mistake to overlook this and order planking stock of the same thickness, without making allowance for backing.

The cross timbers that run in continuous lengths from one side of the boat to the other should be of the best white oak bending stock. If possible, this wood should be cut from the butt logs of vigorous, young, fast-growing trees that show large annual rings with dense, heavy structure. Because these timbers are relatively thin in section, they should bend in easily after steaming and won't be prone to breakage.

Although the stock for the cross timbers was 1/2" thick in the original Dion boat, I am recommending that this dimension be increased by 1/8", to a total of 5/8". This way, 1"-long screws won't break through the inside of the timbers when they're set deep enough into the planking to allow for finish planing and sanding on the outside of the hull. There will even be some room left for puttying over the heads of the screws.

The specifications of these fastenings are: size No. 10; material, silicon bronze. The galvanized iron clinch nails used when the Dion boat was built have not been on the market for many years. Bronze clinch nails may be substituted if any can be found. However, bronze wood screws are available almost everywhere.

They are easy to put in, they pull and hold well and they last. Although they are high-priced, the extra expense is not excessive when compared with other construction costs. Galvanized steel wood screws are not to be considered, for in sizes as small as No. 10s, the galvanizing tends to fill up the threads. As a result, they don't hold as they should, and they soon rust in spite of the galvanizing.

Inwales, thwart knees, stern quarter knees, and the breasthook should be copper-riveted over burrs. Fortunately, copper wire nails and burrs are still available; 2 1/2" nails of 11-gauge wire (English Standard Wire Gauge) would be about right and can be clipped off to the required length for riveting. Where the fastenings must go through planking, timbers, inwales and the top of the thwart knee, 2 1/2" might be a little short, in which case 3", 10d nails would be required.

In spite of the bronze plank fastenings, galvanized drift rod has been specified to pin the stern post to the keel. There is no danger of electrolysis developing in small boats without electrical equipment so long as the diverse metals do not come in direct contact with each other. Galvanized drifts grip the wood better than bronze.

The longitudinal flooring in the Dion boat rests directly on the top of the bent frames and is nailed to them as well as to the top of the keel. This flooring performs an important structural function by supplying the bottom with strength and stiffness, reinforcement which is needed in view of the absence of cross floors and the bottom's extra width. Driving 1 1/2" No. 10 screws into the bent frames would be about right, putting 1 3/4" No. 10s into the top of the keel.

A grown crook for a one-piece breasthook would be a rare find, so this component will normally be made up from two pieces. Needless to say, these two halves must be well fastened together—either with through-bolts, before the breasthook is put into the boat; with a metal plate screwed across the joint on its underside, out of sight; or with both.

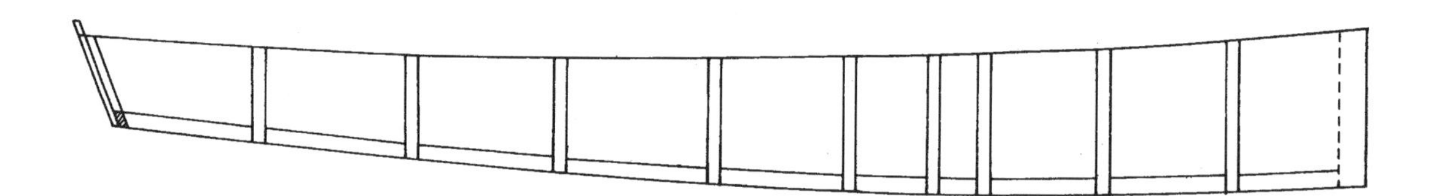

Chapter Five

15-Foot Clamming Skiff

This 15-foot clamming skiff is an authentic example of a type of small fishing boat once common in the waters of Long Island Sound, along Connecticut's south shore.

These skiffs resembled the larger sailing sharpies of the New Haven oystermen, but oars instead of sails were their principal means of propulsion. However, they did carry a small sail that could be raised well forward in the boat to help out when the wind favored from astern. Built without centerboards, such skiffs could not hold a course on the wind. Still, when loaded, the deep draft of their heavily rockered bottoms must have retarded side slip considerably.

In drawing up the lines and the construction details for this account, we have followed quite closely a skiff that was donated to Mystic Seaport Museum by the well-known Captain John Stone of Norwalk, Connecticut, in 1987. He owned the boat for many years, and at one time, when he lived on a neighboring island some distance off the shore, the skiff was used to get to and from the mainland.

It was during this period that the boat was converted to power by the addition of a small, single-cylinder ("one-lung"), make-and-break gasoline engine, since removed. Although this type of skiff dates from the nineteenth century, before gas engines for small workboats had been developed, such boats as Captain Stone's, especially the larger ones of 18 feet and up, are well-suited for small inboard engines.

It was the renowned Stamford, Connecticut, naval architect and schooner-yacht designer, George H. Stadel, Jr., who first called my attention to these clamming skiffs. His recollections of these boats, which made a strong impression on him as a small boy, go back before World War I. In

Captain John Stone of Norwalk, Connecticut, owned this heavily built 15-foot clamming skiff once common along the shoreline in Long Island Sound. The deeply rockered bottom leaves the stem and stern out of the water when the skiff is lightly loaded and helps to move the hull through the water easily when under a load (Judy Beisler photograph, M.S.M.).

particular, he recalls a group of six fishermen, each with his shack on the shore and one of these skiffs. All six went off every day, despite bad weather, for fish, clams, horseshoe crabs, oysters—whatever bounty the Sound offered.

On first impression, this skiff might seem to be an obsolete craft of little practical value today, a boat hardly worthy of serious attention, except as an object of historical curiosity. Yet, on further consideration, it appears there may be a place even now in the fisheries—and elsewhere—for a limited number of such skiffs, or boats much like them.

Heavily built, they are able to carry big loads and to stand up for years under the hardest kinds of use and abuse. Their deeply rockered bottoms have the proven ability to move through the water easily with a minimum of fuss at low and moderate speeds, even when heavily loaded. In fact, they are so substantially rockered that, when light, the bottom of the stem clears the surface of the water and the stern rises well above it.

Construction is quick and simple. Ordinary sawn lumber, nails, good oil-base house paint, and a few hand tools are all that are needed, although power saws, an electric drill and access to a power plane would speed up the process. Almost anyone with modest carpentry experience should be able to put one of these skiffs together in no more than a week. Of course, if there are extras like a centerboard, rudder, spars or an engine installation, the job would take longer.

Various possibilities for current use suggest themselves, and many are quite attractive. These boats would convert for sail quite easily and could support a respectable spread of canvas. While not the fastest craft afloat, they would not be the

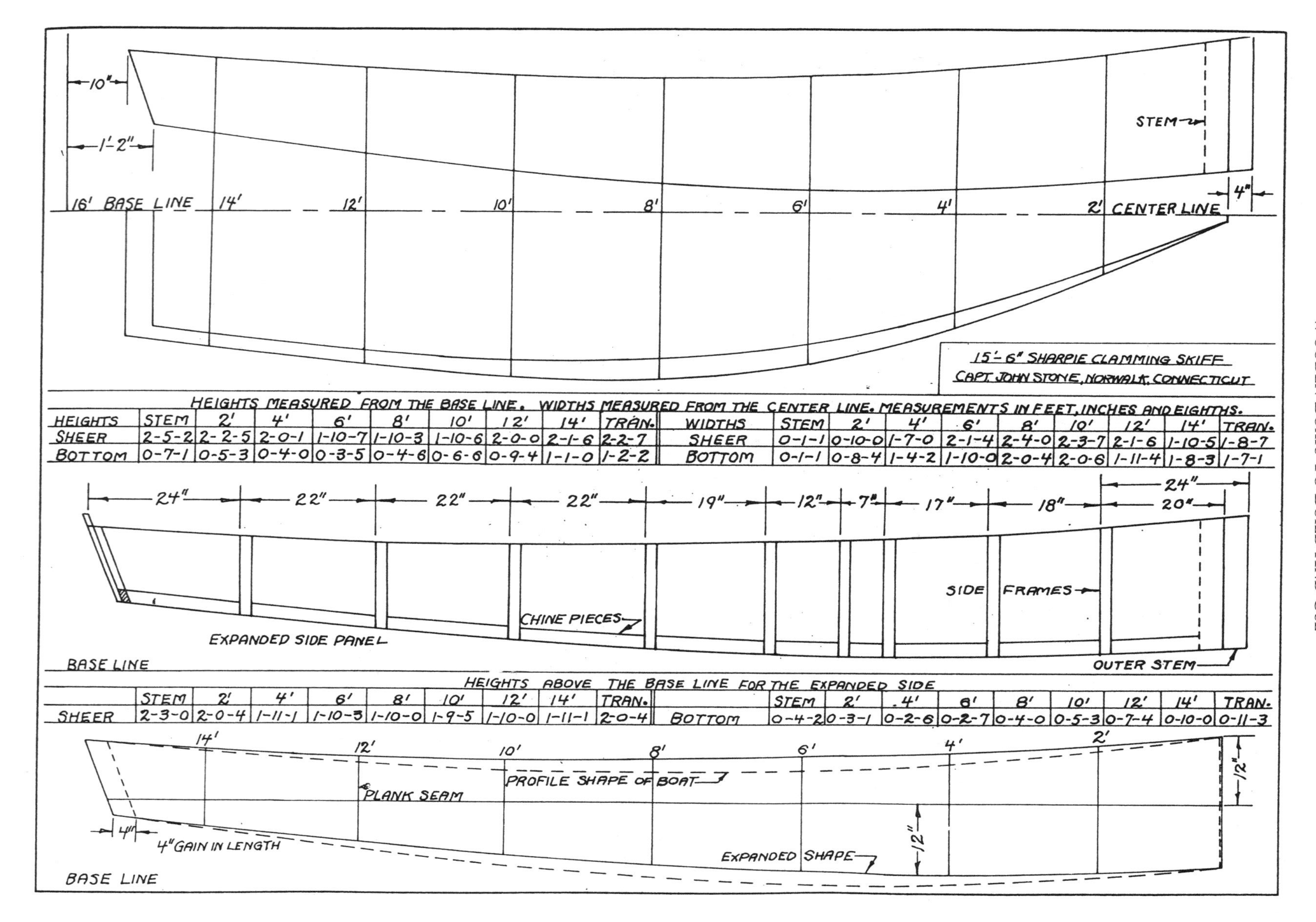

HEIGHTS MEASURED FROM THE BASE LINE. WIDTHS MEASURED FROM THE CENTER LINE. MEASUREMENTS IN FEET, INCHES AND EIGHTHS.

HEIGHTS	STEM	2'	4'	6'	8'	10'	12'	14'	TRAN.
SHEER	2-5-2	2-2-5	2-0-1	1-10-7	1-10-3	1-10-6	2-0-0	2-1-6	2-2-7
BOTTOM	0-7-1	0-5-3	0-4-0	0-3-5	0-4-6	0-6-6	0-9-4	1-1-0	1-2-2

WIDTHS	STEM	2'	4'	6'	8'	10'	12'	14'	TRAN.
SHEER	0-1-1	0-10-0	1-7-0	2-1-4	2-4-0	2-3-7	2-1-6	1-10-5	1-8-7
BOTTOM	0-1-1	0-8-4	1-4-2	1-10-0	2-0-4	2-0-6	1-11-4	1-8-3	1-7-1

HEIGHTS ABOVE THE BASE LINE FOR THE EXPANDED SIDE

	STEM	2'	4'	6'	8'	10'	12'	14'	TRAN.
SHEER	2-3-0	2-0-4	1-11-1	1-10-3	1-10-0	1-9-5	1-10-0	1-11-1	2-0-4
BOTTOM	0-4-2	0-3-1	0-2-6	0-2-7	0-4-0	0-5-3	0-7-4	0-10-0	0-11-3

Allen Storms' nineteenth-century clamming skiff, built from lines published in the March 1988 issue of* National Fisherman, *the same plan reproduced here. Storms expanded the hull to 16 feet, added a samson post, outboard motor well, chine rub strips and flotation (Allen Storms photograph).

slowest either. Perhaps the adaptability of these skiffs for low-speed powering would be appealing. In this case, a number of today's small, efficient inboard engines would be appropriate for a boat of this kind and could be easily installed.

It hardly needs saying that this hull is entirely unsuited for an outboard motor hung over the stern. But, installed in an inboard well, located toward the center of the boat, such a power plant would likely move the boat at sufficient speed to satisfy some requirements. For example, it would serve well as a source of auxiliary power if the skiff were to be sailed, and a small outboard would cost ever so much less than an inboard engine.

The materials used to build these skiffs were few and simple: northern white pine for the sides, bottom, transom, waterways, coaming and seats, with oak for the framing. Of course, in reproducing this boat the builder is not limited to white pine and oak. Cedar, a favorite boat building wood, resists rot better than most other woods and is easy to work. However, it is softer and weaker than pine, does not stand up as well to hard use and soaks up water excessively, picking up extra weight. For a boat that is left in the water much of the time, this can be a real drawback. In the South, the local hard pine, cypress and juniper are possibilities. In the Northwest, Douglas fir, Port Orford cedar and Alaskan yellow cedar would all be suitable.

For framing and for members that must absorb wear and impact—pieces like the outside bottom strip and the gunwale strips—oak is by far the best choice. Oak's ability to hold fastenings better than the softer woods is a prime consideration.

Originally, these boats were nailed throughout; screws were not used, and glue was unheard of. For the most part, nailing is still the easiest, quickest, cheapest and altogether best way to go. The exception would be where the combined thickness of the parts to be fastened together is not great enough to allow nails of sufficient length for good holding power. Examples are the 1"-thick side frames and the short chine pieces to be fastened to the 13/16" side planking,

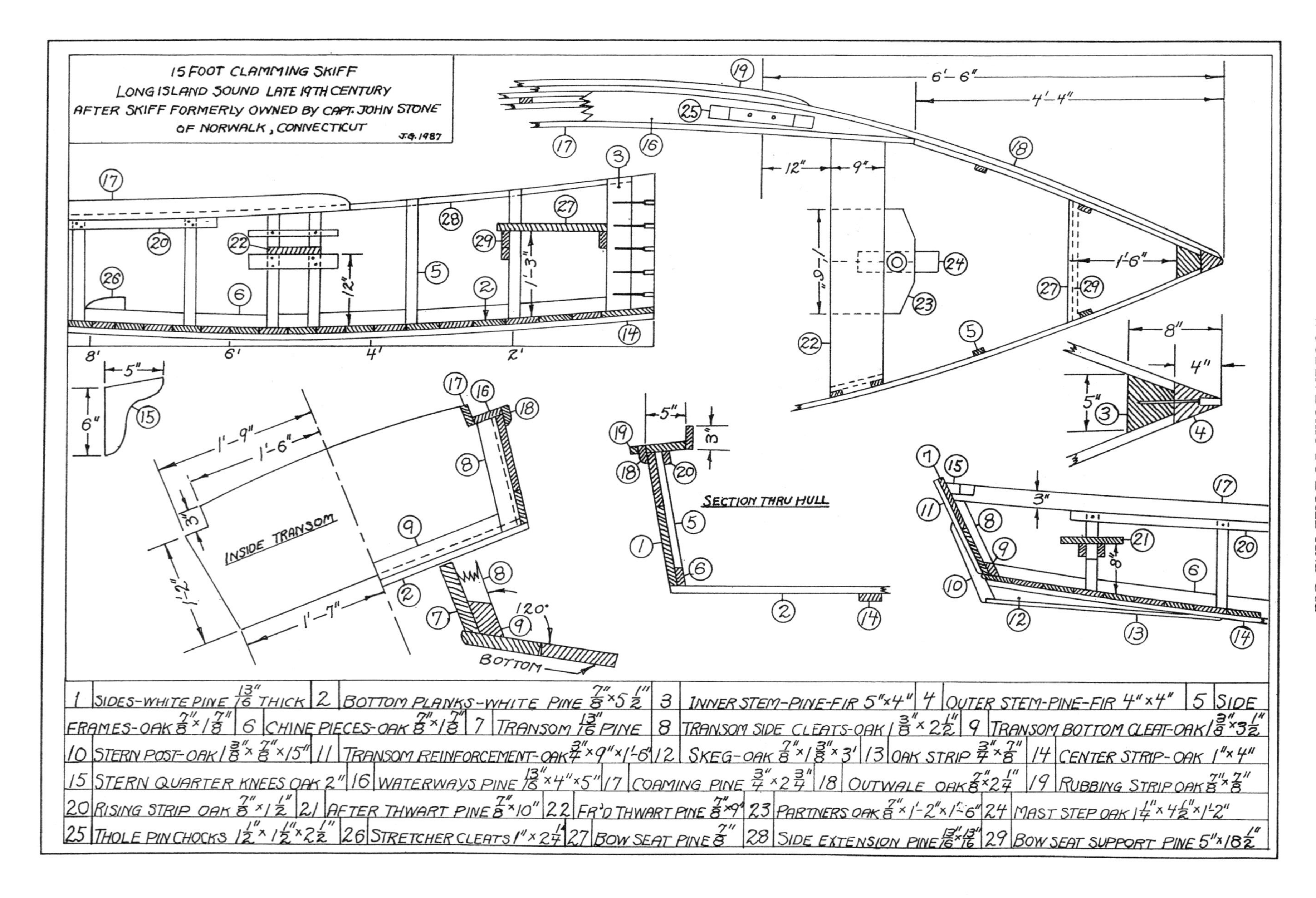
15 FOOT CLAMMING SKIFF
LONG ISLAND SOUND LATE 19TH CENTURY
AFTER SKIFF FORMERLY OWNED BY CAPT. JOHN STONE
OF NORWALK, CONNECTICUT
J.G. 1987
INSIDE TRANSOM
SECTION THRU HULL
BOTTOM
1 SIDES-WHITE PINE 13/16" THICK
2 BOTTOM PLANKS-WHITE PINE 7/8"×5 1/2"
3 INNER STEM-PINE-FIR 5"×4"
4 OUTER STEM-PINE-FIR 4"×4"
5 SIDE FRAMES-OAK 7/8"×1 7/8"
6 CHINE PIECES-OAK 7/8"×1 7/8"
7 TRANSOM 13/16" PINE
8 TRANSOM SIDE CLEATS-OAK 1 3/8"×2 1/2"
9 TRANSOM BOTTOM CLEAT-OAK 1 3/8"×3 1/2"
10 STERN POST-OAK 1 3/8"×7/8"×15"
11 TRANSOM REINFORCEMENT-OAK 3/4"×9"×1'-6"
12 SKEG-OAK 7/8"×1 3/8"×3'
13 OAK STRIP 3/4"×7/8"
14 CENTER STRIP-OAK 1"×4"
15 STERN QUARTER KNEES OAK 2"
16 WATERWAYS PINE 13/16"×4"×5"
17 COAMING PINE 3/4"×2 3/4"
18 OUTWALE OAK 7/8"×2 1/4"
19 RUBBING STRIP OAK 7/8"×7/8"
20 RISING STRIP OAK 7/8"×1 1/2"
21 AFTER THWART PINE 7/8"×10"
22 FR'D THWART PINE 7/8"×9"
23 PARTNERS OAK 7/8"×1'-2"×1'-6"
24 MAST STEP OAK 1 1/4"×4 1/2"×1'-2"
25 THOLE PIN CHOCKS 1 1/2"×1 1/2"×2 1/2"
26 STRETCHER CLEATS 1"×2 1/4"
27 BOW SEAT PINE 7/8"
28 SIDE EXTENSION PINE 13/16"×13/16"
29 BOW SEAT SUPPORT PINE 5"×1 8 1/2"

and the 1" keel strip to be attached to the 7/8" cross-planked bottom.

Formerly, special chisel-point nails would have been used. When their wide, thin ends came through, they were clinched over and turned back into the wood, making an especially strong fastening. The ordinary nails now available do not clinch well. Where length is limited, wood screws make neater fastenings that hold much better than short nails. (It is worth mentioning that the Stone boat's outside keel strip, which was undoubtedly a relatively recent replacement, had been fastened to the bottom with short nails that had started to work loose, and the strip had begun to come away at the bow end.)

Hot-dipped galvanized fastenings, well coated with zinc, will last 10 to 15 years, and sometimes longer. A workboat of this sort—subject to rough, hard use—would likely be worn out by then, and would probably have paid back its original cost several times over.

Of course, the builder could go to bronze screws and ring nails, if the additional cost for bronze were not a consideration. Something the builder really doesn't want are electroplated fastenings. They may look pretty when new, but they last only a little longer than bare steel.

One unusual feature of the skiff's construction is the short sections of chine fitted between the bottom ends of the side frames (or timbers) instead of continuous, full length, bent-in chines. Full length chines are not needed as strengthening members to reinforce the wide side planks, which measure 13/16" to 7/8" thick. The purpose of these short chine pieces is to provide secure nailing for the cross-planked bottom. It would not be enough merely to nail the ends of bottom planking into the bottom edges of the sides.

Inside, the heavy 7/8" thick bottom—without framing, gratings, or the like—provides maximum unobstructed working space. The rowing station is placed well forward, at the point from which the boat handles best, especially when heading into the wind.

Cross planking makes the best of bottoms for such a boat—that is, when it is correctly put on. For a skiff that is going to see hard use and carry heavy loads, a thicker bottom—measuring, say, 15/16" or 1"—might be preferable to one of 7/8". The width of the cross planking here measures $5^{1}/_{2}$", which is about right. Increase this width significantly, and shrinkage opens up the planking seams too much when the boat dries out. Wider planks also tend to cup. Conversely, bottom planks that are much narrower than $5^{1}/_{2}$" lose stiffness. A bottom of narrow planks is more apt to "work" under a heavy load in a seaway and thus take in water.

When the bottom planks are put on, they should fit tight on the inside, yet the seams should open up enough (about 1/8") on the outside to take one or two strands of caulking cotton. This is driven in with a caulking iron and set down hard, but not so hard as to break through on the inside. The seams are packed tight to within about 1/8" of the outside to allow for a final puttying with a general non-hardening, bottom-seam compound. First, however, the raw cotton in the seams is thoroughly saturated with red lead paint or an equivalent primer.

The bottom planking must be wedged up tight when it is fitted in place. One good way is to put it on in several separate sections no wider than can be pulled tight with a couple of door clamps. Slightly wedge-shaped openings the width of one plank are left between the sections; these are filled in last with tapered, wedge-shaped shutter planks that are driven up tight with a heavy hammer. Such planks both tighten and stiffen the entire bottom.

The ends of the bottom planks are nailed to the chine pieces and to the edges of the side planking with 3" nails. If galvanized fastenings are used, these will be ordinary 10d hot-dipped wire nails. For those driven into the edges of the side planking, small diameter lead holes are suggested as a precaution against splitting, but they must not be so large as to seriously reduce the holding power of the nails. (Drill some test holes at the start.) No such lead holes are needed in the short chine sections.

It hardly needs saying that before the bottom planks are put on, the lower edges of the side planking and the short chine sections must be planed true and fair to insure a touching fit for the bottom planking throughout. Also, a caulking seam should be planed onto the lower outside edge of the side planking to provide for driving a thread of caulking cotton after the bottom has been nailed on and the plank ends trimmed flush with the sides of the boat. When caulked, this seam is primed and puttied like the bottom seams.

There is no acceptable substitute for the method described here for caulking the bottom. The various modern flexible seam compounds now widely advertised and used for caulking do not stiffen and strengthen the bottom like well-driven cotton and, therefore, they just don't do a proper job.

It might seem that we have gone to unnecessary lengths and details in describing the proper construction of a cross-planked bottom. Yet a well-planked bottom is critical to the structural integrity of this boat, and such construction is now largely unknown to builders whose experience is limited to the shortcuts of plywood.

Construction begins by getting out the plank sides according to the expanded shape dimensioned in the drawing. This was taken from an accurately made scale half-model of the hull. Best results will be obtained by getting out the sides from two wide, full-length boards each. These will need to be approximately 12" wide at one end but can be several inches narrower at the other, as is often the case with live-edge flitch-sawn lumber. However, three narrower boards can also be used with wholly acceptable results.

The side planks are joined with the 7/8" x 1⁷/8" oak side frames, located as shown on the drawing and fastened with 1¹/2" No. 12 flat-head wood screws driven from the outside. The transom is then framed and fastened together, and the inner stem is beveled as shown. But before these can be assembled, a simple mold to fit the inside of the boat at the 8' station is needed. Nailed together from pieces of rough boards, this mold acts as a spacer to hold the sides apart when they are bent into place.

With this piece fabricated, the sides are securely fastened to the inner stem with 3" nails, bent around the mold at the 8' station, and nailed to the transom. Placed bottom up, this assembly is prepared to receive the bottom planking, as already described. But first, the short sections of chine are fitted between the frames and fastened in place from the outside with 1¹/2" No. 12 wood screws, the same as those used in the frames.

These chine pieces should not be put in until the sides have been bent in place. As the spacing between the sides determines the shape of the boat, it should be checked before the planking is applied, and temporary spreaders or cross spalls should be fitted as needed to hold the dimensions given in the drawings until the bottom is nailed in place.

Measuring 4" wide and 1" thick, the outside center strip of oak is not fastened in place until the bottom has been caulked and puttied, and at least one priming coat of paint has been applied. Just enough fastenings are put in from the outside to hold this strip in place until the boat is turned right side up. Then, it is well-fastened from the inside with three 1¹/2" No. 12 wood screws driven into each of the 5¹/2" bottom cross planks.

This strip greatly stiffens and strengthens the cross-planked bottom and protects its outside surface; for these reasons, it should be extra well-fastened. If a skiff were going to carry heavy loads and be dragged on and off the beach, it would be well to protect the bottom with two additional longitudinal strips, each located halfway between the chine and center strip on either side.

Finishing out the interior is nothing more than simple, straightforward carpentry. No special boat building skill or expertise is required. The drawings are pretty complete, and ordinary common sense should supply what is lacking. Probably no two of these boats were built exactly the same, even by the same builder. More significant than the precise dimensions of one particular boat are the general characteristics of the type it exemplifies.

I hardly need say this to builders, whether experienced or inexperienced. Both have previously proved themselves neither reluctant nor backward in making changes to meet their needs or their own ideas.

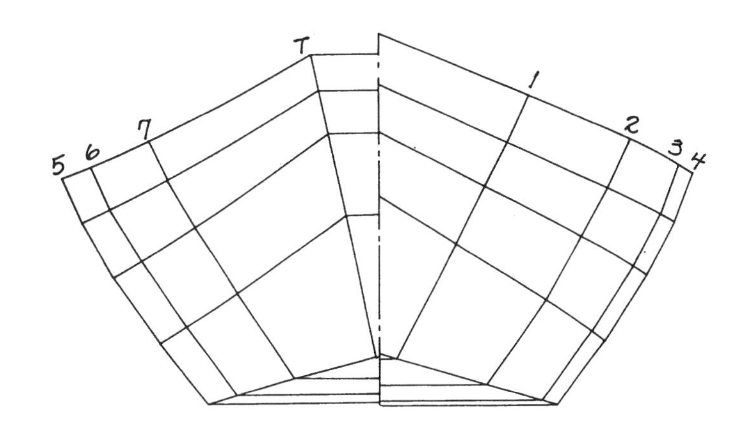

Chapter Six

Nineteenth-Century Working Dory

When I found the old dory, she was quietly reposing in shadowy seclusion in the basement auditorium of the Cape Ann Historical Association in Gloucester, Massachusetts.

Do old boats dream dreams? Pulled out on dry land, snugly tucked away on shore, do their decrepit timbers still shiver faintly, remembering old blasts, the sickening pitch of mountainous seas, the gale's maniacal scream? Who knows?

But ghostly memories do emanate from their ancient bones with power to beguile the viewer, and coming into the presence of this patriarch of dories the visitor is moved.

She is the *Centennial.* In this dory 120 years ago, Alfred Johnson, the intrepid Dane, made the first lone crossing of the North Atlantic. He left Eastern Point, Gloucester, on June 15 of the centennial year, 1876, and docked at Liverpool, his destination, on August 21.

The crossing had not been easy: a week of fog off Nova Scotia's Cape Sable, followed by two gales, in the second of which the dory was hove bottom up with loss of stove and squaresail, and only by fortunate chance re-righted. Newspaper accounts at the time made much of Johnson's encounter with a marauding shark which he drove off with his knife lashed to an oar.

Johnson was the first. Two years later the two Andrews brothers followed in another Gloucester dory of about the same size, and since then there have been other dories and many other single-handed crossings in all manner of craft. But *Centennial* led the way; and it means something to be first.

A more detailed account of the *Centennial*'s feat is to be found in the *National Fisherman Yearbook* for 1976, written by Joseph Garland, Gloucester historian. Garland is the author of *Lone Voyager*, the engaging life of Gloucester's

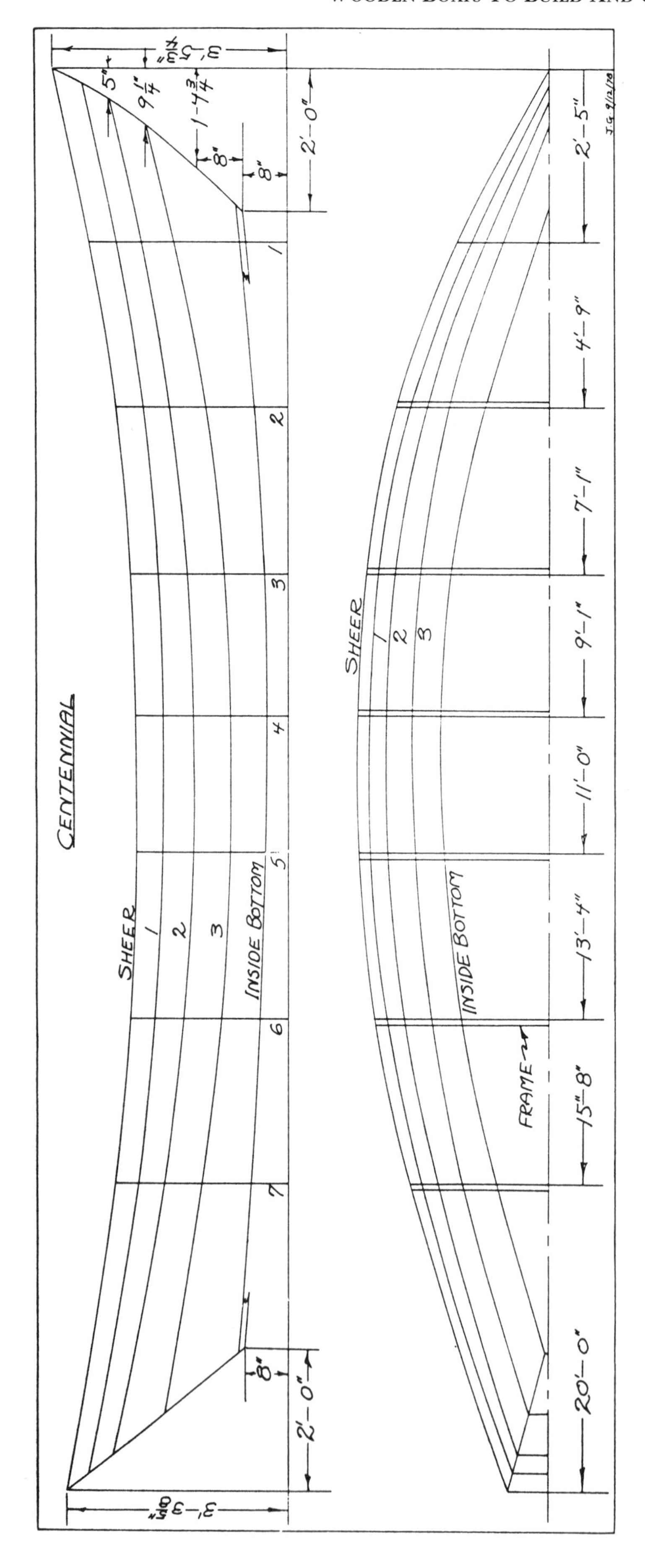

man of iron, Howard Blackburn, who, sans fingers, single-handed the North Atlantic twice in small vessels.

It was Garland who suggested I take off the lines of the *Centennial*, which I have done, and which appear here. I had been aware vaguely that the boat existed, and that it was stored somewhere in Gloucester, but I had not seen it before I drove up from Mystic to Gloucester one Saturday a number of years ago. What had not occurred to me until then was that Alfred Johnson's *Centennial* is the oldest surviving example of an American fishing dory, and hence of particular interest for comparison with present-day dories.

How *Centennial's* dimensions compare with those of the large fishing dories of that time can be seen by consulting Bulletin 21 of the United States National Museum, prepared under the direction of G. Brown Goode, as a catalog of the *Exhibit of the Fisheries and Fish Culture of the United States of America Made at Berlin in 1880*.

Dimensions for the largest American dory at the Berlin Fisheries Exposition are listed as follows: length 15'6" on the bottom, 19'8" on top; width 35" on the bottom, 5'5" on top; depth 22" amidships, 31" fore and 31" aft.

Dimensions for *Centennial*: length 16' on the bottom, 20' on top; width 38" on the bottom, 5'8" on top inside the planking; depth 23" inside amidships, 33" inside bow, 31" inside stern.

Thus, *Centennial* is only very slightly larger than the largest of American fishing dories was in 1880, and

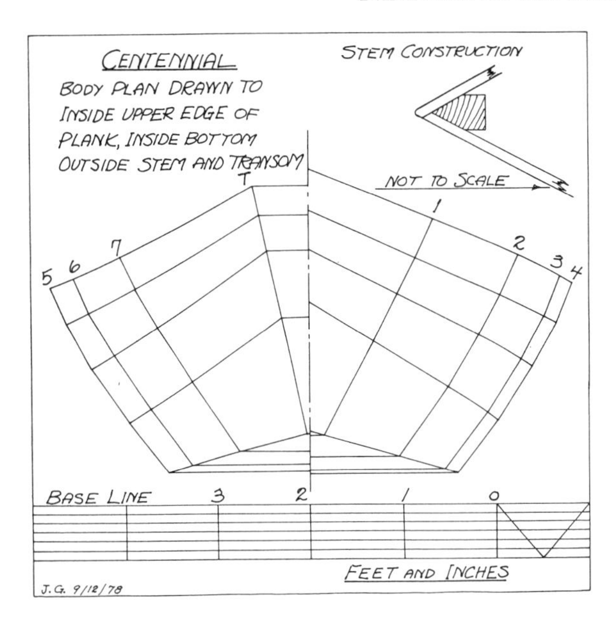

proportionately, closely the same. The large dory described by Goode was planked with pine and framed with oak, and so is *Centennial*. Pine and oak continue to be the standard dory building materials down to the present time.

On this large, $15^1/_2$-foot (bottom measurement) fishing dory, Goode comments as follows: "This size is not used to any great extent by American fishermen, but large numbers are shipped annually to the French at St. Pierre, Miquelon. This dory is built about the same depth and width, as the haddock dory, and very strong; with six, and sometimes seven, pairs of timbers, with a wide band or ribband on the outside to protect the top of the dory. The gunwales, timbers, stern, and about all other materials in this style of dory are larger than are usually put in American dories."

It should be added here that although this dory was the type used in St. Pierre et Miquelon 100 years ago, what we know as the St. Pierre dory today is quite a different boat, much larger, 27' overall against $19^1/_2$', and with more rocker and a greater amount of sheer.

Like the $15^1/_2$-foot dory described by Goode, the *Centennial* has six pairs of oak frames sided about one inch, a bit larger than the 7/8" siding more or less standard for large modern dories. But the molding of *Centennial*'s frames is about what is now considered normal molded width for oak dory frames. Nor are the rest of *Centennial*'s

Captain Alfred Johnson at the helm of Centennial ***(photograph courtesy of Cape Ann Historical Association, Gloucester, Massachusetts).***

OFFSETS CENTENNIAL — FEET, INCHES, EIGHTHS										
	STATIONS	STEM	1	2	3	4	5	6	7	TRAN.
HEIGHTS ABOVE B.L.	SHEER	3-5-6	2-11-3	2-6-7	2-4-2	2-3-3	2-3-0	2-4-1	2-6-6	3-3-5
	1	3-0-4	2-6-4	2-1-6	1-11-1	1-10-3	1-10-4	1-11-7	2-2-7	3-0-0
	2	2-7-5	2-2-1	1-9-0	1-5-7	1-4-7	1-4-6	1-6-5	1-10-0	2-7-4
	3	2-1-1	1-8-2	1-2-4	0-11-3	0-10-2	0-10-2	0-11-7	1-3-2	1-10-2
	BOTTOM	0-9-0	0-8-4	0-5-7	0-4-2	0-3-6	0-4-1	0-5-0	0-6-5	0-8-7
HALF-BREADTHS	SHEER	0-0-0	1-4-0	2-3-0	2-8-2	2-9-6	2-9-6	2-6-6	2-0-5	0-7-2
	1	0-0-0	1-1-5	2-0-4	2-6-1	2-7-6	2-7-5	2-4-5	1-10-5	0-6-3
	2	0-0-0	0-11-2	1-9-5	2-3-2	2-4-5	2-4-4	2-1-4	1-7-6	0-5-3
	3	0-0-0	0-8-1	1-5-5	1-10-6	2-0-1	1-11-5	1-8-7	1-3-2	0-3-5
	BOTTOM	0-0-0	0-1-7	0-11-2	1-5-2	1-7-0	1-6-5	1-3-4	0-9-2	0-0-4
MEASURED TO INSIDE OF PLANKING AND BOTTOM — TO OUTSIDE OF STEM AND TRANSOM. PLANKING 9/16" WH. PINE. PLANK LAP 1 3/8" MEASUREMENTS TO UPPER INSIDE EDGES OF PLANKING.										

scantlings much heavier, if any, than what one would find to be usual practice for fishing dory construction today.

The stem is sided 2 1/8" although I could not get inside under the deck where I could determine the width of the stem's molding. Planking as far as I could tell is 9/16" thick. The decking is 5/8" fairly wide boards that run straight and are joined with tongue and groove. A thickness of 1 1/8" for the oak transom is not at all heavy for a boat of her size.

As for the thickness of the bottom boards, this was difficult to determine with the boat resting flat on the floor, as it was. It could be as thin as 7/8", which I would consider scant for a boat of this size, or it might be as much as one inch. Whether there had once been an outside false bottom or a skeg of any sort can hardly be ascertained with the boat in its present situation.

It is obvious that the structure of the boat has settled somewhat, and that the bottom has flattened out. In addition, some of the garboard fastenings have let go and the garboard on the starboard side has pulled away slightly for a space amidships. Also, as the boat is now resting, the starboard side sags an inch or so below the port. It appears in addition that the hull has taken a slight twist from end to end.

All of this I have attempted to make allowance for and to correct in the finished lines, which I have to say did not give as much trouble as I anticipated and faired out far easier and better than I thought they would.

With Garland's account in the *National Fisherman Yearbook* there is reproduced an old photograph of the *Centennial* posed in profile on the Gloucester dock. This image, shown here, is a broadside view with the camera shooting from about broad amidships and not too high, so that what we get is a pretty fair look at both sheer and bottom rocker. I would judge that at that time she still retained most of her original shape with considerably more sheer curve and fore-and-aft bottom rocker than remains today. In the lines that appear herewith I have endeavored to restore about the same amounts of both.

In addition to resting flat on the carpeted floor, with no chance at all of getting at the underside of the bottom, *Centennial* is decked over, making it difficult to get at most of the interior, and quite impossible to reach some of it. Consequently, measurements for the lines had to be taken from the outside of the hull and planking. To be useful to a builder, however, dory lines, as well as the lines for any clinker boat, must be drawn to the inside of the planking.

First, lines had to be drawn to the outside and then converted to the inside, as they appear here, a process which requires considerable fussing and some personal

Centennial *posed on the dock in Gloucester, Massachusetts, showing some of her original sheer and fore-and-aft rocker. Captain Alfred Johnson, well on in years when this photograph was taken, died in 1927 at the age of 82 (photograph courtesy of Joseph E. Garland).*

judgment in making allowances of lap thicknesses and the like in various parts of the boat.

The fact is that no two surveys in a situation such as this are likely to produce exactly identical sets of lines for the same boat. The structure is too complicated. The same measurement taken twice will often differ slightly, even if the greatest care is taken. Occasional errors will creep in. There is never enough time in a single visit to check everything that needs checking.

The individual judgment of the surveyor enters into the process and molds the results to a greater extent than is generally realized. Taking a set of lines is far from being a completely mechanical and objective process.

Nevertheless, the experienced surveyor can, and should, get lines which are substantially accurate and close, and from which a boat can be built that does not differ in any significant detail from the original. This, I am satisfied, has been accomplished here, and as I have already said, I was pleasantly surprised at how little had to be changed in fairing these lines.

But lines are only a beginning—that is, if one aims to complete a record that would enable the accurate reproduction of a boat in its entirety exactly as it once was. Complete construction details are quite a different matter from a set of lines, and I make no attempt to get into that here.

For a boat in the present situation and condition of the *Centennial*, to faithfully record and reproduce in plan form every detail of deck, internal arrangements, and rig, to the extent that it can now be done, would take days of work with rule and tape, plumb bob and level, not to mention templates and spiling block, and more than one return visit to check and verify, after the drawing-board stage was in progress.

In his *Yearbook* account Garland lists what would have to be included in completely detailed construction plans: "extra frames, centerboard, and iron ballast. She was decked over save for a small cockpit and companion way. Though set up with shrouds and headstays, the mast could be unstepped from a box on deck. Rigging included bowsprit, gaff mainsail, two jibs and squaresail. All running lines leading back to the cockpit. Three watertight compartments."

There was also the rudder of which only a small part remains, and there is possibly more to the interior, as it seemed to me from what I could see through the hatch openings.

While completely detailed plans of *Centennial*'s construction would be desirable as a matter of historic record, as well as of some general interest, I can hardly conceive of anyone wishing to reproduce a replica of *Centennial* for any practical or useful purpose. Certainly not for cruising. She would be horribly uncomfortable. And for a single-handed crossing of the Atlantic, there are much more commodious craft now available.

The bare hull built by standard dory construction according to the lines given here, does have some possibilities, however. The proportions are good. It is withal a rather shapely craft and the very slight curve in the sides does wonders for the appearance. The extra width over that of the standard fisherman's model would be an advantage for recreational use.

Using these lines as the basis, a handy craft in the low-speed range, suitable either for sail or power, or a combination of both, could be built. This hull is appreciably smaller than the 27-foot St. Pierre dory, yet not too small to go to sea in, at least with prudence.

Planked with plywood, a boat built from these lines might even be built light enough to be taken over the road on a trailer. Unless plywood was used, the number of strakes would probably have to be increased to five instead of four, as pine boards as wide as the original garboard (16" to 18") are rarely to be found today.

One final thing to note. In the upper corner of the body section drawing I show a section through the stem. The hood ends of the planking on one side run by the beveled-off plank ends on the other side, and finish without the addition of an outer bent stem piece, or "false stem," so-called. Apparently the outer stem piece which all dories now have was introduced later.

The drawing of a Portsmouth Navy Yard dory of about the same period details a stem construction that is much the same as *Centennial*'s. The same method of finishing off the plank ends was employed on the sharp sterns of the eighteenth-century bateaux used in the French and Indian Wars.

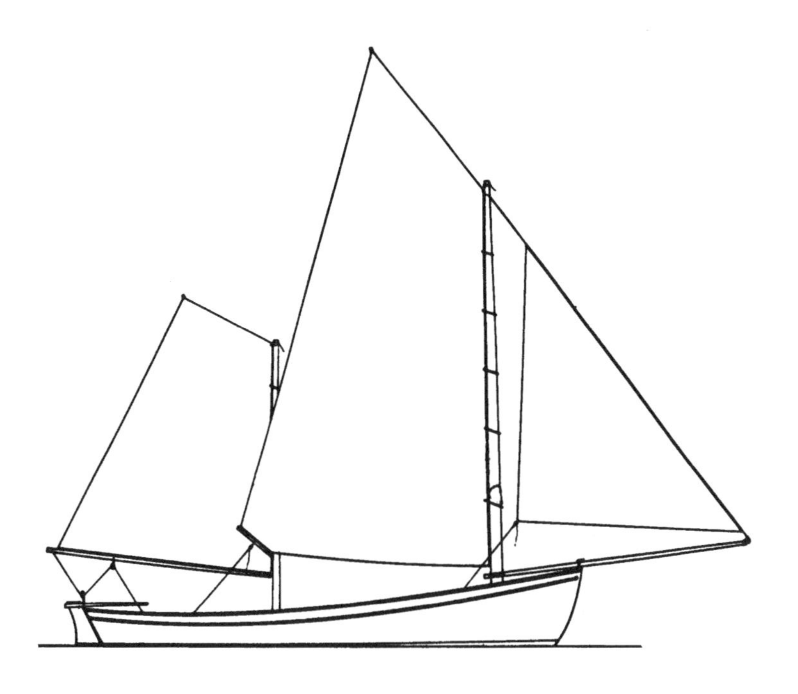

CHAPTER SEVEN

THE MIGRATION OF THE HAMPTON BOAT

From earliest time the inhabitants of New England got much of their living from coastal waters well-stocked with fish. However, it was not until the late nineteenth century that any effort was made to record the boats used by New England fishermen. Among these craft there was a rich diversity of local types and variations.

Attention was first directed to these boats in Henry Hall's *Tenth Census Report, The Shipbuilding Industry of the United States*, published in 1884, and in the United States Fish Commission reports of the 1870s and 1880s by Captain Joseph W. Collins and George Browne Goode. These reports marked a beginning, but other than this, little attention was given for some time to such working craft by historians and the wider boating public.

Martin C. Erismann's study of the Block Island boat, or the Block Island cowhorn as it was sometimes called, and Professor E. P. Morris' *The Fore-and-Aft Rig In America*, published in 1927, were early beginnings in a trend which in recent years has developed into a full-blown movement by maritime historians and others to search out as much as can now be recovered about our heritage of working craft history and design. Because little was written down formally, or previously considered worthy of notice or preservation, there are many gaps in our information to be bridged by conjecture, and these are bound to breed controversy. Yet in spite of all this it is surprising how much information patient and persistent research has so far managed to uncover and piece together.

In 1932 and 1933, *Yachting* magazine fired public interest with a watershed series of 15 articles on native American workboats, for the most part fishing boats, and boats predominantly of New England origin.

Howard I. Chapelle accounted for ten of the 15 articles, dealing in turn with New Haven sharpies, Friendship sloops, Maine pinkies, Cape Cod catboats, Gloucester sloop boats, skipjacks, Bahamian sharpshooters and Bermuda sloops and dinghies. Chapelle's reputation as an authority on maritime history and American working craft design was thereby established, to be further enhanced by the appearance in 1935 of his *American Sailing Craft.*

Consequently, when *Yachting* published "The Hampden Boat," by H. I. Chapelle in July 1938, this article spoke with particular authority and attracted wide attention. It was, however, to elicit a response not anticipated by the author.

Where the name "Hampden" came from, Chapelle declared he had been unable to ascertain. There was a Hampden in Maine, it was true, but the town was somewhat remote from the sea with no discoverable connection with this, or any other, fishing craft. As for the boat itself, there could be no doubt that the type had originated, that is to say developed, in Maine's Casco Bay to meet special local needs, both natural and economic.

Made up of a number of long, narrow coves or bights opening to the southwest, Casco Bay is beset throughout its extent with a multitude of islands, ledges, rocks and flats. Changing and confused tidal currents run strong in the channels and deep bights. Fishing grounds occupying shallow coves and extending among numerous islands and ledges have more than their share of treacherous shoals and submerged rocks. In the wintertime, fishermen were obliged to venture outside into deep water.

To meet such conditions a shoal-draft centerboard boat serves best, provided the model is at once weatherly, seaworthy and burdensome. In addition it must be cheap to build and maintain, simple to rig and within the ability of one or two men to work and handle with ease. Such, according to Chapelle, were the specifications which eventually produced the square-sterned Hampden boat, the forerunner of which appeared to him to have been "the ordinary yawl boat of the coaster and sailing ship." And, gathering assurance further along in the article, Chapelle became quite definite about the prototype. "With the passing years, the Hampden boat slowly developed from the yawl boat to a distinctly different model," he declared, "particularly fitted to pick up lobster pots in heavy seas and gales, to meet a steep chop when beating, to carry sail well and to be generally suitable for the natural conditions she had to meet."

If it is assumed that the Hampden boat originated in this way, how then to account for what in Chapelle's words were a "small number of sharp-sterned Hampden boats, known as Crotch Island pinkies, that were built in Yarmouth in the nineties"? These, in Chapelle's view, had derived from or were influenced by whaleboats, so-called, of an earlier period that were not as large as the standard New Bedford whaleboat, but "probably something like the small Jonesport 'peapods.'"

We are left with the impression that in Chapelle's opinion the double-ended Crotch Island pinkies and the square-sterned Hampden boats were not only distinctly different from the beginning, but were essentially unrelated types.

Chapelle's article was mainly a discussion of lines taken from four builders' half-models offered as representative examples of square-stern Hampden boats. These lines and Chapelle's analysis of their characteristics made an impressive piece of work, yet questions were not long in arising concerning the name Hampden and Chapelle's theory about the type's origin. In the very first issue of the *American Neptune,* dated January 1941, Walter Muire Whitehill, a member of its editorial board of which, incidentally, Chapelle was also a member, questioned the name "Hampden," asking, "Is it possible that the name is derived from Hampton, New Hampshire?"

An answer was not long in coming. In the following issue of the *Neptune,* Charles P. Emerson, who was identified as having formerly owned an old Hampton boat himself, stated he could find nothing locally to justify the spelling "Hampden." Also, fishermen he had talked with in the Portland area agreed that early Hamptons were double-ended and clench built, and they referred him to an 80-year-old South Harpswell man who recalled the old Hamptons as lapstrake double-enders, the first of which had been imported from

Seabrook, New Hampshire, which is only about five miles from the town of Hampton.

In addition, Emerson cited supporting information secured earlier from David Perry Sinnett of Bailey Island which confirmed that the first Hamptons had been imported from Hampton, New Hampshire, and that they were lapstrake double-enders. The first boats built by Sinnett, who is reputed to have turned out over 300 in his lifetime, were likewise lapstrake and double-ended. Later Sinnett went to a square-stern model, and he is credited with having built the first strip-planked Hampton in 1877.

Emerson's comments were not accepted by Chapelle, who replied in both the July 1941 and the July 1942 issues of the *Neptune* with arguments which, it must be said, did little to clarify the issues. He continued to stand by his original contention that the square-sterned boats had been developed from "ordinary yawl boats." And as for the name, that is still "an open question." Chapelle declared, "perhaps either spelling must be accepted for the present." Actually, it was not a mere question of spelling. It was whether or not the prototype of the Hampton boat had come to Casco Bay from Hampton, New Hampshire.

It might seem that questions of origin and name are relatively petty matters and hardly worth controversy, yet they do have their importance. There is likelihood that the stir they occasioned did focus attention on the type, keeping it in view and thus may have contributed to the modest revival that the Hampton boat experienced.

In any case, the continuing controversy came to a head with an iron-bound piece of research by Phelps Soule with assistance from M. V. Brewington and Edward O. Brownlee, which was published in the *American Neptune* in April 1943.

Not only did it settle the question of the name conclusively, proving that the name Hampton had been known and used more or less continuously since the beginning of the nineteenth century, but it added substantially to our knowledge of the type with lines from five builders' half-models taken off and drawn by Brownlee. In addition, there was a sail plan and the interior layout for a typical working Hampton drawn by Brownlee from information supplied by an Old Orchard, Maine, boat builder, and checked by him.

The five half-models were secured from the Crotch Island boat shop of John Pettingill, deceased, formerly a well-known builder of Hampton boats. Two of the models are pinkies with sharp sterns: the rest have square sterns. Only slight and superficial alterations in the lines of these models would be required to transform the square-sterned boats into pinkies, and to give the two pinkies square sterns. In fact, the resemblance of the pinky lines to the lines of the square-sterned boats is so close that this alone is sufficient refutation for me of Chapelle's theory regarding a separate yawl boat prototype for the square-sterned Hamptons.

Chapelle never completely abandoned this theory, but in later years his position softened considerably. By 1951 when he brought out his *American Small Sailing Craft*, he had given up the name Hampden for Hampton. As for his theory of yawl boat origin, he had by this time reformulated it somewhat. "It is probable that the square-sterned boat was not a development out of the New Hampshire boat by way of Casco Bay, but rather that the former developed to the eastward independently and invaded Casco Bay in the 1880s where it first influenced the Crotch Island model, and then gradually replaced it."

By 1960 in his *Catalogue of the National Watercraft Collection*, no reference is made to a possible yawl boat ancestry. In discussing Hampton boats therein, Chapelle wrote: "They received their name from the old double-ended boats of the type originated at Hampton Beach, New Hampshire; they do not, however, resemble the old double-enders, being an entirely different form of boat."

In a sense, this is quite true, of course. Certainly late-model, strip-built, power Hamptons with counter sterns are very different from the earliest clinker-planked, double-ended Hampton whalers said to have been first built in 1805 by Enoch Chase at Seabrook/Hampton, New Hampshire, whence they were imported into Casco Bay. Nevertheless the two are connected,

Elijah Kellogg's double-ended Hampton boat,* Cadet, *fully rigged after restoration at Mystic Seaport Museum (M.S.M. photograph).

standing at the two ends of a continuous line of evolutionary development. There is not one Hampton boat, but a continuing progression of Hampton boats, to be identified and considered according to the time periods in which they were built and used.

By fortunate circumstance one of the early, Maine-built, lapstrake, double-ended Hamptons has survived, and after extensive restoration has been brought back to something close to her original condition. She is the *Cadet*, built by Ebenezer Durgin at Birch Island, Maine, for the Reverend Elijah Kellogg sometime after Kellogg came to the Harpswell church in 1843 and before Durgin's death in 1851. Kellogg, who in his youth had gone to sea for three years prior to his entering Bowdoin College, was an able and fearless sailor who made frequent use of *Cadet* in his ministry to fishermen and their families living on remote islands in Casco Bay. Legend has it he once sailed *Cadet* from Portland, Maine, to Gloucester, Massachusetts, ahead of a northeast snowstorm.

In 1953, after discovering that *Cadet* was stored in a tumbledown building on the old Kellogg place in Harpswell, Chapelle alerted the Marine Historical Association to the possibility of securing *Cadet* and bringing her to Mystic Seaport. After lengthy delays, the transfer was finally effected in 1955, but not before the roof of the barn where *Cadet* was stored had fallen in, with considerable damage to the boat. The donor, aged widow of Elijah Kellogg's grandson, was of the recollection that the *Cadet* had not been in the water for more than 40 years, and she recalled that it had always been spoken of as the "pinky." This last is an important point and of particular significance, as will presently become clear.

Although the *Cadet* came to Mystic in 1955, her spars were not rescued from the wreckage of the old barn until August 1963 when they were pulled free by William T. Alexander, president of the Webb Institute of Naval Architecture, who apparently was summering in his native Harpswell.

"My grandfather lived on the next farm." Alexander wrote, "and was a close friend of Mr. Kellogg. Frank Kellogg, Elijah's son, used the boat for some time, and my father took care of her for him. Consequently, I had a chance to try her out, occasionally, with him...She carried two heavy, unstayed masts with loose-footed spritsails. The jib was carried on a 'crutch type' bowsprit split at the after end to fit down over the stem (head) and resting against the foremast. The bowsprit was held down by an oak pin through the stem and across the bowsprit, which had a curved after end to seat against the mast."

The dimensions of the spars found and identified by Alexander follow. Foremast: length 14'11"; maximum diameter 4 1/2" with sheave for 1/2" line at the head of the mast. Mainmast: 14'2"; maximum diameter 3 7/8", with hole at the top for 1/2" line. Boom for mainsail: 11'2"; 2 1/4" maximum diameter, with three 1/2" holes bored 3" apart horizontally through the outer end. *Cadet*'s overall length measures 23'4"; her beam, 6'6".

A working reproduction of the *Cadet*, completed at Bath Marine Museum's Apprenticeshop in 1977, was sailed that year from Bath, Maine, to Mystic, Connecticut, and back for Mystic Seaport's Small Craft Workshop in June.

According to Phelps Soule in the *American Neptune*, the changeover from sharp stern to square stern by builders like David Perry Sinnett of Bailey Island took place during the period 1875 to 1880. No one seems to have claimed credit for this, which could indicate that no one then considered the innovation important enough.

As already stated, a comparison of the lines of sharp-sterned and square-sterned builders' half-models from the shop of John Pettingill on Cliff (formerly Crotch) Island reveals no marked differences in the basic hull shape. The transition from pinky stern to transom stern would have been easily and simply accomplished, with the aim perhaps of securing more room and working space, as well as more bearing aft for sail-carrying ability. As far as I am concerned, the Pettingill half-models adequately refute any theory of a yawl boat prototype, or the mysterious appearance into Casco Bay of a square-sterned migrant from the east.

A hint of the "double-wedge" hull shape so favorable for conversion to power later on, as Chapelle has explained, was already present in some of the pinky-sterned boats, as can be seen in the Pettingill half-models. With the changeover to the transom stern, the double-wedge characteristic became more pronounced, paving the way for gasoline motors.

A. O. Elden's account of "Power Boating In Casco Bay" in the April 1907 issue of *Yachting* makes the observation that when fishermen first fitted their boats with motors six or seven years previous, they were laughed at. But, since then, Elden continues, power-driven boats proved so successful in lightening labor, cutting working time and in getting fish to market, that at the time of the writing the straight sail-driven Hampton had become a curiosity.

Working sail on fishing boats lasted in Casco Bay until about 1900. During a period of at least 60 years, the sailing rig of the fisherman's Hampton boat changed but little, if at all. This was the same rig as already described for *Cadet*. There were two heavy, unstayed masts, of which the foremast was the larger and set well toward the bow. Both carried spritsails. The foresail was loose-footed, and in some cases, if it overlapped much, it was fitted with a club. The mainsail—somewhat smaller—had a boom, to which frequently the sail was attached only by the tack and clew. A removable bowsprit fitted over the stem head when it was desirable to carry a jib of moderate size, which was set flying. Fittings were the simplest.

This was the standard rig of the working fishermen, but when these Hamptons were taken over for pleasure boats, as a few of them were, the rig proportions were sometimes altered. This is true for the small square-sterned, strip-built Hampton boat, *Cuspidor*, now in Mystic Seaport's small craft collection. *Cuspidor* was built by Sinnett at Bailey Island in 1902 for Dr. Franklin P. Luckey of Paterson, New Jersey, who used to summer at Bailey Island. Only 17'4" overall with 6' of beam, *Cuspidor* carries an oversize foremast that is 16'1" long with a maximum diameter of 5"; the length of the sprit that goes with it is 16'7". Her jib with its long bowsprit is also much larger than those carried by the fishermen. Her

Cuspidor, ***owned by Dr. F. P. Luckey, under sail at Bailey's Island before coming to Mystic Seaport Museum (M.S.M. photograph).***

bottom plank. The other, the *Shamrock*, is a square-sterned boat 23'10" overall with 7'2" of beam and a 21' bottom plank.

In 1900 William G. (Greg) Hall and his son, Will, built seven of these shad boats. The Halls, father and son, were the foremost builders of shad boats in the area, fishing during the 2½-month shad season in the summer, and building boats the rest of the year. There were a number of other builders who produced these boats, but the Halls built the most and were considered to be the best builders. Furthermore, Will had the reputation of being second to none as a sailor and boat handler, and it was for him that the *Ocean Queen* was built.

Although these boats were built to sail, motors were eventually installed, as was easily done, but this change occurred later than the changeover to power in Casco Bay. To make room for *Shamrock*'s single-cylinder, make-and-break Bridgewater engine, the after thwart (through which the mainmast was stepped) was removed, and a

mainsail—or mizzen, if you wish—is, on the contrary, considerably smaller than normal to correspond with a mast only 11'6" in length. Apparently *Cuspidor* sailed well, with her oversize rig suited to pleasure sailing in the summer.

New light has been cast on the sailing Hampton boat and particularly on the relationship between the pinky-sterned boats and their square-sterned counterparts (before the modifications which followed the latter type's conversion to power), by a find in 1979 in Colchester County, Nova Scotia, of three boats unquestionably of Hampton derivation and model. Two of them were built in 1900 for the local shad fishery by Greg and Will Hall in Portapique Village on the north shore of Cobequid Bay at the eastern end of the Bay of Fundy, about 25 miles or so from Truro. One, named the *Ocean Queen*, and known as "the big pinky," is a pinky-sterned boat 25' overall with 8' of beam and a 22'

Shamrock ***fishing after conversion from sail to power. Her mast is stepped through the open cuddy hatch (Harold Cooke photograph).***

This photograph, circa 1893-1895, shows two transom-sterned Bay of Fundy shad boats built by the father-and-son team of Greg and Will Hall of Portapique Village, Nova Scotia. The single spritsail boat,* Myrtle B.*, has Will Hall in the stern and Lorne Cooke, Harold Cooke's father, amidships. The larger boat at right,* MYOB*, carries owner and builder William G.(Greg) Hall and his wife, Jerusha (photograph courtesy of Harold Cooke).

narrow but adequate engine box was built in its place. Forward, a low cuddy was added, as the photograph shows, blocking off the forward mast position but retaining what was originally an alternate third mast position against the forward edge of the forward rowing thwart. Thus, if desired, a sail could still be set on a mast stepped through the cuddy hatch. The centerboard was retained.

The original rig was identical with the standard rig of the Casco Bay Hamptons, an arrangement that had persisted at least from the time of Kellogg's *Cadet* until sail gave way to engine power. This is clearly shown in the old photograph reproduced herewith, which has been dated as early as 1895, and possibly 1893.

The fully rigged boat with sails set, shown on the right in the photograph, is the *MYOB* belonging to Greg Hall, her builder, who is shown with his wife, seated in the boat.

A surprising amount of detail is revealed in this photograph. The slightly longer foremast, which is a full 4" in diameter, is secured by an iron clasp against the heavy beam supporting the after end of the short foredeck. In view is the starboard end of the forward rowing thwart, attached to the forward edge of which is another iron clasp for another mast position, formerly used when the mainmast was brought forward and used alone in bad weather. As already mentioned, this mast position was retained in the *Shamrock* when the cuddy was built on. Though the starboard end of the middle rowing thwart is barely visible, the thole pins which go with it can be seen in the port rail.

The mainmast steps through the after (third) thwart. The foresail is loose-footed, and the main is attached to its boom only at the tack and the clew. The two-part sheet of the foresail is visible, belayed to a removable pin through the starboard rail. There are no washboards on these Bay of Fundy shad boats. In this connection we are told that washboards were not put on Casco

Bay Hampton boats until after 1878, when David Doughty of Great Island, who is credited with introducing them, quit fishing to build boats.

There are no halyards except on the jib, which is set flying, with its halyard rove through a hole in the head of the foremast. The two spritsails do not hoist but are permanently secured to the mastheads with the free ends of their bolt ropes left sufficiently long for this purpose. When the boats were drifting, the sails were tightly furled to the masts with their sprits, and the unstayed masts were then lifted out of their steps and stowed horizontally along the rail. Not shown in this photograph, and possibly something which came later, is a U-shaped iron or crutch rising above the port rail and used for holding the masts with their furled sails in a compact bundle.

The removable plank bowsprit is almost exactly like the one previously described for the *Cadet*. It is slotted to fit over the upper end of the stem head, through which a pin goes to hold it down in place. The after end of this flat bowsprit is made in a crutch shape to fit snugly against the round of the foremast.

The shad boat* Shamrock *looking forward from her stern, showing cuddy and engine box. The clasp for the mast is visible through the open hatch; the crutch for stowing her unstepped mast is on the port rail (John Gardner photograph).

It remains to be seen, after lines have been drawn for these boats, how closely their hull shape resembles that of the sailing Hamptons of Casco Bay, especially the pinky-stern boats. My prediction is that the resemblance will be close. And as for these Bay of Fundy Hamptons, if we may call them that, there appears to be very little difference in the underwater hull lines of the pinky-sterns and the square-sterns. But while these Bay of Fundy boats are remarkably close in shape and rig to the old sailing Hamptons of Casco Bay, there are some differences between the two in construction.

The Bay of Fundy boats are built with plank keels of spruce, birch or sometimes maple 3" thick and 10" or thereabouts wide amidships, tapering toward the ends and perfectly straight for their entire length, which is nearly the length of the boat. Underwater and up beyond the turn of the bilge, the boats are planked carvel with pine as thick as 3/4" or even up to 7/8". The upper topsides are finished out clinker, 1/2" thick. The *Shamrock* has four upper strakes of clinker plank, which I believe was more or less typical, but the *Ocean Queen* has only one clinker plank at her sheer.

One of the older fishermen recalls having been told by his father that the early boats were clinker throughout. The men who made the planking changes are gone, and their reasons for doing so are lost, but we may conjecture. Because of Fundy's 40'-50' tides, these shad boats grounded out on the flats twice each day, and frequently hit bottom when the tide was coming and going. They needed a strong bottom and one with enough flat that they would sit upright in the mud. The heavy, carvel bottom planking was needed here but it was not needed on the topsides, where the thinner clinker strakes were much lighter and quite strong enough. Inside, the bottom was well reinforced with strong, natural-crook floor timbers. Also, the smooth carvel bottom was less noisy than clinker laps would have been, and thus less likely to frighten the shad when drifting. There may have been other considerations, but my guess is these were the principal ones.

Discovering these Fundy shad boats, heretofore completely overlooked, unnoticed and close to the point of extinction, was a lucky find. There is much to be gained from them, as I am sure we shall learn when their lines are drawn, and we have studied them more closely.

Bob Cooper lived in Great Village, Colchester County, Nova Scotia, at the eastern end of the Bay of Fundy. In August 1989, he launched *Jerusha*, his reproduction of the *Ocean Queen*, known locally as "the big pinky." She is the sole remaining example of the double-ended Bay of Fundy shad boats that over 100 years ago dominated the flourishing fishery in Cobequid Bay.

Local history has it that commercial shad fishing got started in the upper reaches of the Bay of Fundy during the third quarter of the last century, when shad fishermen came up from New England, bringing with them their boats and drift nets. In the 2½-month shad season of those days, the waters at the upper end of the bay teemed with these much sought-after fish, yielding a rich harvest to the experienced fishermen from the States. These men split and salted their catch, packing it in barrels that they shipped to Boston and other distribution centers to the south.

It was not long, however, before the Nova Scotians who lived on the bay became aware of the shad fishery's financial rewards and saw how the fish were caught. Soon the natives took over, and in so doing they adopted—at least to begin with—the methods and equipment that had been brought up from the States. The early boats of which the *Ocean Queen*, now at Mystic Seaport Museum, is the last, were nearly identical to Casco Bay Hampton boats of the period before the introduction of the gasoline engine. They were also strikingly similar to other early double-ended New England fishing craft, including the Block Island boat, the Noman's Land boat, the Isle of Shoals boat and some of the larger, so-called Colonial whaleboats.

In time, as sails gave way to the new internal combustion engine (and very likely for other reasons, as well), the double-ended pinky stern was replaced by the square, transom stern. Although this transition did not occur abruptly, it must have been very nearly complete by 1910 or shortly thereafter, when sails had also been replaced on Casco Bay Hampton boats and most other small commercial fishing craft.

In 1900, there were 20 shad boats fishing in Cobequid waters. That same year, William G. (Greg) Hall and his son, Will, the foremost shad boat builders in the area, turned out seven new ones. Sometime previously, they had built the *Ocean Queen* for Will, who undoubtedly preferred sails to a motor in view of his reputation as one of the ablest sailors and boat handlers in the Cobequid Bay area. The pinky-sterned *Ocean Queen* is 25' overall with an 8' beam. Instead of the common plank-on-edge keel

In the summer of 1979, the only shad boat still fishing in the Cobequid Bay area of the Bay of Fundy was this smaller, later model owned and fished by Russell Cooke. Note the pinky stern and rudder post (John Gardner photograph).

The Ocean Queen *shortly after arrival at Mystic Seaport Museum. The ketch-rigged, double-ended* Ocean Queen, *built about 1900 by Greg and Will Hall in Portapique Village, Colchester County, Nova Scotia, is 25' 1 1/4" x 7' 4 1/4", and carvel planked on sawn frames except for the sheer which is lapped (Maynard E. Bray photograph, M.S.M.).*

of her New England prototypes, she has the heavy, flat plank bottom adopted by the Bay of Fundy shad boats for the reasons already noted.

Both the sailing shad boat's two unstayed masts carried spritsails. In addition, there was a jib, which was set flying with its halyard rove through a hole in the head of the foremast. The two spritsails were permanently secured to the mastheads, with the free ends of their bolt ropes left long for this purpose. When the boats were drifting with their nets, the sails and their sprits were tightly furled to the unstayed masts, which were then lifted out of their steps and stowed horizontally along the rail.

Sometimes in bad weather the mainmast (with its spritsail) was brought forward and was used alone. In the photos sent to me by the late Bob Cooper, the *Jerusha* is sailing under foresail alone, and Bob's letter, which follows in part, clearly indicates he was well-pleased with the boat's sailing performance.

"When I last wrote, it looked like an early summer [1989] launching, but it was not until August before *Jerusha* (named after the wife and mother of the shad boat builders, Greg and Will Hall) took her first bath. ...it was a typical Bay of Fundy launching—the boat floated off the drag at high water. The small children had a tour of the harbor, and the adults, champagne!

"We had about six outings with her under sail in the fall. Perhaps I was prepared for the worst, but I couldn't be more pleased with her sailing ability. No one would buy her for her windward performance, of course, but in this regard she is adequate. She will clear the little harbor in three tacks, the same as I have done with my 12' Edson Schock dinghy. To date, we have not used the rest of her rig (mainsail and jib) except for a short run with mainsail added on the day we took the photos, haulout day in late November. As for speed, we close reached across the bay in 40 minutes, an average speed of 6 1/2 knots.

"We used about 400 lbs. of grader blade for ballast. You will note that more is needed farther aft to trim her, although she seems to sail about right as she is with two people in her stern sheets.

"She differs in two ways from the *Ocean Queen.* Her topside planking has three lapped strakes instead of only one, and she has perhaps 4" more freeboard aft, the suggestion of a local fisherman whose knowledge of shad boats is second to none. Otherwise, *Jerusha* is a close replica of the original.

"With the exception of her planking and bottom plank, the timber used in her construction was got out with a chain saw mill. Planking is white pine; stem and stern post, floors and knees are hackmatack; breasthooks, apple; frames, white oak; bottom, yellow birch. Fastenings are galvanized boat nails, with copper clinch nails for the laps and copper rivets for the gunwales."

After the completion of the *Jerusha*, Cooper generously donated the *Ocean Queen* to Mystic Seaport Museum's collection of historic boat types. This was an important acquisition, representing, as it does, the migration of a boat type from New England to the Canadian Maritimes. She is but one instance of the close relationship—historical, ethnic, economic, social and cultural—of these neighboring regions, which, except for their separate governments, might well be considered parts of the same country.

Bob Cooper sailing Jerusha under foresail alone. She is carrying 400 pounds of ballast and has more freeboard aft than* Ocean Queen *(photograph courtesy of R. Cooper).

Launch day on the Bay of Fundy for* Jerusha, *the shad boat reproduction built by Bob Cooper of Great Village, Nova Scotia (photograph courtesy of R. Cooper).

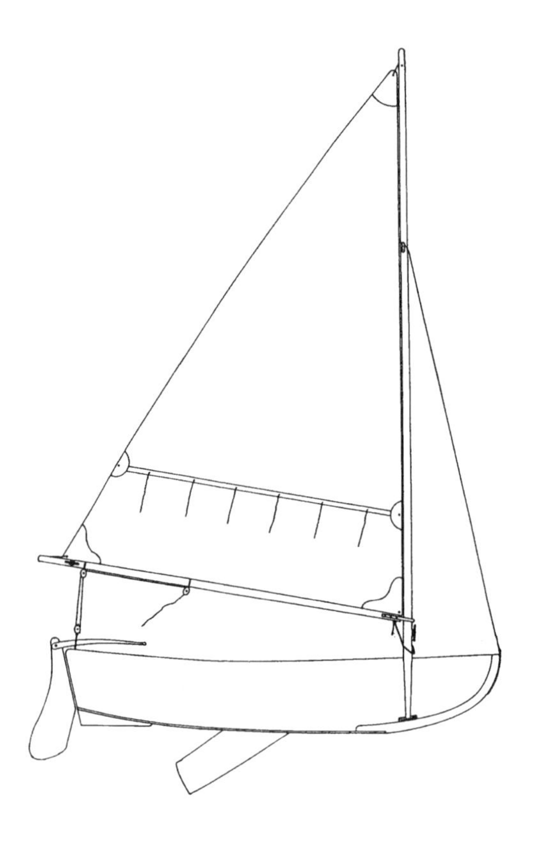

Chapter Eight

13-Foot 7-Inch Swampscott Sailing Dory Skiff

One of the hundreds of small islands scattered along Maine's jagged coastline, McGee Island lies in the upper reaches of Muscongus Bay about $2^1/2$ miles west of Port Clyde and $3^1/2$ miles south of Friendship. Nearly round except for the indentation in its southern side from the close proximity of Barter, a smaller island that once may have been joined to McGee, it is slightly less than half a mile across, and covers somewhat more than 100 acres. Much is now thickly wooded with a growth of towering, mast-size spruce of which occasional dead ones along the ledge-bound shore provide nesting sites for ospreys.

Some time early in this century McGee Island was acquired by the Erickson family of Swampscott as a retreat where they could bring their numerous brood of children to rusticate and run wild during July and August when for two brief, idyllic summer months such favored spots along the Maine coast are nature's paradise.

And it was on one of those somnolent August afternoons when the wild raspberries hang heavy with ripening fruit, and lupine, knee-deep, purple and blue shading to pink, cover the slope leading up from the beach, that we came to the island from Mystic Seaport to measure the boats.

The Ericksons were boating people. Much of their stay at McGee in the old days must have been spent in and around boats, judging from the number and variety, ranging from yachts to dories and small tenders, that we found stored on two floors of a large boathouse.

For me the most noteworthy boat on the island was not one of those we found stored away in the boathouse, but was still

The 5' 3" beam of the 13' 7" dory skiff is accentuated in transom view in this idyllic scene where John Gardner works on the beach before the cookhouse and the lupine-covered shoreline (Ben Fuller photograph).

very much in use on the waterfront, a small utility skiff that went by the name of "Fat Boat" and was so called because of its exceptional width of beam in proportion to its length. It had been built many years before to Mr. Erickson's specifications by Guy Gardner of Swampscott, George Chaisson's competitor, and builder of Sam Brown's far-famed Yankee Dory. They must have been gratified by the result, and perhaps just a little surprised, for despite her wide beam, *Fat Boat* sits easily and lightly in the water and moves briskly under oars even when weighted down with a heavy load. Naturally exceptionally stable, *Fat Boat* handled herself well in a chop. In addition to widening her out, Gardner had sharpened her entrance somewhat so that she doesn't pound like some dory-built boats with more upright stems that carry their flat bottom well up into the bow. In spreading her out, her shape had been transformed into what is virtually a round hull boat, and Gardner had found it possible to do this and still retain the four plank sides. Four planks, if laid out right, go on ever so much quicker and easier than the seven or eight or more usually required for a round hull boat as wide as this one.

The upshot was that we pulled *Fat Boat* up on the beach where I could measure her and take off her lines. After our return to Mystic, I laid out the lines and drew up the construction details for inclusion in the Museum's collection of small craft plans available to the public for study and construction.

The shape and proportions of this small rowboat intrigued me. Enlarged and with some alterations, a modified *Fat Boat* would do well, it seemed, as a small sailboat. With her ample beam, *Fat Boat* could carry sail and still be stiff and stable, with a comfortable amount of room for occupants to move about in. With sail down she could be rowed with rowing stations for two. And with an outboard motor, she would move along at moderate speed, although her rockered bottom, so essential for sailing, would disqualify her as a speedboat.

After laying out several sets of lines in which I introduced various changes, I cut according to my final revision, a solid half-model, scale $1^1/2$ inches to the foot, changing it and fairing it in several places, before it satisfied me. Finally it was from the finished and corrected half-model that boat lines were taken that appear here. It is my conclusion, after long experience, that more is to be learned about a design from a solid half-model than is ever revealed in a draft of the lines. At least that is the old way, the way that the designs for thousands upon thousands of able and beautiful vessels from rowboats to clipper ships came into being.

They tell us that very soon it will all be done by computer, and better and quicker than ever it was done in bygone times when designs were modeled by hand and eye. It may well be so, but that day is not here yet, not quite. There are still a few of us left who like to do it the old way, and get enjoyment doing it. There is satisfaction in turning a block of wood into a fair and shapely half-model, shaping it shaving by shaving with keen edged tools. Without making any exalted claims, cutting a model in this way is nothing less than a form of sculpture by which beauty is created as functional utility is achieved. There is no better example than Captain Nat Herreshoff's elegant half-models that furnished the designs for the many incomparable fine yachts built by the Herreshoff Mfg. Co. But even lowly utility craft like *Fat Boat*, while making no claims to elegance, should line out true and fair to present a pleasing appearance on the water. Boats of whatever sort when fair, neatly lined and properly proportioned, always bring a pleasure to the sailor's eye. Misshapen hulks are an abhorrence.

The boat we are considering here is an old-fashioned boat. The design is traditional and it is built largely of wood, the traditional building material per se. But wait—the wood is glued together, and with epoxy, no less. This is enough, I am afraid, to degrade and disqualify it in the eyes of the deep-dyed purists. So much the worse for them, I say. This boat is

Before a backdrop of majestic spruce, John Gardner records measurements of* Fat Boat*'s classic lines. Despite her beam and rounded shape, Guy Gardner was able to plank the lapstrake hull in only four strakes (Ben Fuller photograph).

traditional and right in the things that count, in its shape, looks and the way it behaves on the water. What holds the wood together is hidden and out of sight, and if glue does that job better than metal fastenings, then, glue by all means.

There had to be a time once long ago when metal fastenings were new, and I suppose there may also have been a few purists in that far off time that would have none of metal fastenings at first. But most working boat builders are practical mechanics who are not loath to take up new things when they come along if they prove good.

An example that comes to mind is the gimlet-pointed wood screw that became available sometime in the 1840s. Without such screws in small sizes the classic Adirondack guide-boat might never have achieved the degree of perfection it attained. And as for plywood, even though L. Francis Herreshoff's caustic condemnation 50 years ago of "laminated wood," primitive forerunner of modern marine plywood, is still remembered, I daresay Herreshoff would now be specifying Bruynzeel and similar high-grade marine plywoods if he were designing yachts today.

This boat can still be built by the die-hard purist without glue, if he can find boards of natural sawn lumber wide enough to make the planks, good quality white oak bending stock to replace the glued-up frames and a grown crook of the right shape for the stem. This substantial stem piece can be spliced from two or more pieces without glue, or it can be steam-bent, but special equipment would be required to bend a piece of oak large enough in section.

Select quality cedar or cypress boards wide enough to plank this boat are extremely difficult to obtain nowadays, and naturally bring premium prices. It is much the same for white oak, a specialty item of very limited availability, and what is offered as white oak often turns out to be of indifferent quality and unsuitable for bending stock. More than one builder of traditional boats from my plans has complained to me of late that almost as much time and effort was expended in locating and assembling materials as went into the construction of the boat. And more than one would-be builder has given up when he found it too difficult to get the required materials.

The boat that is offered here was designed to avoid such difficulties, and it does this by utilizing two easily obtained and moderately priced modern materials, namely marine plywood and thermoplastic (epoxy) glue. It is worth repeating that in what matters—i.e., shape, appearance and performance on the water—this is a traditional boat, and a wooden boat built with wood throughout. Only a few prejudiced purists will consider it anything less than a traditional boat because of its utilization of plywood and glue.

This is a boat that is simple and inexpensive to build without any sacrifice of structural strength. Indeed the use of glue throughout appreciably increases structural strength. No steaming is required, eliminating the need for a steam box. Because the frames and the stem are built up by gluing together a number of short pieces, these pieces can be cut from ordinary straight-edged boards with very little waste. Spruce has been specified for much of the internal structure because spruce lumber of suitable quality is plentiful and inexpensive. Spruce is strong, light in weight, holds fastenings well, glues well and is not unduly susceptible to rot, provided care is taken in its use, as will be explained. One excellent source of spruce for this boat are spruce ledger boards widely used in building construction for staging and the like. Boards can be picked out that are relatively free from knots. A few small knots don't matter anyway.

Not a heavy boat, it is not an ultra-light boat either. When the ultimate in speed is desired, a boat can hardly be built too light, but this is no specialized racing machine. A moderate amount of weight is a virtue in a sea boat, providing better handling because of it, and permitting construction that stands up well to wear and rough use. Too often scantlings are slimmed down to the utmost, when there is no need for it, resulting in such common failures as broken frames and split planks. Out where the big ones roll, I feel more comfortable in a boat that I know has been put together solid enough to take a pounding without springing a leak or starting to fall apart.

As has already been said, this is an old-fashioned boat for people who like the old ways, who want a traditional boat that is safe, comfortable, handy and smart.

This is a general-purpose boat, intended to do it all and do it acceptably well, a family boat for all generations, a boat that is safe enough for youngsters to learn to sail in, a boat that is stable enough and roomy enough for their grandparents to relax in in comfort.

Not a slug by any means, this little skiff with its 85 square feet of canvas should move smartly under sail and be plenty for young sailors to handle when the breeze picks up. For those who might want a larger rig, it would be simple to lengthen the mast several feet with additional canvas added to correspond. The 5'3" beam would afford ample stability to support an enlarged rig. Under oars, with either one or two persons rowing, the boat will not lag. And while I have to admit I don't like outboards, either for their noise or their pollution, it is hardly possible nowadays not to make provision for at least a minimum outboard on a general-purpose boat. A motor greatly extends the boat's working range. There are times when the wind fails or blows from the wrong quarter, others when the need to get from here to there as expeditiously as possible indicates the faster motor instead of oars. Also, for someone who does not sail or who finds rowing for any considerable distance unduly taxing, an older person with health problems, perhaps, who has a mind to go fishing, or to take the boat for a leisurely afternoon on the lake or river, a small motor greatly extends the recreational possibilities offered by this boat. Three to five horsepower should be ample. With large motors the boat will squat with little or no gain in speed, but with increased noise, pollution and fuel consumption.

With the motor removed and tucked away under the stern sheets or hidden forward of the mast, this is a boat to get close to nature in, what little there is left, a boat in which to explore small creeks too shallow and winding for larger craft to penetrate. Much natural beauty still remains along our waterways for nature lovers to feast upon and luxuriate in, for environmentalists to cherish and protect, beauty that large power craft charging along at frantic speeds, spreading noise and pollution, miss entirely and never see. This is the sort of craft whose presence the small creatures of fur and feather will tolerate to some extent and accept, notwithstanding that out where the rollers are when the wind blows and the sail is set, this boat can hold its own in a rough and tumble with the waves.

Boats of this sort are rarely, if ever, to be found at commercial boat shows. Unsuited for cloning by the thousands from thermoplastics or aluminum, they do not offer the high profit margins that modern mass marketing demands. Part of a way of life that is passing, demand for them has diminished to a very low ebb. In various parts of the country a limited few are still being turned out in small commercial shops, operated more often than not by one or two boat builders struggling to make a living producing a quality product mostly by hand work with hand tools, and doing it that way because they find fulfillment doing it and because it is the only way it can be done and done properly.

Even though such motivated builders are content to work long hours for what amounts to a meager return, the cost of the custom boat commercially built in this way will not be cheap because of the amount of hand work required, and hand work is expensive.

For those who cannot afford or do not wish to pay for a custom-built boat, there is another way, and that is to do their own hand work and build their boats themselves. It is the old way. One hundred years ago and more, it was a common thing along the North Atlantic coast for inshore fishermen to build the boats they fished in. Young fishermen starting out learned to build by building. Some took to it better than others. Wilbur Morse, for instance, a young Muscongus Bay fisherman who

The simplicity of* Fat Boat*'s interior belies her ease of handling and versatility. Fitted with patent swivel row locks, she was set up with two rowing stations, with ample room for carrying gear between the stern sheets and the aft thwart (Maynard Bray photograph).

built a sloop to work in, sold it before it was finished to another fisherman, built another, and sold that one too and kept on building Friendship sloops for the rest of his life. Sam Crocker, a young Cape Codder, who back in the Great Depression started a boat to go fishing in that he finished up as a yacht, went on to become the leading designer of shoal-draught yachts of the pre-war generation.

In George S. Wasson's *Sailing Days On The Penobscot*, published by the Marine Research Society of Salem, Massachusetts in 1932, we read how fishermen at Isle au Haut and other Penobscot Bay locations built their own boats, how a boat might be worked on for years, off and on, and then one fine day the finished boat would emerge from a woodshed, an old hen house or some other unlikely place, complete and ready to go. Farther down the Bay at Ashe Point, local fishermen turned out their wherries and small double-enders built to their own individual specifications, and according to their personal requirements, in much the same way. At about the same time up at Canton, New York, not far from the St. Lawrence on the way to Montreal, in an empty barn in the winter of 1873, an undersized 30-year-old, an unemployed store clerk with a hacking, consumptive cough, built his first boat, a small rowboat, using borrowed patterns. He built it to take him into the woods and waterways of the Adirondacks to recover his health, but before he had barely finished he had sold the boat and had an order for another. He kept on building boats until he died in 1906, becoming one of this country's foremost builders, designers and manufacturers of planked sailing canoes and other fine small pleasure craft in the latter part of the nineteenth century. His name was J. Henry Rushton.

It would be easy to go on citing examples of notable boat builders who started out on their own as unskilled novices. Somehow in past generations, when boats were needed, the necessary skill and the means seem generally to have been found to produce them, as happened with the Adirondack guide-boat. More than 100 years ago, locked away in the remote seclusion of the Adirondack wilderness, untaught woodsmen with simple hand tools and a few patterns to go by, working in cramped quarters by the dim light of kerosene lamps, built 16-foot guide-boats that weighed no more than 72 pounds, complete with oars and ready to go, boats the builder-guides could carry through woods on their backs, and that would last a lifetime of use given proper care. These were boats whose construction was so delicate and precise that it has been compared to that of the violin. How these north-woods guides, lacking training and working under such unfavorable conditions, ever managed to do what they did–and not a few, but many working independently and alone–is an unexplained mystery. It is almost as if the essential skills lay latent and waiting in their hands, waiting to be awakened and brought forth.

And so it is today. In recent years not a few would-be boat builders have come to the boat shop at Mystic Seaport seeking instruction, complete novices who had never planed a board, or held a sharp chisel in their hands. Yet after brief exposure to the art they went home and successfully built boats to enjoy and be proud of, as their numerous letters and photographs offer no uncertain testimony.

It is the same for those who build from books. This is going on all the time, not only close by, but as far away as Alaska, New Zealand and Africa.

Nor do age, gender or occupation present any barriers. Seniors and adolescents, women as well as men, doctors, secretaries, insurance salesmen, carpenters, clergymen and so on *ad infinitum* can build boats. The lightning can strike anywhere, and as Pete Culler used to say: "Experience starts when you begin." The only way to learn to build a boat is to go ahead and build it regardless of drawbacks and difficulties. The road ahead has its ups and downs, its peaks of elation, and at times a need for the crying towels or the moaning chair mentioned by H. I. Chapelle.

Building a relatively simple yet highly functional small boat like *Fat Boat* opens up an area of experience that is closing for many of us, affording scope for creativity and individual expression we seem to be increasingly denied. In the construction of such small, simple boats there is leeway for builders to strike out on their own, to take charge, freedom to express themselves and throw off constraints imposed by others. In building a boat in this way, the builder builds himself into it in a very real sense, making it his boat in a way it never could otherwise be.

Technology is rapidly outgrowing human capacity to benefit from it. More and more our world is becoming a shadow world conjured out of the depths of the television tube, or appearing and disappearing from the computer screen. To a greater and greater extent we are losing physical contact with the real world in lives that are programmed for us, and although we may not be consciously aware of the loss, it registers nonetheless, taking its toll in various subtle ways, in feelings of emptiness, of something missing, of vague longings unfulfilled.

Humans are tool-making, problem-solving animals; by nature we are self-reliant, independent and free. We are linked to our physical environment by bonds that reach back to the beginning of time. In fact we are a part of the natural environment and draw our strength from it. In manipulating that environment we change it and change ourselves. We think with our hands as well as with our brains. Only when artificial restraints are removed to allow hands and brain to work together can the individual function at peak potential. That is why building a small traditional boat by hand with hand tools in the traditional way is such a satisfying, fulfilling experience. In the present-day world in which we find ourselves, there are too few opportunities left for hands and brain, harmoniously meshed together, to create tangible goods in which utility and beauty are conjoined.

But one that does remain is to build a small, traditional boat.

For some, building a traditional boat can be a release, a renewal of inner strength, a way to cope. There was a doctor in Cleveland, Ohio, who belonged to a local organization of boating enthusiasts in that city. They knew him as a prominent physician and hospital administrator, a very busy man who attended their meetings infrequently, but who could be depended upon for a generous donation. What they did not know about him was that he was secretly building a trimaran under lock and key in the spacious attic of his residence. After a punishing day at the hospital when his nerves were frazzled and he felt himself coming apart, he would disappear, slip away up into the attic, lock himself in and go to work on the trimaran, and soon he would be at peace with himself and the world, driving nails and fitting this piece to that.

Boats have been built in attics, and quite successfully in some instances. When the impulse to build strikes, a place in which to do it will be found. Like a bird seeking a place to nest, some odd, unlikely locations may be considered before the would-be boat builder finds a spot where he can settle in and go to work.

Back at the beginning of this century when gasoline engines for pleasure boats were new and experimental, a Connecticut lad not yet out of his teens set out to construct one of Bill Hand's early power launches in the attic of the family dwelling. At least one winter was spent sawing and hammering in the confines of the attic before the job was finished and the launch was ready to emerge into the light of day; but then it was discovered that the boat would have to be partly taken apart before it could be extracted from under the rafters. This was done, the boat was lowered to the ground, then reassembled and the motor installed. Finally the boat was ready for the water, where after a few trial runs the new launch proved to be all and more than its young builder had ever hoped for.

Was it then that a wild idea popped into his head, or had it been lurking there from the beginning? He had relatives living in Digby, Nova Scotia. Why not set course for Digby and run down to visit them in his new power launch? Just because no one had ever before taken a power launch outside for such an extended ocean cruise was no good reason why it couldn't be done. He would do it! Imagine how his family and friends must have implored, begged, threatened and forbade him to even consider such a foolhardy stunt. Their warnings and imploring went unheeded. He set out and made it, and it was a relatively fast and uneventful trip, at the end of which he found himself an international celebrity. The New York and London newspapers were full of it. The event seized the public's imagination. It was the first time anyone had dared venture out into the open ocean in a small power launch for any extended cruise. His exploit had opened a new era in power boat cruising.

The lad's name was George Bonnell, and the launch, *Old Glory I*, was the first of a continuing line of *Old Glory*'s that got larger and larger and more expensive as time went on. But the first was the only one built by George himself. It should be no surprise to learn that George Bonnell was one of the founding members of the Cruising Club of America. Business success came to him as a pioneer tax accountant with offices in Rockefeller Center and clients like Heinz of Heinz Pickles. Yet one goal he had set for himself eluded him to the end. He always intended to build another boat, but somehow he never managed to get around to it.

It had been his practice over the years when business was particularly good to divert some of the proceeds to the purchase of boat-building materials—lumber, paint, tools and the like—to be set aside for future use, and so it was that when Bonnell finally retired to his estate in Essex, Connecticut, unable by that time because of old age and ill health to build boats, the place, nonetheless, was loaded with an accumulation of everything a boat builder could ever want or need. Teak, mahogany, cedar and pine were stacked outside in covered piles for protection against the weather. Some of this lumber had been there so long that the ends of a few of the planks that stuck out had started to rot. Unopened cans of paint lined shelf upon shelf, some of the cans all but rusted through with age. There were tools of all sorts in the original boxes and unopened packages as they had come from the manufacturer, kegs of galvanized boat nails, gross upon gross of bronze wood screws, and in addition, a boat shop waited fully equipped with power machinery.

All throughout his long life it would appear that George in his secret thoughts had dreamed about recapturing those experiences of long ago that had been such a high point in his life. Sadly, it was not to be. He had waited too long.

On the day after Christmas 1887, the bark *Aquidneck*, home port, New York, loaded with Brazil wood and proceeding across the bay at Guarakasava, Brazil, about 400 miles down the coast from Rio, misstayed and stranded broadside on a hidden sandbar. There she lay for three days, helpless, pounded and raked fore-and-aft by the crashing waves until her back was broken.

All aboard got off safely, the captain and owner, Captain Joshua Slocum, his wife, their young son, Garfield, aged six, Victor, Slocum's older son and mate, full grown at 14, and "as strong as a windlass," and the crew of ten.

After salvaging all that could be removed in safety from the stricken vessel, Capt. Slocum sold the hulk, enabling him to pay off the crew, with a small moiety remaining, but not enough for him to engage passage home for himself and his family. There they were stranded thousands of miles from home on the edge of the Brazilian wilderness virtually penniless. They had no intention of begging passage back. Capt. Slocum, who was one of the ablest shipmasters of his day, who had sailed in all quarters of the globe as master of some of the finest ships afloat at the time and who had experienced shipwreck before, was not wanting in resourcefulness and courage. He was not long in coming up with a plan. There was one thing they might do, if they dared try, which was to build themselves a boat and sail back. That was what they did, covering 5,510 nautical miles in $53^{1}/_{4}$ days at sea in a 35-foot boat, seven and a half feet wide with three feet depth of hold, modeled after Slocum's recollections of Cape Ann dories with additional features taken from a photograph of a "very elegant Japanese sampan" which Slocum happened to have with him. He rigged his boat like a Chinese sampan.

Directly after the first of the new year, Slocum and his family set to work with what remnants and odds and ends they had been able to save from the wreck. It was a meager stock, indeed, but in barely more than four months time, despite a number of setbacks, they were rigged, provisioned and ready to sail. Launched on the 13th day of May, the day the slaves in Brazil had been set free, their little vessel was named *Liberdade* (freedom).

The building of *Liberdade* is certainly one of the most remarkable feats of boat building ever accomplished by anyone anywhere. It is cited here not to imply any similarity of *Liberdade* to *Fat Boat*, although it is true that both are dories of sorts, but as a prime example of what can be accomplished by unpracticed, but resolute and ingenious boat builders in the face of towering difficulties.

For some notion of the extraordinary manner in which *Liberdade* was contrived and constructed, Slocum's own words from *Voyage of the Liberdade*, published by Roberts Brothers of Boston in 1894, can hardly be improved upon.

"To begin with, we had an axe, an adze and two saws, one 1/2 inch auger, one 6/8 and one 3/8 auger-bit; two large sail-needles, which we converted into nailing bits; one roper, that answered for a punch; and, most precious of all, a file that we found in an old sail-bag washed up on the beach. A square we readily made. Two splints of bamboo wood served as compasses. Charcoal, pounded as fine as flour and mixed in water, took the place of chalk for the line; the latter we had on hand. In cases where holes larger than the 6/8 bit were required, a piece of small jack-stay iron was heated, and with this we could burn a hole to any size required. So we had, after all, quite a kit to go on with. Clamps, such as are used by boat builders, we had not, but made substitutes from the crooked

The 35-foot* Liberdade *which sailed 5,510 nautical miles from Brazil to the coast of South Carolina in 53 1/4 days in 1888 with Joshua Slocum, his wife and two sons aboard. From* Voyage Of The Liberdade *by Joshua Slocum, Press of Robinson & Stephenson, Boston, Massachusetts, 1890.

guava tree and from massaranduba wood.

"Trees from the neighboring forest were felled when the timber from the wrecked cargo would not answer. Some we did not use the saw upon at all, it being very hard, but hewed it with the axe, bearing in mind that we had but one file, whereas for the edged tools we had but to go down to a brook hard by to find stones in abundance suitable to sharpen them on.

"The unskilled part of the labor, such as sawing the cedar planks, of which she was mostly made, was done by the natives, who saw in a rough fashion, always leaving much planing and straightening to be done, in order to adjust the timber to a suitable shape. The planks for the bottom were of iron wood, 1 1/4 x 10 inches. For the sides and top red cedar was used, each plank, with the exception of two, reaching the whole length of the boat. This arrangement of exceedingly heavy wood in the bottom, and the light on top, contributed much to the stability of the craft.

"The iron wood was heavy as stone, while the cedar, being light and elastic, lent buoyancy and suppleness, all that we could wish for.

"The fastenings we gathered up in various places, some from the bulwarks of the wreck, some from the hinges of doors and skylights, and some were made from the ship's metal sheathing, which the natives melted and cast into nails. Pure copper nails, also, were procured from the natives, some ten kilos, for which I paid in copper coins, at the rate of two kilos of coin for one kilo of nails. The same kind of coins, called dumps, cut into diamond-shaped pieces, with holes punched through them, entered into the fastenings as burrs for the nails. A number of small eyebolts from the spanker-boom of the wreck were turned to account for lashing bolts in the deck of the new vessel. The nails, when too long, were cut to the required length, taking care that the ends which were cut off should not be wasted, but remelted, along with the metal sheathing, into other nails.

"Some carriage bolts, with nuts, which I found in the country, came in very handy; these I adjusted to the required length, when too long, by slipping on blocks of wood of the required thickness to take up the surplus length, putting the block, of course, on the inside, and counter-sinking the nut flush with the planks on the outside; then screwing from the inside outward, they were drawn together, and there held as in a vise, the planks being put together 'lap-streak' fashion, which without doubt is the strongest way to build a boat.

"These screw-bolts, seventy in number, as well as the copper nails, cost us dearly, but wooden pegs, with which also she was fastened, cost only the labor of being made. The lashings, too, that we used here and there about the frame of the cabin, cost next to nothing, being made from the fibrous bark of trees, which could be had in abundance by the stripping of it off. So, taking it by and large, our materials were not expensive, the principal item being the timber, which cost about three cents per superficial foot, sawed or hewed. Rosewood, ironwood, cedar or mahogany, were all about the same price and very little in advance of common wood; so of course we selected always, the best, the labor of shaping being least, sometimes, where the best materials were used.

"These various timbers and fastenings, put together as best we could shape and join them, made a craft sufficiently strong and seaworthy to withstand all the buffetings on the main upon which, in due course, she was launched."

And then on the 24th of June, as Slocum relates, "after having sailed about the bay some few days to temper our feelings to the new craft, we crossed the bar and stood out to sea," on a voyage that was to last 127 days until at 8 p.m. on October 28, 1888, 13 days from Mayaquez and 21 days from Barbados, they dropped anchor in Bulls' Bay off the coast of South Carolina with Cape Roman Light clearly visible to the north.

On the voyage back fortune smiled on the little vessel and its crew. Winds and currents favored for the most part. They were spared the buffeting of violent storms. Slocum's knowledge of the coast, a lifetime of experience with boats of every sort, his skill as a helmsman, his proficiency as a navigator (his compass, charts, sextant and chronometer, he had managed to save from the wreck of the *Aquidneck*), all of these things stood them in good stead, and in ports of call along the way Slocum was well known and had many friends, which also helped a lot.

The sailing ability of *Liberdade* says much for her hull design and her rig. The average for the 53 days spent at sea was nearly 104 nautical miles a day, although on her first day in the current of the Gulf Stream she logged 220 miles. It was there, bowling along at breakneck speed, she hit a floating spar that tore off her false keel, although she went on none the worse without it. Once she was bumped by a whale and lifted part way out of the water. The sharks that little Garfield had worried about never showed up.

It might seem to some, perhaps, that we have gotten off the track and wandered aimlessly afield in recounting the *Old Glory I* and *Liberdade* episodes; but I think not, for both have relevance in considering the construction of most owner-built boats, including *Fat Boat*.

Old Glory I strikes a cautionary note—if you have a mind to build yourself a boat, don't hesitate, delay and put it off too long. And when it comes to finding a place to build, that always was a problem, but never to the extent that it is today. Yet the really serious would-be builder will not be deterred; eventually he will find a place. There was the man who built a Whitehall in the family living room. Let us hope his wife liked boats as much as he did, and that he still had a wife when the boat was launched. And then there was the young boating enthusiast who went back to Manhattan from one of Mystic Seaport's boat building classes to put together a pram in a Brooklyn high-rise apartment, umpteen floors above the ground.

The saga of *Liberdade* is a heroic tale of what can be accomplished by dint of pluck and elbow grease with a minimum of mechanical equipment. It is a good antidote for popular television features detailing the construction of such simple wooden objects as footstools, plant stands or magazine racks in spacious shops exhibiting every conceivable kind of expensive and unnecessary woodworking machinery—advertising come-ons for the power-tool manufacturers.

In enlarging and adapting *Fat Boat* for sailing, while at the same time maintaining a capability for rowing as well as for mounting a small outboard motor, almost no changes in the original basic hull conformation were made, except to carry back the lower end of the stem and sharpen the entrance, replacing the flat of the dory bottom at the forward end with a V-section. This V-section should largely eliminate the often objectionable pounding that flat-bottom dories are notorious for when heading into chop, as well as to make for an easier-driven boat.

The overriding objective in reworking *Fat Boat*'s design was versatility, coupled with adaptability—that is to say, a little boat that would meet the diverse needs of many different owners, as well as a boat that builders could modify to suit individual need or taste without spoiling it. The basic dory design used here is an exceptionally forgiving one. *Fat Boat* is easily adaptable to a number of different sailing rigs, the spritsail or the sliding gunter would permit masts short enough to be stowed in the boat. Other options might be to add a jib, or to lengthen the mast by a couple of feet and increase the area of the main to correspond. The length of the hull could be increased simply by inserting an additional several feet of length amidships. Everything else including the ends would remain exactly the same. Then again, the beam could be increased a moderate amount merely by widening the bottom. If this increase were to take the form of a narrow triangle or thin wedge with its sharp end forward, the increase in width at the stern would help to support additional sail area or a larger outboard motor, although the boat may not row quite as easily with the wider transom.

In planning the construction of *Fat Boat*, account was taken of the current tendency to build boats as light as possible not only to increase their speed, but to facilitate getting them in and out of the water when they are trailered. Increasingly, boats are being transported over the road on trailers. Opportunities to use them are greatly extended thereby, and more and more boat owners are keeping their boats at home because it is cheaper and there is less chance of vandalism.

Many boaters with moderate incomes who find themselves unable to afford the expense of larger boats, are discovering they can have just as much fun at a fraction of the cost with smaller boats in the easily trailerable range, like *Fat Boat*. Clearly the boat should be built light, as light as is advisable, but if it is to last, remain tight and withstand the severe strains that hard sailing occasionally puts on a boat, it should be solidly built. There should be no skimping on materials or undue downsizing of scantlings. And it is here that experience and judgment come into play.

Lumber Recommended For *Fat Boat*

For ultra-light boats in *Fat Boat*'s size range, some builders are using 1/4-inch plywood for planking, and claim they are finding it adequate, but over the long haul and for a boat to stand up under hard use I should not want to go under 3/8 inch, certainly not for this boat. Besides, those who use 1/4-inch frequently cover it with fiberglass which itself is a heavy material, with the result that the combination weighs more than 3/8-inch plywood, not to mention the considerable expense of the fiberglass, and the labor and the extra trouble applying it. As planking material for outboard skiffs widely used as fishing workboats in Alaskan waters, 3/8-inch marine plywood has proved adequate, passing the most stringent tests.

In constructing *Fat Boat* considerable weight is saved by substituting spruce for oak where oak is generally used. Spruce is the strongest wood for its weight of any timber grown in North America. It glues well and takes and holds fastenings well, is widely available and moderately priced, that is with the exception of Sitka spruce, and possibly some of the other western varieties. The species I am recommending here is eastern red spruce, widely distributed for use as ledger boards, staging planks and the like.

Instead of using spruce for the thwarts, stern benches, stern sheets, top of the centerboard trunk and the like, northern white pine is a good choice, and there are also a number of local lumber species native to different parts of the country that make acceptable substitutes. One Tennessee boat builder recommends sassafras. Whatever is used, it should not be too heavy.

No steam bending is required in the construction of *Fat Boat*. The curved stem and frames are made up of short sections cut from straight boards, laid out to avoid any weakness from short, or cross grain, pieced together and glued. Butts must be well staggered, from side to side.

In selecting the plywood for the planking, decking, waterways, transom and bottom, pains should be taken to get first quality marine grade. The best is none too good. A little extra spent here is no extravagance. Bruynzeel stands out as one of the superior brands.

In addition to the plywood for the boat itself, two 4' x 8' sheets of ordinary exterior plywood, good one side, will be required to provide the surface for drawing the full-size laydown of the lines, and for laying out and splicing the planks later. These two panels should be not less than 1/2 inch thick, and must be perfectly flat and unwarped. Plywood that has taken a twist from improper piling in storage should be rejected. After these panels have served their purpose for this, they can be salvaged for use elsewhere. They will not have been cut or otherwise damaged except for a few small nail holes that are easily puttied and painted over.

The 4" x 6" joists that form the sides of the ladder-frame should be select stock, well-seasoned and perfectly straight. If they are not straight, it will be difficult to make a true and accurate setup of the boat. After the ladder-frame has served its purpose, it is taken apart and the joists are used to make a form on which the mast is assembled and glued. Later on they can be salvaged undamaged except for a few nail holes.

For making the molds, the cross cleats for the ladder-frame and the two horses upon which the ladder-frame rests, ordinary 1-inch spruce or pine boards will do nicely.

Mahogany is specified for the rudder because it is a moderately hard, close-grained wood that works well and holds its shape. As it will be removed when the boat is launched, or taken out of the water, the extra weight of the mahogany is of no consequence.

The mast and boom are spruce, pieced together and glued. The centerboard and the centerboard trunk are plywood with the exception of the headledges and bed logs, which are oak. Skeg, stern quarter knees, transom knee, breasthook and boom jaws are also oak, but altogether there is not enough oak in these small pieces to increase the weight by any significant amount. In considering the overall weight of the boat, the additional weight of the epoxy adhesive used throughout should not be overlooked. Epoxy is a heavy material. That is the reason, aside from extra cost, that covering the boat inside and out with epoxy instead of paint, as is frequently done, is not recommended. The use of epoxy as an adhesive, however, wherever it can be applied, is indispensable, adding immensely to the structural strength of the boat.

Everything considered, the materials needed for the construction of *Fat Boat* would seem to offer no particular problems, either as to availability or cost. As for cost, while the materials specified should be readily affordable for most prospective builders, they will not come cheap. There is no such thing as cheap lumber any more.

Tools And Related Equipment

What remains to be taken into account are the tools and various related equipment. Considerable effort has been made to make the construction of this boat as simple and as straightforward as possible. The builder, if he has to, can manage quite well with a minimum of power tools, say an electric drill and a small table saw, although the addition of a small band saw would expedite the job and make it much easier. A small electric handsaw for cutting plywood would be handy.

It is more or less taken for granted that anyone undertaking to build *Fat Boat* will have the usual assortment of hand tools and some experience in using them. These should include a block plane either by Stanley or Record for beveling the laps. For best results it should be a low-angle block plane. The low-angle plane is no longer manufactured, but they are still occasionally obtainable secondhand at flea markets, yard sales, tool auctions and the like, and are worth making an effort to obtain. At one time Sargent manufactured an excellent low-angle block plane of which surviving examples sometimes can be obtained.

At least one larger plane will be needed, and if the builder is restricted to one plane, it should be a jack plane which will do it all sufficiently well, including rounding the mast and boom. Jack planes of acceptable quality are still put out by Stanley, Record, and Lee Valley in Canada, but these do not quite measure up in quality to those formerly produced by Stanley, which can still be obtained secondhand. In addition to planes, a spokeshave will be needed to clean up curved surfaces. The spokeshave can also be used for beveling the planking laps.

While this is not the place for instruction in the use and care of tools, it still needs to be emphasized that too many inexperienced woodworkers do not know how to adjust hand planes to get the best results, or how to sharpen them properly. This is something to look into by those who may have doubts about their proficiency. Good work cannot be done without sharp, properly working tools, planes especially, and this means having the right sharpening equipment, a grinding wheel that does not overheat and burn the blades, and oil or other sharpening stones for honing the final edges.

A sharp handsaw is also a requirement. For a handsaw which will see much use, a 10-point 22-inch cross-cut panel saw would be a good choice, 12 points would not be too fine, and for cutting plywood, excellent. Of course it would have to be sharp with just a bit of set. Seven points is much too coarse. So long as there is a power table saw there will be little need for a hand ripsaw, but the kit should include one just the same.

It has been said that a boat builder can never have too many clamps, but no more than a half-dozen 6-inch and 4-inch C-clamps ought to suffice here. A couple of larger Jorgensen hand screws might also prove useful. One such hand screw, when clamped flat to the top of the bench with a C-clamp, makes an effective substitute vise for holding plank in a vertical position for planing the edges. Some sort of work bench with a vise is a necessity, and if need be, there are any number of ways of improvising one. How to do it will have to be left to the initiative and ingenuity of the boat builder.

If the spruce used is obtained from ledger boards and the like, it will come rough, and will need to be run through a surface planer to size and smooth it. Arrangements can generally be made with a local woodworking shop to make one job of this for a reasonable cost. In the old days builders would have had to have done this by hand with a plane, as still could be done today, but it is a lot of work.

The builder is advised to make photocopies, enlarged if possible, of the drawings and the materials list from the book. Mounted on separate sheets of cardboard, these will be readily available for easy reference during construction without having to turn each time to the book, saving time as well as the book.

Parts are identified in the building directions by parentheses enclosing the number of the part preceded by the number of the drawing(s) or sheet(s) in which it appears, thus the first reference to "Bottom," part No **5**, keyed on Sheets **2**, **5**, and **6**, is followed by **(2, 5, 6-5)**.

Laying Down And Fairing The Lines

Before construction starts, the lines are laid down full-size and faired. Because of the simplicity of dory lines, and because *Fat Boat*'s lines were accurately faired prior to scaling the offsets, this process can be considerably simplified here. It will suffice to draw the full-size outline shape

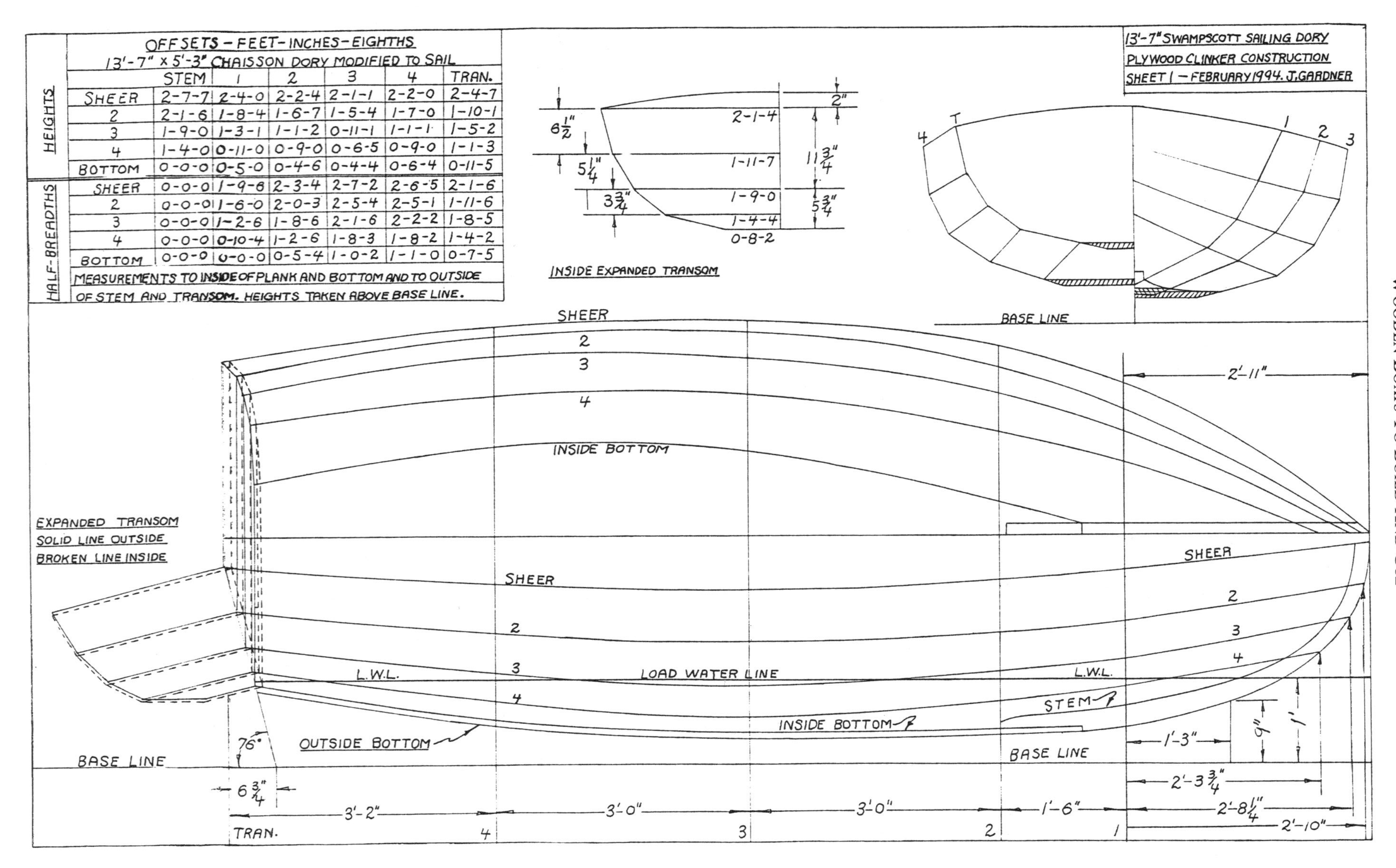

OFFSETS – FEET – INCHES – EIGHTHS

13'-7" x 5'-3" CHAISSON DORY MODIFIED TO SAIL

		STEM	1	2	3	4	TRAN.
HEIGHTS	SHEER	2-7-7	2-4-0	2-2-4	2-1-1	2-2-0	2-4-7
	2	2-1-6	1-8-4	1-6-7	1-5-4	1-7-0	1-10-1
	3	1-9-0	1-3-1	1-1-2	0-11-1	1-1-1	1-5-2
	4	1-4-0	0-11-0	0-9-0	0-6-5	0-9-0	1-1-3
	BOTTOM	0-0-0	0-5-0	0-4-6	0-4-4	0-6-4	0-11-5
HALF-BREADTHS	SHEER	0-0-0	1-9-6	2-3-4	2-7-2	2-6-5	2-1-6
	2	0-0-0	1-6-0	2-0-3	2-5-4	2-5-1	1-11-6
	3	0-0-0	1-2-6	1-8-6	2-1-6	2-2-2	1-8-5
	4	0-0-0	0-10-4	1-2-6	1-8-3	1-8-2	1-4-2
	BOTTOM	0-0-0	0-0-0	0-5-4	1-0-2	1-1-0	0-7-5

MEASUREMENTS TO INSIDE OF PLANK AND BOTTOM AND TO OUTSIDE OF STEM AND TRANSOM. HEIGHTS TAKEN ABOVE BASE LINE.

of the boat in profile together with the plan views or half-breadth shape at the sheer and of the bottom. To get both views on a 4-foot width of plywood it will be necessary to superimpose one view over the other. The baseline used for the profile view will serve as the centerline for laying out the half-breadth views.

One foot and eleven inches above the base line and parallel to it, a line is laid down representing the top of the ladder-frame stringers. Where this line crosses the stem and the stern transom in the profile view should be marked for reference when the boat is set up for planking. This line squared across the sectional view indicates the location of the top of the cross spalls on the planking molds. When the molds are set up for planking, the tops of the cross spalls will rest on the top of the ladder-frame.

Begin by snapping the baseline for the profile view with a chalk line, stepping off the sections on it, and squaring up the section lines, which will serve for both profile and plan views. The section lines are most accurately erected perpendicular to the baseline using trammels as beam compasses.

The body plan or sectional views from which the planking molds are made can also be superimposed in the upper center of this lines laydown. There are not enough lines from the other views to be confusing. Measurements for the body plan are taken from the offsets. The height of sheer and bottom are checked from the profile laydown, and the widths of sheer and bottom are checked from the half-breadth laydown.

Instead of superimposing the body plan over the profile and half-breadth views, some might find it less confusing and more convenient to lay it down on a separate piece of plywood.

Making The Molds

In making the planking mold for each section in turn, both sides of the section are first laid down on the lay-out surface. Their shape is then picked up and transferred to the pieces of 3/4" pine boards from which the mold will be made. These are cut out and laid back on the lay-out surface. Any adjustments that are required to get exact conformance to the original sectional outline are made, and the pieces of the mold are temporarily tacked in place to prevent them slipping out of register while they are being securely fastened with glue and 1½-inch wood screws. Before the mold is taken up, its vertical centerline is marked on the cross spall. This will later line up with the centerline on the ladder-frame.

There are several methods for picking up the shape of lines laid down on the mold loft floor. One is to lay small thin threaded nails or tacks at intervals along the line with one side of their heads partly embedded in it. The other half of the head sticks up, and when a soft wood board is placed over them and briskly tapped, the projecting part of the heads sink in enough to mark it. By running a batten through these indentations, a line can be drawn which reproduces the original line exactly. In this way the sides of the planking molds are marked to be cut out prior to their assembly.

Equipment needed for laying out and fairing the lines includes a straightedge, 6 to 8 feet in length, which is best made of dry, well-seasoned white pine 3/4" thick, a carpenter's framing square, rules and tapes, pencil dividers, trammels, battens of various sizes, a chalk line and chalk, several sharp awls, pencils and a hammer. Some 2d nails to hold the battens in place hardly need to be mentioned.

Sub-Assemblies Prior To Setup

Prior to setting the boat up for planking it will be necessary to cut out and shape a number of parts and to make up several sub-assemblies. These will consist of the ladder-frame and horses, the section molds, the transom, the stem and the inner layer of the bottom, serving as a pattern for the outer layer which is cut out at the same time.

Making Up The Bottom

The bottom **(2, 5, 6-5)** is made up of two layers of 3/8" plywood glued together. In laying it out, the forward end runs one foot forward of Station 2 and is 2⅝" wide to correspond to the siding of the stem. The two layers are cut out at the same time, using one as the pattern for the other. The

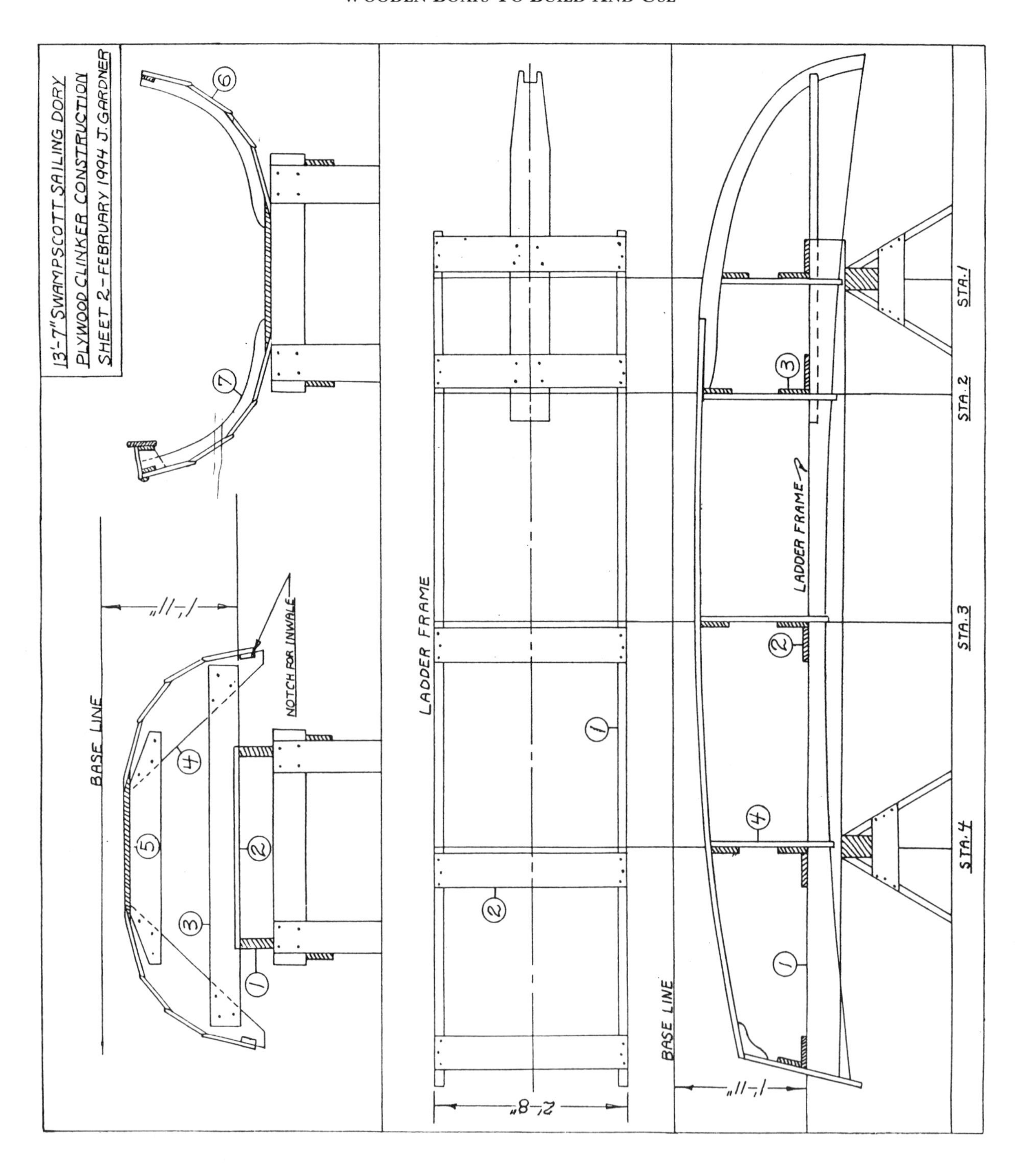

centerboard slot laid out according to the dimensions given on Sheet 6 is mortised through both, taking care to keep it slightly under 1¼" in width to insure a snug fit for the headledges. The six bottom cleats **(5-15)**, 1½" x 1½" spruce, spaced 17" on centers, are glued and nailed in place with 1½" No. 10 annular nails on the inside of the bottom. Their ends stand 1½" back from the chine to allow passage for water to drain around them. The cleats are not cut away for the centerboard trunk until it is installed later.

Making Up The Transom

The transom **(6-41)** is also made up of two layers of 3/8" plywood glued together. For economy the inside layer is made in two parts cut from scrap. These center on the outside of the 10"-wide, 1½"-thick vertical reinforcement **(5-34, 6-43)** for mounting the outboard motor and are glued and nailed to it with 1½" No. 10 annular nails. The lower end of this reinforcement should be cut long enough to bevel for a tight fit against the inside of the bottom. This bevel, which is also the bevel of the inside of the transom, is picked up from the profile view in the lines laydown.

To glue the two layers of the transom together, the outer layer, outside face down, is placed on the flat surface of one of the plywood panels used for the lines laydown, and its inner and upper surface is given a thin coating of epoxy adhesive. The outer surface of the inner layer is likewise coated with adhesive, and the two layers are pressed together and tacked through to the underlying plywood panel with a couple of thin nails to hold the two layers in register. Around the edges are placed thin strips of pine, in short lengths through which thin resin-coated nails, spaced a couple of inches apart, are driven down tight and into the plywood panel underneath. These will supply enough pressure to bring the two glued surfaces in firm contact, which is all that is needed. When the glue has set the fine strips are split away and the nails are pulled. In addition to these nails a number of heavy objects, say several cement blocks, placed on the assembly would help to bring the gluing surfaces in firm contact throughout.

Making Up The Stem

Making up the stem **(3, 4, 5-20)** may not be quite as easy as might first appear. It is recommended that a mold of the stem shape be made from scrap plywood, taking up the shape from the lines laydown by the method using the heads of nails or tacks, as already explained. If a piece of plywood large enough for this mold is not available, two or three smaller pieces could be glue-spliced together to make it. After the shape of the mold has been marked and it has been cut out (on the band saw, if there is one) it is trimmed to the line with the spokeshave, and laid back on the lines laydown. It should fit exactly.

The stem is glued together from three lifts of 7/8" spruce. It would be a good idea if the first lift, which is made up from two pieces, be glue-spliced together with a 4-inch tapered scarph. When the glue has set and the lift has been trimmed with the spokeshave, if necessary, to conform to the stem mold, its inner surface is coated with glue and the pieces that make up the center lift are also coated with glue, both on the surface that will contact the lift below and on their ends, these for the second time, a first coat on the meeting ends having been applied sometime before and allowed to soak in and dry, sealing them. This done, and when the ends have been drawn together tight, the two lifts are nailed together with 1 1/2" No. 10 annular nails about half an inch from the inside edge where they will not be in the way of the chisel and plane, when the stem is beveled. As soon as this is done, the third lift is added in exactly the same way. C-clamps can then be applied to the assembly at intervals to pull it together and supply pressure, or it can be nailed from either side with 2 1/2" double-headed staging nails to substitute for C-clamps. These are easily pulled when the glue has set.

The Setup For Planking

The ladder-frame for setting up the molds for planking is clearly diagrammed on Sheet **2**. The centerline put on with the chalk line is clearly marked on the cross cleats **(2-2)**. When the planking molds are put in place, their centerlines marked on their cross spalls **(2-3)** should fall exactly on the fore-and-aft centerline of the ladder-frame. The setup must be level athwartships. This can be checked by resting a spirit level on each of the cross spalls. On the planking molds the setup need not be level fore-and-aft.

In making up the ladder-frame, the cross cleats at Stations 3 and 4 must be located 3/4", or the thickness of the cross spalls, aft of the station lines, precisely as shown in the drawing, but at Stations 1 and 2 they should be 3/4" forward of the station lines. So located, the frame molds will not require beveling.

When the station molds are in place and have been centered and leveled, their cross spalls are fastened to the ladder-frame cleats with several staging nails. Several short diagonal braces can be fitted to hold the molds in an upright position, and they can be further secured by a temporary

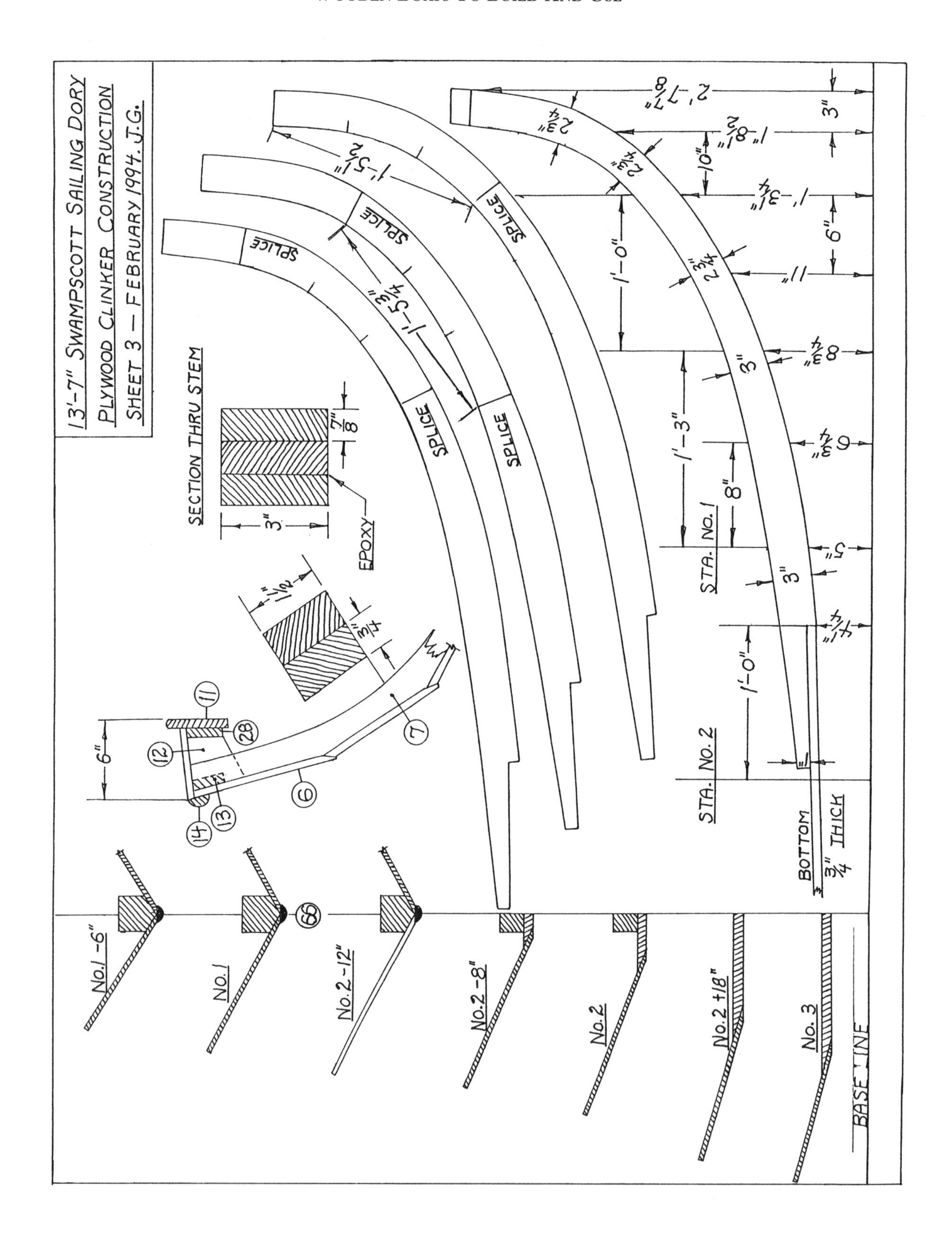
13'-7" SWAMPSCOTT SAILING DORY
PLYWOOD CLINKER CONSTRUCTION
SHEET 3 — FEBRUARY 1994. J.G.
SECTION THRU STEM
EPOXY
SPLICE
STA. No. 1
STA. No. 2
BOTTOM
3/4" THICK
No. 1 -6"
No. 1
No. 2 -12"
No. 2 -8"
No. 2
No. 2 +18"
No. 3
BASE LINE

batten running fore-and-aft and tacked across the bottoms of the molds.

At this point the sides of the setup are checked to make sure they are fair, prior to installing the bottom assembly. A fairing batten, approximately 3/8" thick and 1 3/4" to 2 1/4" wide and long enough to span the molds with some to spare at the ends, is run lengthwise around the sides of the molds more or less in the same direction the planking will take. The batten should be fair without showing bulges or flat spots. If any unfairness is evident, adjustment will have to be made to correct it, otherwise it will show up in the finished hull.

Prior to installing the bottom assembly, which comes next, it is advisable to temporarily tack a full-length batten to the molds on either side to hold them rigid.

Installing The Outer Bottom Layer Of Plywood

The bottom assembly, which consists of the stem, the inner bottom layer of 3/8" plywood with the six bottom cross cleats attached to it and the transom, the reinforcement for the outboard motor, and the stem bottom knee, is bent on centered on the centerline marked on the ladder-frame, and temporarily fastened in place at either end. This done, the bottom assembly is ready to receive the outer bottom layer of plywood. The joining surfaces of the two layers are coated with epoxy adhesive. If put on too thick the excess will only be squeezed out, making a mess when the two surfaces are pressed together.

Start by nailing the forward end of the bottom where the stem shoulders are to receive it. One and one-half inch No. 10 annular nails are used for nailing the outer bottom layer. Proceed by working back toward the stern, bending the plywood down, clamping it at intervals and nailing through into the bottom cleats, already fastened to the inside of the bottom. At the stern slightly longer nails could be used in the bottom of the transom, the reinforcement for the motor and the stern bottom knee.

How To Make And Use Leatherheads

If there is an abundance of clamps, clamps set up at intervals along the sides, to be left in place until the glue hardens, should be enough to draw the two layers together in a close fit. If not, 1" No. 8 flat-head wood screws put in 4" to 6" apart around the outside, will do the job. They will come out easier after the glue sets if soaped when driven. Also, what I shall call "leatherheads" might be substituted, consisting of nails with washer-like pieces of leather under the heads. Leatherheads function much the same as small double-headed staging nails. When they are driven in tight against a surface, the leather applies pressure without marring it, as well as holding the head of the nail out far enough to be grasped with carpenter's tongs for easy extraction. They are easily made from small pieces of thick leather pierced for the nails with an ice pick, and can be used over again several times. When *Fat Boat* is planked they will be used to fasten the planking temporarily to the planking molds.

Beveling The Stem And Bottom

As soon as the glue holding the bottom together has set, the bottom, stem and transom are ready to be beveled for planking. The bevel angles must be right so that the planks lie flat and in touching contact with the molds when the planks are bent on and glued in place.

A centerline is gauged on the stem with lines marked on either side, parallel to it and 1/4" out. The stem is beveled in as far as these outer lines. The angle of the changing stem bevels is taken by means of a fairing batten 5/16" to 3/8" thick, 2" to 2 1/2" wide, and long enough to reach from the stem to Station 2 and a foot or two beyond. This batten, which simulates the planking, is applied in the direction the planking runs. Wood is cut away from the stem until the batten lies flat from the inside line to the outside of the stem. The planking will lie flat to this bevel when it is applied.

Using the fairing batten in the same way, the transom is beveled from the outside to its inside edge.

Beveling The Bottom

Beveling the bottom to receive the garboard requires a different procedure. First, a batten 5/16" to 3/8" thick, and running from stem to stern, is nailed down tight to the molds with leatherheads just inside the location marked for the upper inside edge of the garboard. This batten will support the outer end of a bevel stick, a straight piece of 3/4" pine board approximately 18" to 20" long and 2" wide. On the underside of the inner end of the bevel stick is fastened a spacer, a flat piece of pine 4" to 6" in length and exactly the same thickness as the batten on which the outer end rests. In beveling the bottom, the bevel stick is applied at right angles to the curve of the bottom, with its outer end resting on the batten and the inner end, with spacer attached, resting on the outside edge of the bottom, which is cut away in conformity to the slant of the spacer until the inside corner of the bottom is reached. The beveled edge of the bottom will touch the under surface of the spacer throughout.

The angle of the bottom bevel, together with the length of the beveled surface, changes progressively as the twist of the garboard approaches the stem and stern, lifting the outer edge higher and higher. With the bevel stick this change is allowed for automatically.

Planking With Strakes Made Up In Two And Three Sections

In planking the boat with strakes made up in two and three sections as shown on Sheet 9, it is possible to cut these sections out of standard 4' x 8' plywood panels with very little waste. Ability to spile these strakes accurately is of critical importance. The spiling technique is explained in standard boat building manuals by H. I. Chapelle, Bob Steward and others. I have to say I like my way of doing it, as set forth in the appendix of my *Building Classic Small Craft, Vol. 1,* which was first published by International Marine Publishing Company of Camden, Maine, in 1977.

If this boat were timbered out with all its frames in place before planking, it might be easier to fasten on the planking one section at a time, splicing and gluing the strakes when they were put in place on the boat, but because there are no intermediate frames between the molds as yet around which to bend the plank section and to support their scarphed ends for glue splicing, the planking is best put on the boat in full-length strakes reaching from stem to transom. Obviously this requires that splices be scarphed and glued prior to fastening the strake on the boat. This can be a critical operation, which, although not difficult, calls for exactitude. The slightest slippage when a splice is glued can alter the shape of a strake enough to spoil its fit completely.

One tried and true method is to make a full-length spiling of the strake and transfer its shape to the mold loft floor. Using the same spiling, get out the two sections of the strake that are to be spliced together. Scarph the ends for the splice, and lay out the sections on the mold loft floor. They should fit exactly the outline of the strake already marked there. Tack the two sections in place with leatherheads, coat the splices with glue and then bring the scarphed surfaces together with enough pressure to establish firm contact. The weight of one cement block should be ample. Be sure there is a sheet of paper under the splice to prevent any glue that might squeeze out from sticking to the floor. If the spiling was accurately made the strake should fit perfectly when it is bent on the boat, and it will serve as a pattern for the corresponding strake on the opposite side.

Installing The Inwales

Before the planking goes on, the inwales **(3-13)** are put in place. They run the length of the boat in the cutouts made for them in the upper ends of the planking molds **(2-4)**, and are fastened to the stem and the stern. These cutouts should not be too deep in order for the inwales to project about 1/8" above the sheer line to allow for planing later on when the sheer is faired to receive the waterways and decking which land on the inwales and are glued and nailed to them. Small dabs of glue attach the inwales to the planking molds to hold them in place until the boat is planked. After the sheer strakes have been bent on and glued and nailed to the inwales, a thin saw will separate the inwales from the planking molds quite easily.

Laying Out The Plank Bevels

Finally, prior to hanging the garboards in place on the boat, a uniform bevel 7/8" wide for the lap with the plank above is planed

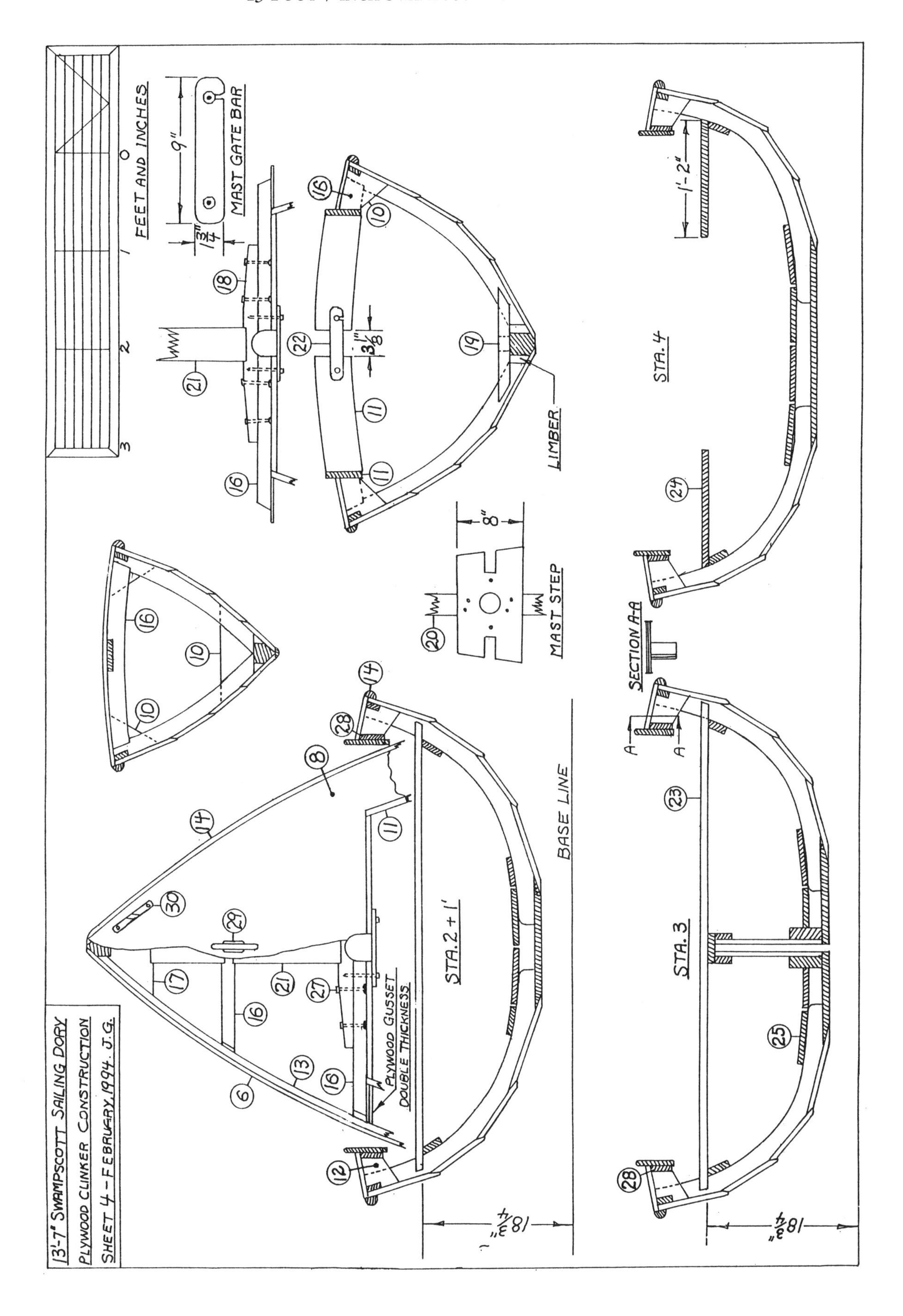
13'-7" Swampscott Sailing Dory
Plywood Clinker Construction
Sheet 4 - February, 1994. J.G.
Feet and Inches
0
1
2
3
Mast Gate Bar
9"
Mast Step
8"
Limber
Section A-A
Base Line
Sta. 2 + 1'
Sta. 3
Sta. 4
Plywood Gusset Double Thickness
1'-2"
18¾"

on the upper side leaving the inside edge 1/16" thick. The garboards are glued in place and nailed to the stem and transom with No. 10, 1 1/2" annular nails, although No. 10 bronze wood screws could be used instead. Leatherheads, later to be pulled and plugged, are driven through the plank into the molds to hold the plank in place, and No. 6 round-head screws are driven through the bottom lap wherever needed to pull it tight, also to be removed and plugged when the glue has set. Epoxy fairing gel can be used to fill the screw holes.

The succeeding planks, with exception of the sheer plank, have a uniform 7/8" lap bevel on the upper side, the same as the garboard. But on the lower inner side the overlapping bevel will change along the length of the boat, as the planking angle changes, and must be made accordingly.

The usual procedure is to take off the bevel angle at each mold station with a small bevel square or planker's bevel, and to transfer it to its respective station on the plank, where the wood is pared away with a sharp chisel until the plank shows the correct angle of bevel at that spot. After this has been done for each mold station in turn, the wood between stations is planed away with a sharp plane or spokeshave set fine until the bevel angle at one station merges gradually and imperceptibly into the bevel angle at the next.

Fitting And Fastening The Planks

The plank above the garboard is put on with epoxy adhesive thick enough not to leak out of the lap. Starting at the bow, its forward end is nailed or screwed to the stem as was done with the garboard, and it is bent around in place and temporarily nailed twice with leatherheads at each mold station. It is first nailed down tight through the lap into the mold and then through the plank into the mold just below the upper bevel. At the stern either 1 1/2" No. 10 annular nails or No. 10 flat-head bronze wood screws fasten the plank to the transom. The planks are got out a trifle long at the transom to be trimmed off later. Along the lap wherever necessary to draw it tight, 1" No. 6 round-head screws are driven to be removed later when the glue has set and the screw holes plugged with an epoxy fairing gel or filler forced in with a limber putty knife.

The two remaining planks, one of which is the sheer plank, go on in the same way. As already mentioned, the sheer plank is also fastened to the inwale with glue and 1" No. 12 annular nails.

In fitting the laps, should the lap surface on the upper outside edge of a plank be a little short at any place due to the angle of the plank knuckle, the overlap of the plank above should not be shortened, even if it projects below the closure of the lap as much as 1/8". The very slight gap that results will be hardly detectable, and should be filled with epoxy gel.

For appearance the outer edges of the laps may be rounded slightly and any bits of excess glue scraped off. The protruding bottom edges of the garboards are planed off flat and continuous with the bottom, and because when plywood is planed on a bias end grain is exposed, the surface of the edges is given a coat of epoxy to sink in and seal it. Later the bottom may receive protective reinforcement along the sides in the central portion of the bottom if it is deemed advisable. These reinforcements may be oak 5/16" thick, a foot or more wide at their widest part, with their inside edge straight and parallel to the centerline of the boat in order not to drag water. The fastenings are 7/8" No. 8 bronze wood screws.

Fastening On The Skeg

Before the outside is given a priming coat of any sort, it would be well to fasten on the skeg **(5, 6-40)**, which is 2'4" long, molded 6" wide at its after end, tapering to 1" at its forward end. The skeg is got out of 7/8" oak tapered to 5/8" at its lower edge. Put on with glue, it is fastened from the outside through the forward end into the next to last bottom cleat with a 3" No. 14 bronze wood screw. Two 1 1/2" No. 12 or No. 14 wood screws are put in on the centerline through the bottom from the inside. A 4" common galvanized spike is driven on an angle through the after end of the skeg up into the bottom knee. Later when the boat is turned over, 4" spikes are put in through the aftermost bottom cleat and the bottom knee. Lead holes are bored for them to give a tight driving fit.

Making Up And Fitting The Frames

After the skeg is fastened on, the boat can be turned right side up and set on horses or other level supports at a convenient height for working. The next operation is to cut out the frames **(2, 3, 5-7)**, glue them together and fit them in place beginning with the frames located between the mold stations.

Frames are glued together in two layers made up of shorter pieces of 3/4" spruce joined with butt joints. In this way it is possible to get curved frames out of straight-edge boards, but there should be as few pieces to a frame as possible, and the butts from side to side should be as widely spaced as possible.

Probably the best way of going about shaping and fitting these frames is to work from a template that can be made up from scrap plywood, and that will serve with only slight alterations for several frames.

In fitting the frames to the sides of the boat, this pattern is used to take off the shape, which changes very little from station to station. Frames are made from two lifts glued together, each of which is composed of two parts butted together. The location of the butt joints in the two lifts should be separated by at least 6", and even somewhat more is better.

Although the finish molding of the frames is 1½", the four parts from which each frame is made should be molded 2" when first cut out to allow plenty of extra wood for possible slippage in gluing and for fitting. First, after the pieces for a frame have been sawed out, the ends that will be butted together are fitted for a tight contact and are then separated and given a preliminary coating of epoxy, allowing plenty of time for the glue to sink in and dry.

In gluing up a frame, its pieces are first assembled dry on a flat surface, bench top or floor. The top lift is removed, without disturbing the under lift, then the meeting surfaces of both lifts, including the ends, are coated with epoxy, and the lifts are reassembled. Starting at the outer end of the short part of the top lift, 1¼" No. 10 annular nails are driven 3" to 4" apart and about 7/8" from the side to be fitted to the boat. When the butt joint is reached it is pulled together as close as possible (cabinetmaker's dogs would work well here) and nailing continues to the end of the long part of the under lift. The frame assembly is turned over, and the short part of the under lift is nailed, but not before the butt joint has been drawn together tight. The frame assembly is set aside to cure, while another just like it is put together for the other side. If timbering starts in the center of the boat it would be safe to begin by making up six or eight identical frame assemblies to begin with.

Using the template, the frames are fitted against the inside of the boat. A molded width of 1½" is gauged and marked on one side of the frame, the excess wood is cut away (a band saw does this best) and the inside of the frame is cleaned up and its corners slightly beveled with a spokeshave. The frame is ready to be installed. Glue is applied and the frame is nailed from the outside with No. 10 1½" annular nails through the laps, and with an additional nail midway between laps, so that each plank is nailed twice to the sides of each frame. Two nails through the bottom will be enough, but because the bottom is thicker, these nails should be 2" long.

Bronze wood screws could be used instead of nails, but bronze annular nails are cheaper and quicker and easier to drive, and just as good here. Note that with the exception of the two foremost, the frames are not set square across the boat, but are allowed to assume their natural position against the sides of the boat, reducing beveling to a minimum. If the builder wished to keep down weight as much as possible, he might get his frames out of stock 5/8" thick instead of 3/4", thereby reducing the finished siding of his frames to 1¼", but I don't believe the total saving in weight would be five pounds, if it were as much.

As soon as frames have been put in on either side of a planking mold, the mold can be removed, pulling the leatherheads on the outside that are holding it, and twisting the mold sideways enough to clear the inwale. It is replaced by a frame which is glued and nailed in like the others.

The two forward frames are put in square across the boat to line up with the deck beams **(4-16)** which cap them in order

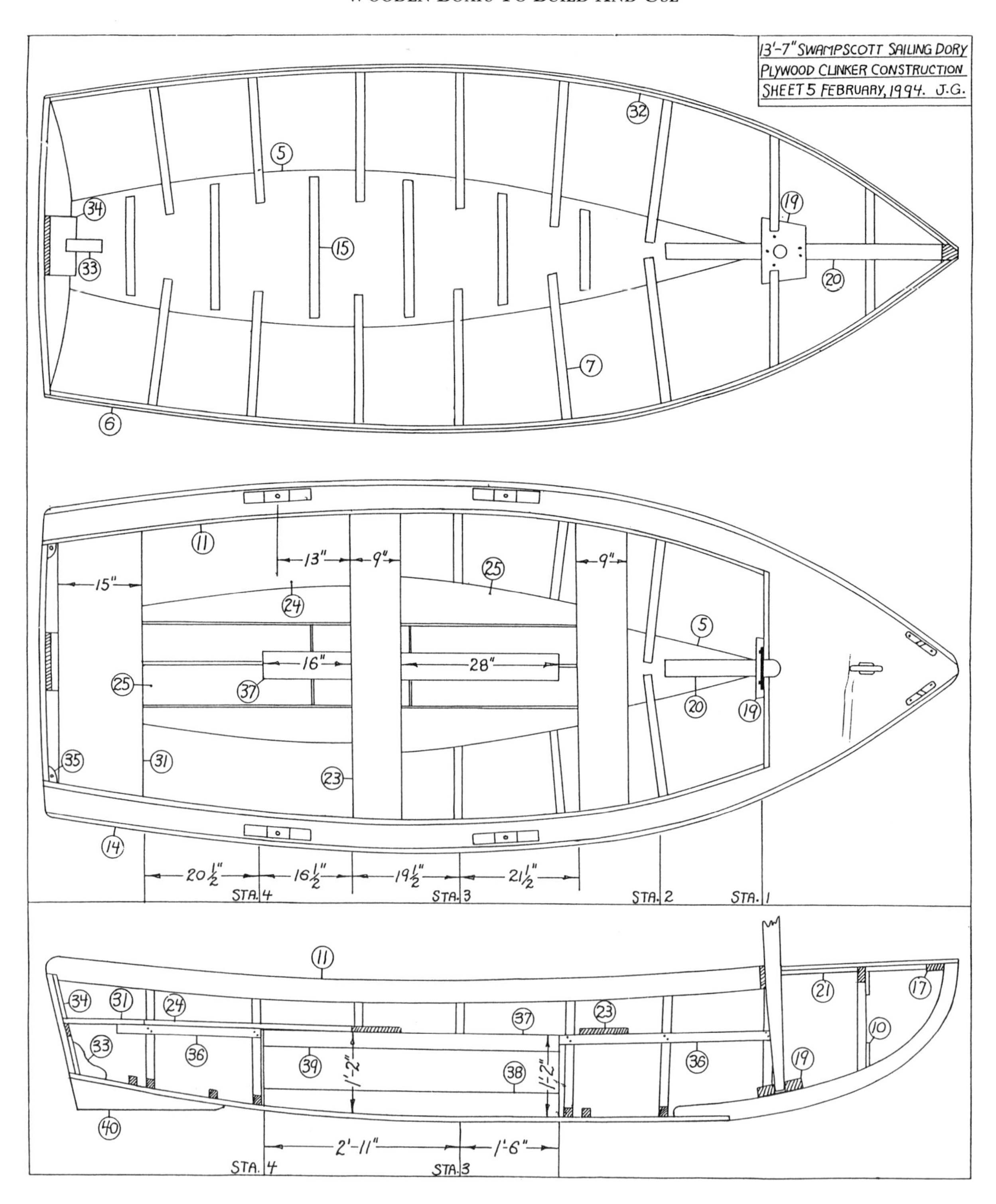
13'-7" SWAMPSCOTT SAILING DORY
PLYWOOD CLINKER CONSTRUCTION
SHEET 5 FEBRUARY, 1994. J.G.
15"
13"
9"
9"
16"
28"
20 1/2"
16 1/2"
19 1/2"
21 1/2"
STA. 4
STA. 3
STA. 2
STA. 1
1'-2"
1'-2"
2'-11"
1'-6"
STA. 4
STA. 3

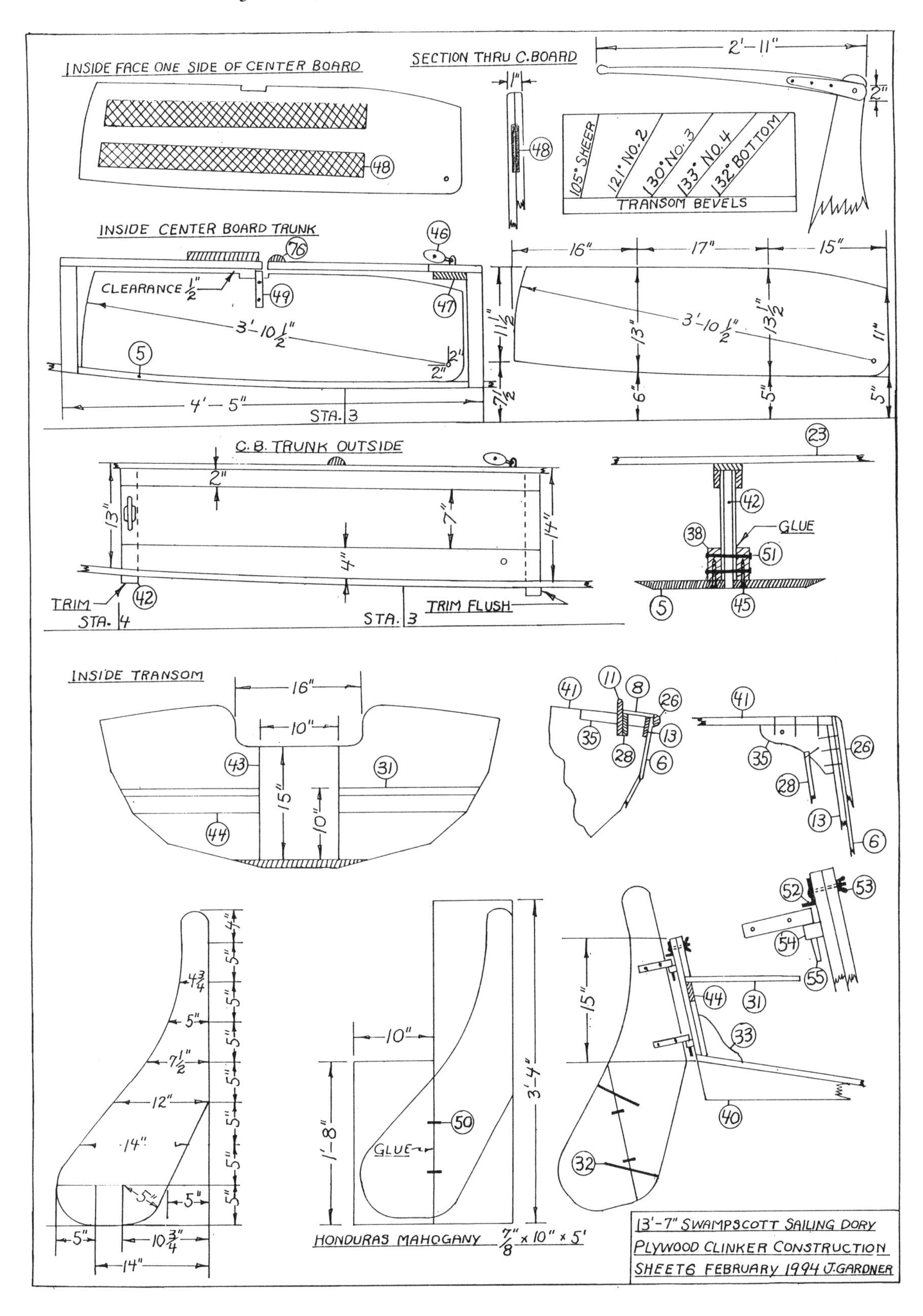
INSIDE FACE ONE SIDE OF CENTER BOARD
SECTION THRU C.BOARD
2'-11"
TRANSOM BEVELS
105° SHEER
121° NO.2
130° NO.3
133° NO.4
132° BOTTOM
INSIDE CENTER BOARD TRUNK
CLEARANCE 1/2"
3'-10 1/2"
4'-5"
STA. 3
C.B. TRUNK OUTSIDE
TRIM
STA. 4
TRIM FLUSH
GLUE
INSIDE TRANSOM
HONDURAS MAHOGANY 7/8" x 10" x 5'
13'-7" SWAMPSCOTT SAILING DORY
PLYWOOD CLINKER CONSTRUCTION
SHEET6 FEBRUARY 1994 J.GARDNER

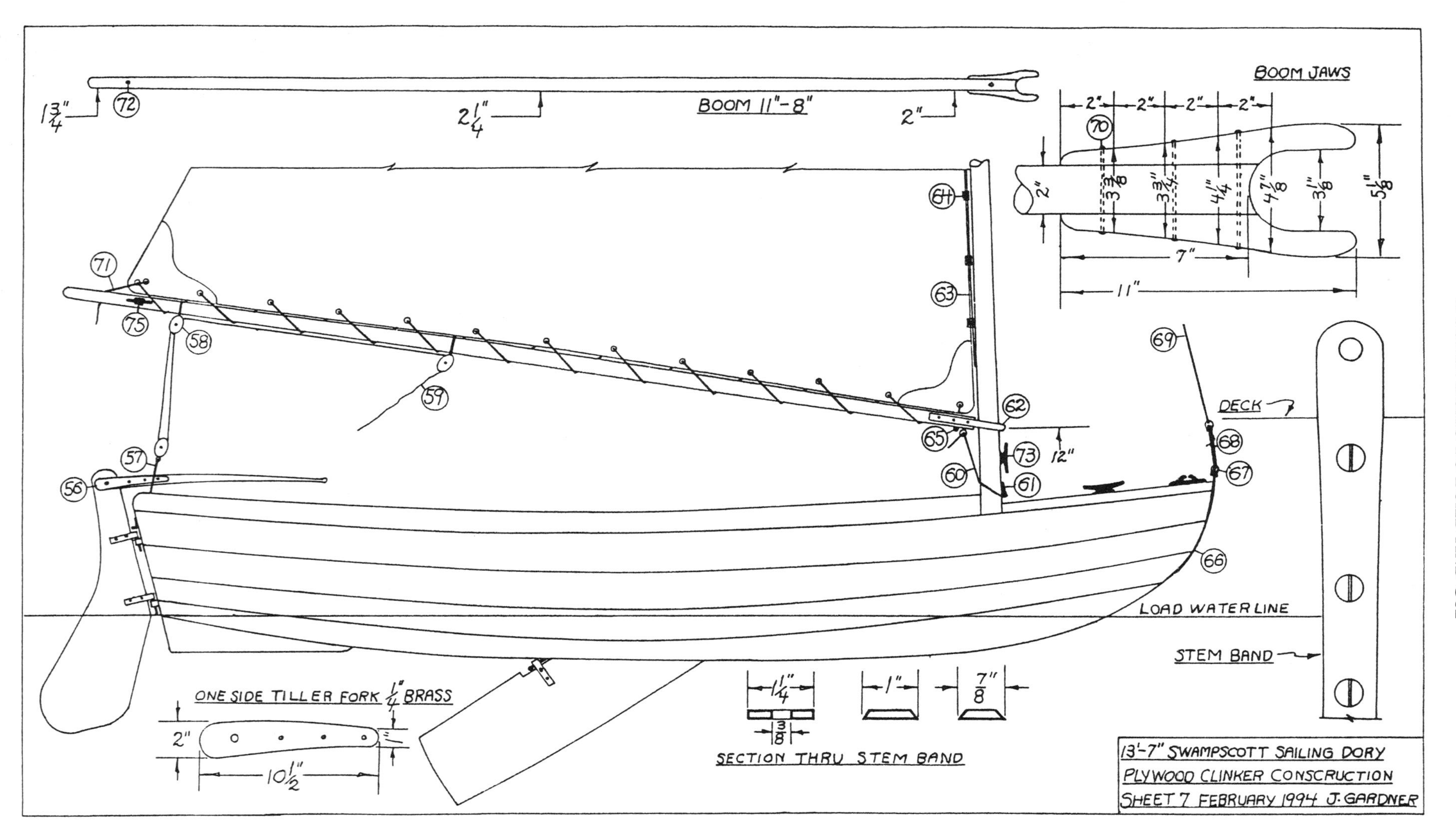
BOOM 11"-8"
BOOM JAWS
DECK
LOAD WATER LINE
STEM BAND
ONE SIDE TILLER FORK 1/4" BRASS
SECTION THRU STEM BAND
13'-7" SWAMPSCOTT SAILING DORY
PLYWOOD CLINKER CONSCRUCTION
SHEET 7 FEBRUARY 1994 J. GARDNER

that the two can be joined together at the ends with triangular plywood gussets glued and nailed in place. This ties the ends together and helps to reinforce the sides of the boat against the sideways thrust of the mast. Additional reinforcement is provided by the plywood decking, which goes on later.

Centerboard Trunk And Centerboard

At this point a number of roads stand open for the builder to take, one being the construction and installation of the centerboard trunk and centerboard, followed by the seat risers, thwarts, stern benches and stern sheets. Dimensions and details for the trunk and centerboard are fully laid out on Sheets **5** and **6** of the plans, but some direction on how to proceed will be in order. The slot for the centerboard, which has already been cut through the bottom, should finish a scant 1¼" in width and 1/2" shorter than the overall length of the trunk so that the outside of the bottom ends of the headledges will shoulder out 1/4" on the inside of the bottom. The bottom ends of the headledges which go into the centerboard slot in a tight fit are cut back 1/4" in width to allow for this.

The three bottom cleats located in the way of the centerboard trunk are cut back 2⅛" from the slot to allow for the combined thickness of the 1¾" bed log and the 3/8" plywood side. One of the sides is cut out and fitted to the inside curve of the bottom, serving as a pattern for the other side. To these sides the 4" wide bed logs fitted to the curve of the bottom are glued and well nailed with 1½" No. 10 annular nails, and the two inch wide capping strips at the top are also glued and nailed with 1" annular nails.

When the glue has set and the excess, if any, has been cleaned off, a "dry" assembly of the trunk is made in the boat with the headledges in place, the sides clamped to them and the assembled trunk pressed down firmly against the inside of the bottom. Several screws are driven, 1¼" No. 10 bronze, to hold the trunk together. Three through the plywood on either side at the ends will suffice, and the trunk is lifted out of the boat. Holes for the 5/16" bronze bolts through the bed logs and headledges are bored, while it is easy to get at them, but the bolts will not be driven until later.

The hole for the 3/8" bronze carriage bolt that the centerboard will pivot on is also bored at this time before the trunk is disassembled for gluing. It is located 2¼" up from the bottom, and 2½" aft of the after side of the forward headledge. The screws previously driven through the plywood into the headledge are removed, the trunk is disassembled, glue is applied and the pieces are put back and reassembled in the boat. The screws that were removed are replaced, lining up the pieces of the trunk as previously assembled. More screws are added. The bolts for which holes were bored are driven and tightened. A 1/4" bolt is put in at either end through the capping strips and headledges. Finally, the bed logs are pulled down tight and fastened through the bottom from outside with 3" No. 14 bronze wood screws, spaced 4" on centers. If possible, some means should be devised for hauling the trunk down tight against the inner bottom until it is fastened in. One way would be with pipe clamps that have an adjustable jaw on the outer end that can be removed to allow the pipe to pass down through the bottom slot, and then replaced and slid up into contact for clamping.

The inside of the trunk will have to be painted. A plastic foam brush with an extended handle will serve for this. Before lead paint was outlawed, I would have recommended a coat of red lead primer followed by a coat of anti-fouling. Some might want to fiberglass the plywood sides on the inside before the trunk is fastened together. Others might want to coat the inside of the trunk with epoxy.

The centerboard is made up, as shown in the drawing, by gluing together two pieces of 1/2" marine plywood, that have been pocketed on the inside for two strips of 1/2" brass to add extra weight. I would prefer lead if it can be obtained because it is heavier, and where it is completely enclosed and sealed with epoxy I see no reason not to use it. In cutting the pockets, a power router will not be needed. A hand router and/or a sharp chisel will do quite well for this. The hole for the pivot bolt is bored in the lower forward corner 2" in from the forward edge and 2" up from the bottom. It should be 7/16" in diameter.

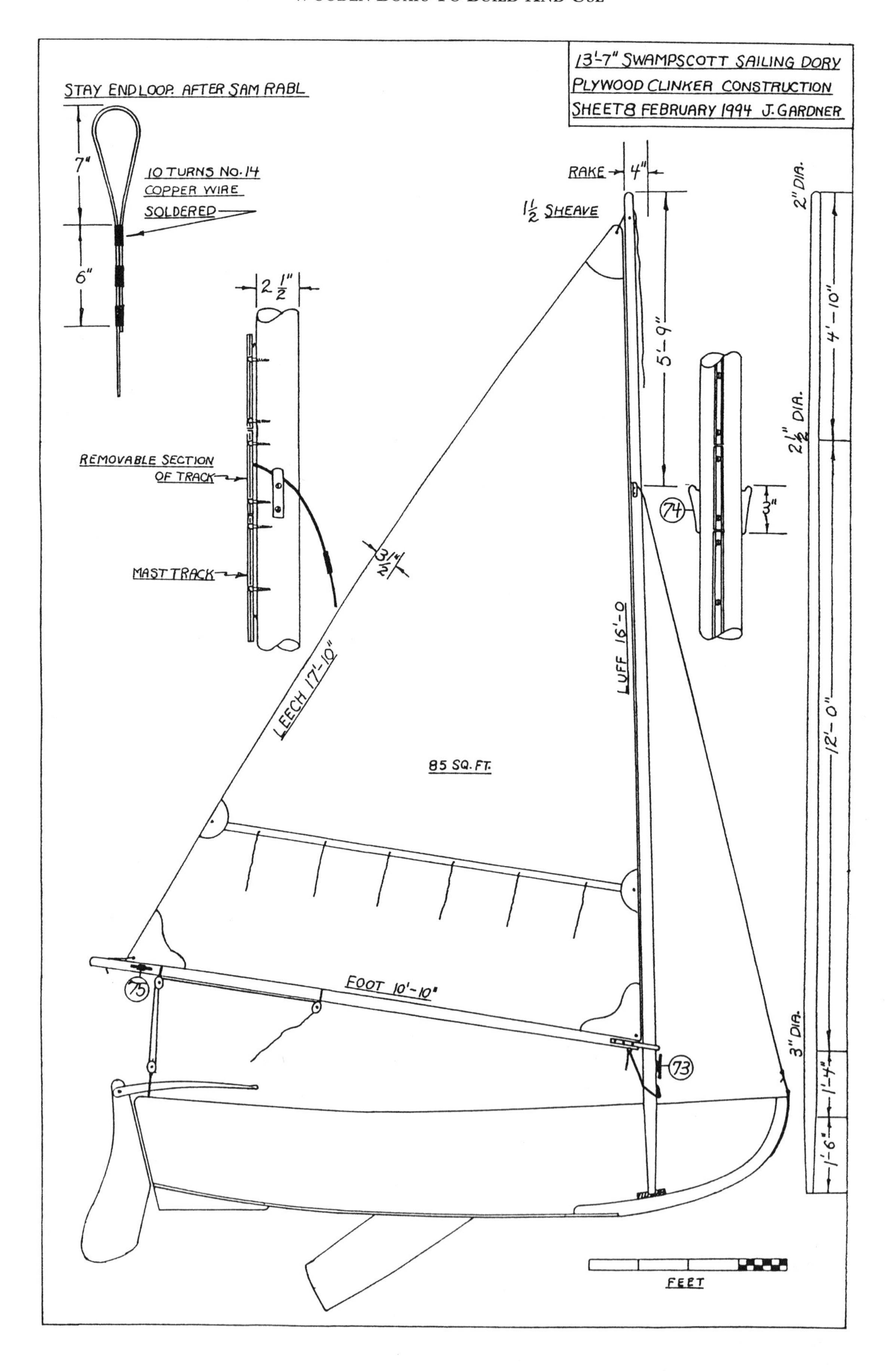
13'-7" SWAMPSCOTT SAILING DORY
PLYWOOD CLINKER CONSTRUCTION
SHEET 8 FEBRUARY 1994 J. GARDNER
STAY END LOOP. AFTER SAM RABL
7"
6"
10 TURNS NO. 14
COPPER WIRE
SOLDERED
2 1/2"
REMOVABLE SECTION
OF TRACK
MAST TRACK
RAKE 4"
1 1/2 SHEAVE
5'-9"
3"
74
2" DIA.
4'-10"
2 1/2" DIA.
12'-0"
3" DIA.
1'-4"
1'-6"
3/4
1/2
LEECH 17'-10"
LUFF 16'-0
85 SQ. FT.
FOOT 10'-10"
75
73
FEET

The centerboard trunk is capped with 3/4" pine $3^{1}/4$" wide put on in two pieces that butt just forward of the forward edge of the after thwart. The location of this thwart is given on Sheet 5. It rests on top of the capping strip and levels across to give the height for the seat risers which run in an easy continuous sweep from the second frame from the bow to the last frame at the stern. The seat risers are 3/4" spruce 2" wide, and are fastened to the frames with two 2" No. 8 annular nails or 2" No. 10 bronze wood screws.

Thwarts And Stern Sheets

The after section of the capping strip is fastened down with 2" No. 10 bronze wood screws, and the after thwart is put in place over it and fastened both to the capping strip and the seat risers, also with 2" No. 10 bronze wood screws. The forward thwart goes on next, and is fastened to the risers. The two 9" wide thwarts should be spruce no less than 7/8" thick.

For the stern sheets and the benches on either side that run from the stern sheets to the after thwart, pine 7/8" thick is the preferred material. To get the 15" width specified for the stern sheets it will undoubtedly be necessary to glue two boards together, and if pipe clamps were available for hauling down the centerboard trunk, they are just what is needed here.

A two-piece cleat of 3/4" pine 2" wide supporting the after edge of the stern sheets is glued and nailed across the inside of the transom. The ends of the stern sheets are supported by the after ends of the seat riser. Fastened to the cleats and the risers with glue and 2" No. 10 bronze wood screws, the stern sheets, in addition to providing a place to sit, become an important structural member reinforcing the transom against the strains imposed by the outboard motor.

In all likelihood the 14"-wide side benches that connect the stern sheets with the after thwart will also have to be glued together from two 7/8" pine boards. These side benches are joined at their after ends to the stern sheets and at their forward ends to the after thwart with 6"-wide cleats of the same material put on underneath with glue, and screwed through from the upper side with $1^{1}/2$" No. 10 bronze wood screws. To stiffen and support the side benches, a vertical post bottoming on the next-to-the-last bottom cleat is put in on either side. It should be located far enough back from the edge of the side bench to be out of the way. I don't believe the stern sheets will need any additional support, but if this should prove to be the case a post can be raised from the last bottom cleat midway of the boat and far enough back to lie out of sight.

Installation Of The Two Forward Frames

It has already been noted that the two forward frames are put in place plumb with the baseline and square with the fore-and-aft centerline. The second frame receives the sideways thrust of the mast. Its after side lines up with Station 1 and is located forward of it. The location of Station 1 can be marked on the sides of the boat before the planking mold is removed, and the bevels of the mold with the sides of the boat should be taken at this time to be applied in fitting the frame.

It is essential that the sides of the second frame line up square with the centerline of the boat for the gussets that connect its two ends to the deck beam running between them to lie flat against the surfaces of both the deck beam and the frame. This requires that the frame be beveled to fit against the sides of the boat, and to allow for extra wood for beveling the molding of the frame sides, when they are got out, is increased to $1^{3}/4$".

After the two sides of the second frame have been beveled, glued and nailed in place, their upper ends are cut in a half inch to receive the ends of the deck beam. Note that their lower ends (Sheet 4) are cut back from the stem about an inch to provide for limbers. Later they will be connected by the mast step, and must line up for the mast step to fit in place.

Getting Out And Installing The Deck Beam At Frame No. 2

The deck beam which is sided $1^{1}/2$" and molded $2^{1}/4$" is cut out to the curve of a beam mold laid out to a rise of 4" in 6'. This beam mold will need to be longer than the beam of the boat when it is used to give the outward pitch of the narrow side decks or waterways, and the tops of the inwales are dressed off to receive the side decking.

The ends of the deck beams are shouldered into the recesses cut for them in the upper ends of the side frames, and are glued and toe-nailed into the timber heads. Joining the ends of the side frames and the deck beam, on both the forward and after sides, are triangular gussets of 3/8" plywood. These gussets measure seven inches both on the outside of the frames and on the under side of the decking. Glued in place and well-nailed with 1 1/2" No. 10 annular nails, these gussets are important structurally in contributing support to the deck when the cutout is made for the mast.

Fitted between the ends of the gussets on the after side of the deck beam is a spacer of 3/8" plywood the same width and shape as the deck beam. Its function is to fill in between the gussets to afford a continuous level surface on which to attach the forward length of coaming later. On the forward side no spacer between the gussets is necessary as the deck beam reinforcement is glued and bolted directly to the deck beam. It is 2 1/2" wide in the center, tapering to 1 1/2" at the ends with the taper starting 6" from the center. The aperture is not cut until the decking is on.

The forward frame is also put in square with the centerline with the ends of the deck beam shouldered into the timber heads and reinforced with 3/8" plywood gussets 6" on the sides, glued and nailed in place. The gusset connecting the bottom ends of the frames is 4" in depth.

Framing The Forward Deck

The breasthook (**4, 5-17**), spruce or oak approximately 1" thick with a fore-and-aft width of 6", is fitted to the stem and inwales (**3, 4, 6-13**), glued and nailed to them. The 4"-wide center strip supporting the deck and bracing the forward thrust of the mast is made of 3/4" spruce. It is notched into the deck beams and breasthook and is glued and nailed to them. This completes the framing for the forward deck, but before the deck is put down the framing for the 5"-wide side decking or waterways is fitted and fastened in place.

On one side of each frame a 3/8" plywood gusset, notched around the inwale, fits against the inside of the sheer plank extending down from the sheer line 4 1/2". The top edge of the gusset is cut to match the camber of the beam mold, and 4" inboard from the inside of the sheer plank a vertical line is plumbed marking the inboard side of the gusset. Three inches down on this line is located the bottom outboard corner of the carlin band, later to be bent in and fastened, and from this point a line is drawn to the outside bottom corner of the gusset, marking the bottom edge of the gusset.

The 3/8" plywood gusset is glued and nailed with 1 1/2" No. 10 annular nails to the side of the frame. A piece of spruce frame stock 1 1/2" thick with its grain running vertically is cut to fill in between the frame and the inboard end of the gusset. It is glued to both the frame and the gusset, nailed to the gusset with 1 1/2" No. 10 annular nails and to the frame with one 3" galvanized common nail. The combined 1 7/8" thickness of plywood and filler piece will give good nailing for the carlin band and the coaming, and on top there will be ample width for supporting and fastening down the side decks, as well as nailing butt splices when short lengths of deck are joined.

At the stern the stern quarter knees (**6-35**) support the after ends of the side decks. These knees are oak, plus or minus an inch, 5" long on the inwales and 8" on the transom. Their inboard ends which project inside the side deck and coaming will be bored later to take the ends of the rope horse.

Carlin Bands

Before the side decking can be put down, the carlin bands (**3, 4, 6-28**) go on to support the inboard edge of the decking and take nailing. These bands are 5/8" spruce, 3" wide and run continuous from the stern quarter knee to the deck beam at Station 1, in a fair, smooth sweep without any bumps or hollows. A saw cut here or a thin shim there may be needed to achieve this. The bands

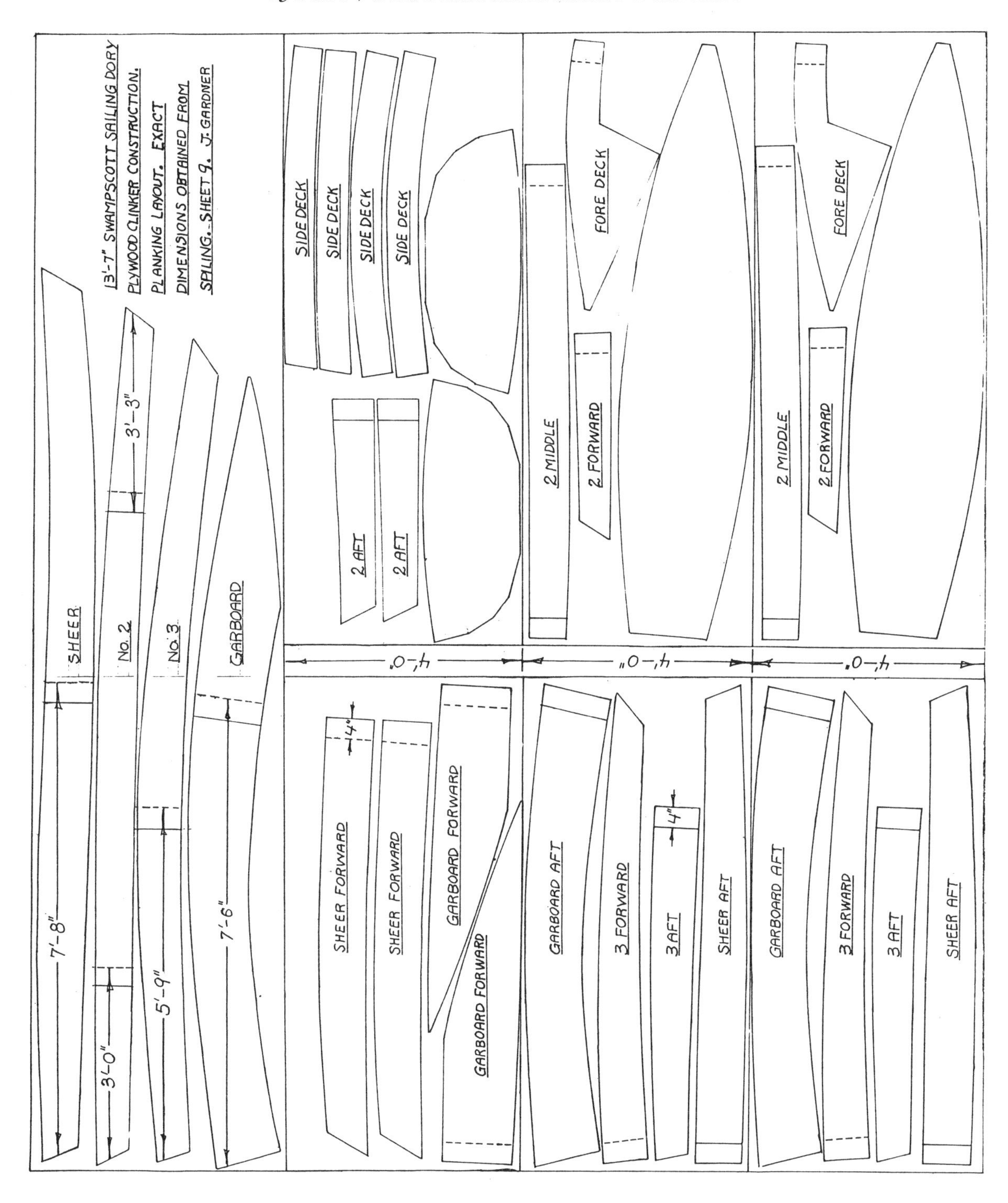

are put on with glue and 2" No. 10 annular nails.

It may be necessary to remove a few shavings with the jack plane before the surface is ready for decking. Final testing is done with the beam mold. It is better to keep the carlin bands slightly high when they are put on than to allow them to sag even a little, the extra is easily planed off before the decking is installed.

Putting Down The Plywood Deck

Decking is 3/8" marine plywood put down with glue, generously nailed with 1$^{1}/_{4}$" No. 12 annular nails, puttied and faced with epoxy gel and painted. Forward of Station 1, the deck is put on in two identical halves that join on the centerline. Their outboard ends are carried aft the width of the side decking one frame to butt against the forward ends of the side decking midway of

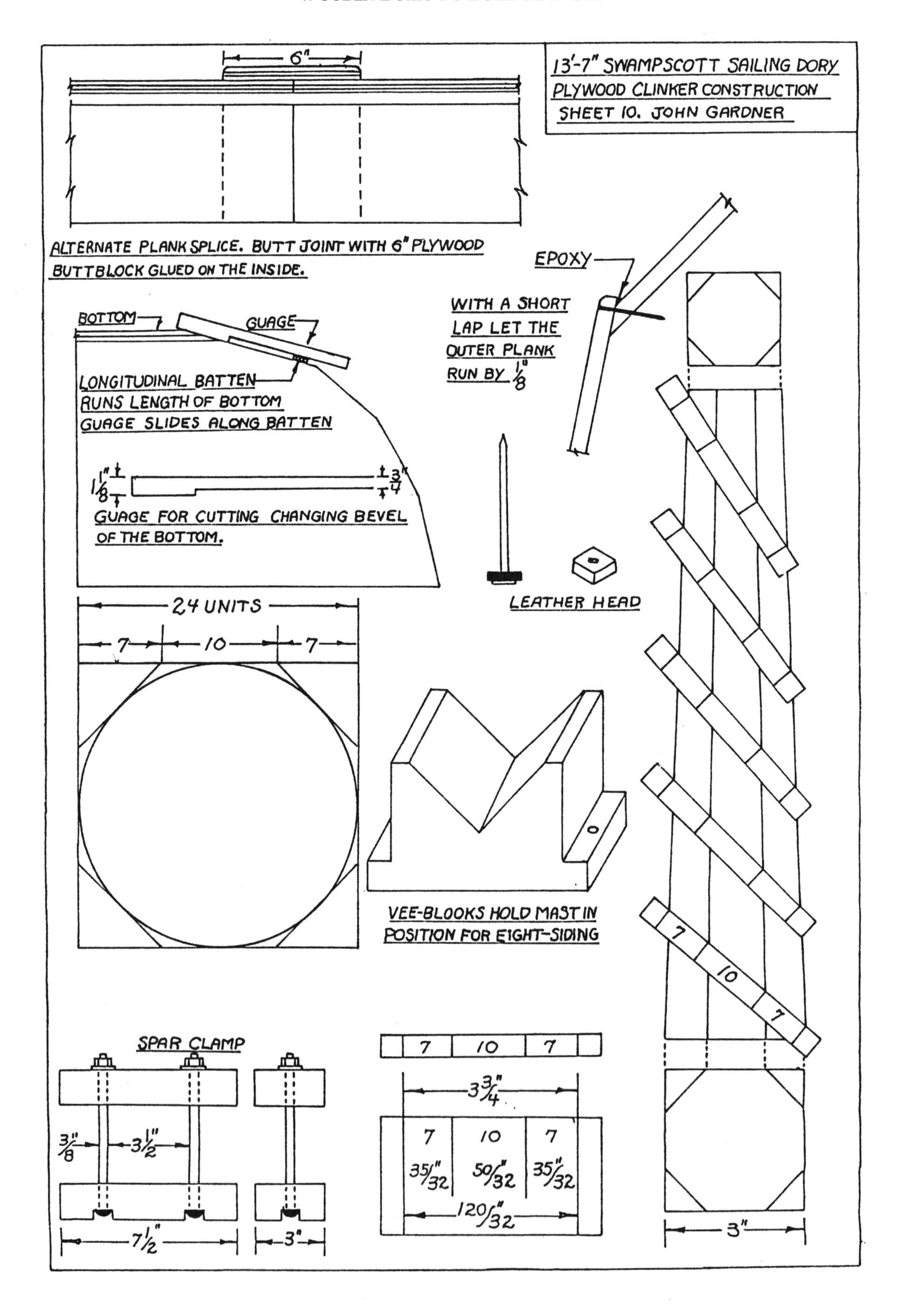
13'-7" SWAMPSCOTT SAILING DORY
PLYWOOD CLINKER CONSTRUCTION
SHEET 10. JOHN GARDNER
6"
ALTERNATE PLANK SPLICE. BUTT JOINT WITH 6" PLYWOOD BUTTBLOCK GLUED ON THE INSIDE.
BOTTOM
GUAGE
LONGITUDINAL BATTEN RUNS LENGTH OF BOTTOM GUAGE SLIDES ALONG BATTEN
1 1/8"
3/4"
GUAGE FOR CUTTING CHANGING BEVEL OF THE BOTTOM.
EPOXY
WITH A SHORT LAP LET THE OUTER PLANK RUN BY 1/8"
LEATHER HEAD
24 UNITS
7
10
7
VEE-BLOOKS HOLD MAST IN POSITION FOR EIGHT-SIDING
SPAR CLAMP
3/8"
3 1/2"
7 1/2"
3"
7
10
7
3 3/4"
35/32"
50/32"
35/32
120/32"
3"

the first frame extension. From there aft, the side decks are put on in two or even three shorter lengths butted on the frame extensions.

On the outside the side decking finishes flush with the outside of the sheer plank and is nailed to the inwale. On the inside it finishes flush with the inside of the carlin band to which it is nailed as well as to the frame extensions and the stern knee. Put down with epoxy adhesive and nailed before the glue sets, this is extremely strong construction which stiffens the sides of the boat immensely.

At the forward end of the boat the two deck sections are well-nailed along the centerline, to the deck beams, the breasthook and the inwales. The breasthook provides solid fastening for the bow chocks. The 6" cleat midway of the foredeck should bolt through the deck beam.

The half oval trim **(4-26)** which covers the outer edge of the plywood deck can be put on now. As there is a chance it could get banged up later on and need replacement in part, it is best to set it in paint instead of epoxy. For fastenings, 1 1/2" No. 12 bronze wood screws bunged 1/2" would be about right. While spruce or pine would do for the trim, oak, mahogany or some other hard wood would be better.

Locating And Cutting The Hole For The Mast

Before leaving the deck, it would seem logical to cut the hole for the mast. As this must be located precisely, special care must be taken to get it right. First, the boat must be leveled both fore-and-aft and athwartships, and secured in place. The center of the deck is found and marked on the after edge of the deck. A plumb bob dropped from this mark should hit the center of the stem, or essentially the same operation can be carried out with a long mason's or carpenter's level, plumbing up from the stem. In all likelihood the two will not correspond exactly, but will be close.

In that case check the boat again in several places to make sure it is level athwartships. Perhaps an adjustment will need to be made. Finally, plumb up from the center of the stem and mark the edge of the deck. Square in from that mark and draw a line. Measure in on that line 1 3/16" and draw another line parallel to the edge of the deck. The point where these two lines intersect is the center of the 3 1/8" diameter hole to be cut for the mast as diagramed on Sheet 4.

The 2" hole for the bottom end of the mast is not cut in the step until the step has been fitted, but not fastened in place. The step is taken up to do this, then put back and fastened. The hole in the step should center 3/4" forward of the hole through the deck for the mast. This will produce somewhat more than the four inches of after rake specified for the mast, but the forestay will haul the mast forward enough to offset this.

Making The Mast

The mast will be spruce, glued together in four 7/8" lifts, each of which will be spliced from two or three pieces. The location of these splices must be carefully distributed throughout the mast so that they are well separated when the lifts are glued together. Scarphs are cut twelve times as long as the thickness of the material, in this case 10 1/2" long. In planing the taper for the scarphs, the ends are left about 1/16" thick. This will be planed off smooth after the glue has set.

Care should be taken to line the lifts up straight when they are spliced, and although the finished diameter of the mast is slightly more than three inches at the deck, it would be well if the width of the pieces from which the lifts are spliced were approximately 1/2" wider.

After the four lifts are glued and their splices have been dressed off with the plane, a straight line representing the center of the mast is snapped down the center of one of the lifts, and along this line the widths for the taper of the mast are laid out. At the bottom, spots 1 1/16" out are marked on either side; up 1'-6", spots 1 5/16" out; up 2'-10", spots 2 9/16" out; up 8'-10", spots 1 7/16" out; up 14'-10", spots 1 5/16" out; and up 19'-8", spots 1 1/16" out. A fairing batten run through these spots and marked will outline the mast as it is tapered.

A batten long enough to reach the entire length probably won't be available, nor is one necessary. By overlapping with a shorter batten the job can be done just as well. Check the line for fairness by eye, and make any minor adjustments required.

After the taper is marked, the excess can be trimmed off with the drawknife and planed down to the line on either side, being careful to keep the edge square with the surface. When one of the lifts has been cut to shape, it is used as a pattern for the others. Of course the lifts could be sawed to shape with a handsaw, should one be available, and in the old days the builder would have thought nothing of trimming off the excess by hand with a sharp ripsaw.

The next step is to glue the four lifts together. As they have been cut from the same pattern, they should register perfectly when they are laid one on top of the other. To clamp them until the glue has set, special clamps will be required, as well as a bench on which to make the assembly. The top of this bench must be perfectly straight and flat from end to end.

As previously mentioned, a bench can be improvised by taking the ladder-frame apart and splicing together the 4" x 6" stringers that formed its sides to make the bench top. This can be supported on a couple of horses at a comfortable height for working, and braced where needed with posts to stiffen it and keep it from sagging.

Clamps For Gluing The Mast

Clamps for gluing the mast can also be improvised. They are easily made and inexpensive. Not more than a dozen or so will be needed. As can be seen in the diagram, they consist of two 6"-long strips of 7/8" spruce board, some of the same strips that were used for the mast. The bottom piece is bored for a tight fit for two 3/8" carriage bolts located $3^1/2$" apart on centers. The top piece is bored with half-inch holes, $3^1/2$" apart also, so it will drop down easily on the upright bolts. Pressure is applied by tightening the nuts on the upper ends.

The clamps are placed at regular intervals along the bench with the bolts upright and the top pieces off. Three of the lifts coated with glue are assembled between the bolts on the bottom part of the clamps; $2^1/2$" staging nails or leatherheads are driven through the lifts to hold them in line and keep them from slipping out of register. These nails must be long enough to go well into the bottom lift, but not through it. A Jorgensen Clamp or hand screw is applied to the sides of the lifts to hold them in position as each nail is driven. Five nails equally spaced will probably be enough, starting at the foot and ending at the head. This done, the assembly is turned over, the nail heads will come between the clamps and the fourth lift is put in place with glue and nailed with $1^1/2$" leatherheads. The tops of the clamps are dropped on over the bolts, and nuts with washers are threaded on and tightened to give firm, uniform pressure throughout the length of the mast.

When the glue has set, the clamps are loosened and removed, the nails are pulled, any glue that has leaked out of the seams is scraped off and the tapered surfaces are given a final brushing with the smoothing plane.

Tapering And Eight-Siding The Mast Preparatory To Rounding

The stick, already tapered one way, must now be tapered the other way, that is, on the other two sides. When this is done the stick will be square in cross section.

In laying out the taper on the untapered sides, the glue line at the center of the stick is used for the centerline from which the measurements for the taper are made. These are the same as were previously used in tapering the lifts, and the taper is lined out in the same way. When one side is done, the stick is turned over, and the taper is lined out on the opposite side. The excess wood can now be cut away, producing a stick that is uniformly tapered on all four sides.

Preparatory to rounding the mast, it is eight-sided or cut octagonal in section. Lines running the length of the square stick that show how much wood to remove are marked with a special gauge that the builder can make himself and that is detailed on Sheet 9.

To hold the stick in position for its corners to be removed to make the mast eight-sided or octagonal, it is set in V-blocks, which hold it with its corners uppermost. This is also diagrammed on Sheet 9. When the excess wood at one corner has been removed with the drawknife and dressed down with the plane to the lines that were previously marked, the stick is given a quarter turn, and the process of removing the excess wood is repeated and so on around the stick, making it octagonal in section.

Sixteen-Siding And Final Smoothing

The next step is to make the mast 16-sided. With a mast as small as 3", a few full-length shavings with a sharp plane set moderately fine, taking off each corner in turn, should do it. It should finish a plump 3" at the deck; 3 1/16" would not be too large. Keep turning the mast in the V-blocks as you plane until each corner has been taken down in turn and you have gone around the stick. The final planing is done with a very sharp plane set to take off tissue-thin shavings as the mast is slowly rotated in the V-blocks. Check frequently with calipers to make sure the mast is not getting too small. Very little wood is removed in the process, yet enough and from the right places, so the mast comes out round and smooth enough that little finish sanding is required.

In the bygone days of wooden ships before there were modern abrasives, spars were given their final smoothing with planes.

Making The Boom

The procedure for making the boom is exactly the same as that for the mast. The boom jaws are constructed as detailed on Sheet 7. They are best made of seasoned white oak or black locust. They are riveted together over washers either with 3/8" copper rod or soft, malleable bronze. Hard bronze will not head over. Bronze can be softened by heating to a dull red, and quenching in cold water. It behaves just the opposite from steel. When bronze is heated too hot, to a temperature approaching a white heat, it will break and crumble when it is pounded.

The Rig

The rig is simplicity itself. To remove it, the sail is dropped, the stopper knot in the end of the halyard at the head is untied, the boom downhaul loosened and the sheet unrove from the horse. The whole business can then be lifted clear and wrapped around the boom to which the sail is laced. The halyard remains rove through the sheave at the head of the mast with one end cleated to the halyard cleat. The other end is also made fast to it. Another cleat could be added on the side of the mast for this, although it is not shown in the drawing.

To bend on the sail again the process is simply reversed and without disturbing the mast or loosening the forestay. The use of mast track and sail slides makes it possible to do this without disturbing the stay, which was the reason for specifying mast track in spite of extra cost and possible difficulty in obtaining it.

Mast hoops could be used but they would be clumsy and less efficient. They would have to be large enough to pass over the loosened forestay and forestay thumb cleats as well as the halyard cleats at the base of the mast, and the mast would have to be unstepped and lowered before the sail with the boom attached could be removed.

Also if such large, sloppy, mast hoops were used, the sail probably would not draw as well.

The 19'-8" spruce mast, 3" at the bury and tapering to 2" at the head, weighs with mast track and loosened forestay attached a little over 20 lbs, making it possible even for youngsters to raise or lower it. In raising the mast, the heel, 1 1/2" in diameter, is inserted in the step, and with the mast gate **(4-22)** open the mast is pushed up, as a ladder is raised, until it stands upright and the gate can be closed. In lowering the mast this operation is carried out in reverse.

With its 85 square feet of canvas in a jib-headed sail that is 16' on the luff, this 13 1/2-foot boat should sail smartly, although it was not designed for racing. If more speed is desired, the mast could be lengthened by two feet or so with the sail cut to correspond. The superior efficiency of the high-aspect Marconi rig is well established. Of course the forestay would have to be lengthened and carried up higher on the mast, but this is a small matter. With its more than five feet of beam and ample bearing on the water *Fat Boat* could handle the heightened rig without any loss of stability worth considering. Just the same, if the boat is to be sailed by youngsters and novices, the more conservative of the two rigs is the one to be recommended.

If the builder should decide to go with the taller mast and a larger rig, he might want to consider adding single side stays or shrouds. This would be easy to do. These could be made up exactly like the forestay and positioned on the mast conveniently below it, leading down to chain plates bolted through from the outside to the first frame aft of the mast. Side stays would be set up with lacings to the chain plates. Other rigs might be considered for *Fat Boat*, the sliding gunter in particular. The gunter is popular and widely used in small boats in *Fat Boat*'s size range. Its two-part mast is easy to handle and can be taken apart and stowed in the boat. Moreover only very slight interior alterations would be required if *Fat Boat* were to be rigged with a sliding gunter, whereas with a sprit rig the foredeck would be superfluous and had best be omitted.

The forestay (7-69) is 4-mm stainless steel wire rope, the upper end of which is looped to go around the mast. The lower end is looped around a thimble holding a ring through which the lashings pass that connect the stay with a shackle or ring in the upper end of the stem band. The lashings, 3-mm polyester braided rope or the equivalent, are laced three or four turns, set up tight and made fast with a couple of half hitches around the standing part.

To secure the loops at either end of the forestay the method first described by Sam Rabl in *How to Build Twenty Boats*, back in 1915, and later recommended by Rabl in *Boat Building In Your Own Back Yard*, is both easy to do and amply strong. In making a loop the wire rope is turned back on itself six inches or more, and the two parts are tightly wrapped in three places with ten to twelve turns of No. 14 soft copper wire and soldered. In making the wire servings the last turn, instead of going around the two parallel sections of wire rope, is hauled up between them and pulled tight before being cut preliminary to soldering, or the serving may be soldered while the wire is still under tension before it is cut.

In case it should be found that more rudder would be desirable with the larger rig, it would be quite easy to provide for this by dropping the rudder deeper in the water. This could be done very easily by lowering the gudgeons. To make provision for such an eventuality, the upper end of the rudder could be made several inches longer.

Enough foam flotation should be provided to float the boat and its occupants even with a small outboard motor aboard. This is desirable in any case, but it is imperative when the boat is to be used by youngsters and novice sailors. A considerable amount of foam can be packed away out of sight under the stern sheets, and on both sides under the stern benches, and a corresponding, or even greater, amount can be placed forward in the eyes of the boat under the bow deck. The wooden structure of the boat itself will furnish a considerable additional amount of flotation.

Final Instructions

A final brief word to those who may read the foregoing with the idea of building *Fat Boat*. A lot of thought, planning and boat building experience is packed into the foregoing text, including the drawings that accompany it. Prospective builders, in particular novices, are strongly advised to study this material thoroughly before starting out. It is not a bad idea to first build the boat in one's head following the itinerary of operations set forth here, before putting saw and hammer to wood.

You may want to make changes. Fine. Most builders do, and sometimes they come up with genuine improvements. This is your boat, your opportunity for creative self-expression, but make sure you think through fully and exhaustively what the outcome of any prospective changes is likely to be. Make sure you are not about to implant any sleeping time bombs.

There is joy in building a boat along with plenty of hard work. It has its ups and downs. There are moments of triumph and moments of despair. Chapelle has said, "Be sure to have a few crying towels." You will make mistakes, everybody does. However, as they used to say, the test of a good mechanic is his ability to get out of his own mistakes. So you use your head, buckle down, stick with it, give it your best shot and the result should be a good little boat that will take you out and bring you back safely, one that should be easy to handle, a joy to look at and an enduring source of pleasure and wholesome fun.

Parts List

(1) Ladder-frame stringers **(2-1)**; Douglas fir; joists, or the equivalent, seasoned, dry and straight, $1\frac{1}{2}$" x $5\frac{1}{2}$" x 12'.

(2) Ladder-frame cross cleats **(2-2)**; pine, spruce or fir; 3/4"-1" x 5" x 2'8".

(3) Cross spalls **(2-3)**; pine; 3/4" x 5" wide.

(4) Planking molds **(2-4)**; 3/4" pine; fastened together with $1\frac{1}{2}$" steel wood screws.

(5) Bottom **(2, 5, 6-5)**; two layers of 3/8" marine plywood glued together.

(6) Planking **(2, 5, 6-6)**; 3/8" marine plywood.

(7) Frames **(2, 3, 5-7)**; spruce; sided $1\frac{1}{2}$", molded $1\frac{1}{2}$"; made up of two, two-part layers glued together with the joints in the two layers well staggered.

(8) Deck and side decks or waterways **(3, 4, 6-8)**; 3/8" marine plywood.

(9) Not used.

(10) Gussets **(4, 5-10)**; 3/8" marine plywood

(11) Coaming **(3, 4, 5, 6-11)**; pine or spruce; 5/8" x 5"; stands $1\frac{1}{4}$" above the decking.

(12) Support for waterways or side decking and backing for carlin band **(3, 4-12)**; spruce; $1\frac{1}{2}$" thick; fitted between carlin band and frame and against inside of gusset and glued and nailed in place.

(13) Inwale **(3, 4, 6-13)**; spruce; 3/4" x $2\frac{1}{2}$"; continuous from stem to transom; can be glue-spliced.

(14) Half oval trim **(3, 4, 5-14)**; pine or oak; $1\frac{1}{4}$" x $1\frac{3}{4}$" in section.

(15) Six bottom cleats **(5-15)**; spruce; $1\frac{1}{2}$" high, $1\frac{3}{4}$" wide; 16" on centers; ends $1\frac{1}{2}$" from chine; nailed and glued to inside of bottom.

(16) Deck beams **(4-16)**; spruce; sided $1\frac{1}{2}$", molded $2\frac{1}{4}$"; deck beam camber laid out to a rise of 4" in 6'.

(17) Breasthook **(4, 5-17)**; spruce or oak 1" thick; fore-and-aft width 6"; fitted between the inwales and against the stem; slightly crowned to correspond to the crown of the deck; glued and nailed in place.

(18) Deck-beam reinforcement **(4-18)**; oak; sided $2\frac{1}{2}$" tapering to $1\frac{1}{2}$" at the ends; molded to fit the curve of the deck beam.

(19) Mast step **(4, 5-19)**; oak; $1\frac{1}{2}$" thick; 8" fore-and-aft measurement; forward end 12" wide, after end 14" wide.

(20) Stem **(3, 4, 5-20)**; spruce; glued up in three 7/8" layers, and finished to dimensions given in Sheet 3.

(21) Fore-and-aft deck support **(4, 5-21)**; 3/4" thick, 4" wide; in two pieces, from breasthook to deck-beam reinforcement.

(22) Mast gate bar **(4-22)**; galvanized steel or bronze; 3/16" thick, $1\frac{1}{2}$" wide.

(23) Thwarts **(4, 5, 6-23)**; pine or spruce; 7/8" thick.

(24) Stern bench **(4, 5-24)**; pine; 7/8", 15" wide.

(25) Walkways **(4, 5-25)**; pine; 3/4" x 7" wide, tapering to 4" outside, forward ends.

(26) Half oval trim **(6-26)**; same as **(3, 4, 5-14)**.

(27) Bolts, 3/8" galvanized **(4-27)**; deck beam to deck-beam reinforcement.

(28) Carlin bands **(3, 4, 6-28)**; spruce; 5/8" x 3" wide, supporting the inside of the waterways or the side decking.

(29) Cleat **(4-29)**; 6".

(30) Bow chock **(4-30)**; 6".

(31) Stern sheets **(5, 6-31)**; pine; 7/8" x 15" wide.

(32) Bronze drifts **(6-32)**; 5/16" diameter.

(33) Stern bottom knee **(5, 6-33)**; oak; sided $1\frac{1}{2}$", 6" on bottom, 6" up on inside of transom reinforcement.

(34) Vertical reinforcement of the transom for mounting the outboard motor **(5-34)**; spruce; $1\frac{1}{2}$" x 10" wide; the same as **(6-43)**.

(35) Stern quarter knees **(5, 6-35)**; oak; 1" thick; extends 8" on the transom, 5" on the inwale.

(36) Seat risers **(5-36)**; spruce; 3/4" x 2" wide; continuous from second frame from the bow to the after most frame; glue-spliced if necessary.

(37) Centerboard trunk **(5-37)**.

(38) Bed logs of centerboard trunk **(5, 6-38)**; oak; sided $1\frac{3}{4}$", molded 4"; bottoms cut to the curve of the bottom; 4' 5" long.

(39) Capping strips either side, top of centerboard trunk **(5-39)**; pine; 3/4" x 2" wide.

(40) Skeg **(5, 6-40)**; oak; 2' 4" long, 6" molded width after end, 1" at forward end.

(41) Transom **(6-41)**; two layers of 3/8" marine plywood glued together.

(42) Headledges **(42-6)**; oak; 1¼" x 2" wide.

(43) Vertical reinforcement of the transom for mounting the outboard motor; spruce; 1½" x 10" wide; same as **(5-34)**.

(44) Cleat supporting stern sheets **(6-44)**; pine; 3/4" x 2" wide.

(45) Bronze wood screws **(6-45)**; No. 14, 3" long, attaching centerboard trunk bed logs to the bottom; 4" on centers.

(46) Swivel block for centerboard tackle **(6-46)**.

(47) Reinforcement **(6-47)** for fastening **(6-46)**; oak; 4" wide.

(48) Lead or brass inserts for increasing centerboard weight **(6-48)**; 1/4" thick, 3" wide, 2' 6" long.

(49) Strap bail for attaching the centerboard pennant **(6-49)**.

(50) Dowels **(6-50)**; 5/16" bronze rod.

(51) 1/4" bronze bolts through bed logs and headledges from side to side **(6-51)**.

(52) Rudder stop **(6-52)**; 1/4" bronze angled 90 degrees; can be turned to clear rudder pintle when wing nut is loosened.

(53) 1/4" diameter rudder stop bolt **(6-53)** with wing nut.

(54) Gudgeons **(6-54)**; bronze.

(55) Pintles **(6-55)**; bronze.

(56) Tiller fork **(7-56)**; 1/4" brass or bronze; two pieces, one either side, through-riveted with 1/4" copper or soft bronze rod.

(57) Rope horse **(7-57)**; 1/2" Dacron.

(58) 3" sheet blocks **(7-58)** for 3/8" Dacron.

(59) Sheet **(7-59)**; 3/8" Dacron.

(60) Downhaul **(7-60)**; 1/4" Dacron.

(61) Thumb cleat **(7-61)**; oak; 3/4" x 2½".

(62) Boom jaws **(7-62)**; oak; 1⅛".

(63) Mast track **(7-63)**; 3/4".

(64) Slides for mast track **(7-64)**.

(65) Stopper knot **(7-65)**; forward or tack end of boom lacing.

(66) Stem band **(7-66)**; 3/16" bronze.

(67) Shackle or ring **(7-67)** for lashing stem band to forestay.

(68) Lashing **(7-68)**; rope diameter is 2.5 - 3.0 mm.

(69) Forestay **(7-69)**; 4-mm stainless—4 x 19 or 7 x 19.

(70) 1/4" malleable bronze or copper rod **(7-70)** rivet-headed over washers.

(71) End of boom lacing serves as outhaul **(7-71)**.

(72) Bee hole **(7-72)** for outhaul.

(73) 6" halyard cleat **(7-73)**.

(74) Thumb cleats **(7-74)** for forestay loop; 3" x 1⅛".

(75) Outhaul cleat **(7-75)**; 3½".

(76) Section of 2" half round **(7-76)**; thwartship capping centerboard trunk.

Alphabetical List Of Parts
(Name, Sheet, Part Number)

Bee hole, Sheet 7, No. **38 (7-38)**
Boom jaws **(7-62)**
Bow chocks **(4-30)**
1/4" Bronze bolts, bed logs **(6-51)**
3/8" Galvanized bolts **(4-27)**
Bottom **(2, 5, 6 -5)**
Bottom cleats **(5-15)**
Breasthook **(4, 5-17)**
Bronze drifts **(6-32)**
Capping strips **(5-39)**
Carlin bands **(3, 4, 6-28)**
Centerboard trunk **(5-37)**
Cleat 6" **(4-29)**
Cleat, stern **(6-44)**
Coaming **(3, 4, 5, 6-11)**
Cross spalls **(2-3)**
Decks & Side deck **(3, 4, 6-8)**
Deck beams **(4-16)**
Deck-beam reinforcement **(4-18)**
Deck support **(4, 5-21)**
Dowels, rudder **(6-50)**
Downhaul **(7-60)**
Forestay **(7-69)**
Frames **(2, 3, 5-7)**
Gudgeons **(6-54)**
Gussets **(4, 5-10)**
Half oval trim **(3, 4, 5-14**; same as **6-26)**
Half round 2", centerboard trunk **(7-76)**
Headledges **(6-42)**
Half round, 2", centerboard trunk **(7-76)**
Inwale **(3, 4, 6-13)**
Lashing **(7-68)**
Ladder-frame stringer **(2-1)**
Ladder-frame cross cleats **(2-2)**
Lead or brass strips **(6-48)**
Mast gate bar **(4-22)**
Mast step **(4, 5-19)**
Mast track **(7-63)**
Outhaul **(7-71)**
Outhaul cleat **(7-75)**
Pintles **(6-55)**
Planking molds **(2-4)**

Planking **(2, 5, 6-6)**
Rivet, bronze or copper rod **(7-70)**
Rudder stop **(6-52)**
Rudder stop bolt **(6-53)**
Rope horse **(7-57)**
Sail slides **(7-64)**
Seat riser **(5-36)**
Shackle or ring **(7-67)**
Sheet **(7-59)**
Sheet blocks **(7-58)**
Skeg **(5, 6-40)**
Stem **(3, 4, 5-20)**
Stem band **(7-66)**
Stern bench **(4, 5-24)**
Stern bottom knee **(5, 6-33)**
Stern quarter knees **(5, 6-35)**
Stern sheets **(5, 6-31)**
Stopper knot **(7-65)**
Strap bail **(6-49)**
Support for waterways **(3, 4-12)**
Swivel block **(6-46)**
Swivel block reinforcement **(6-47)**
Tiller fork **(7-56)**
Thumb cleat **(7-61)**
Thumb cleats forestay **(7-74)**
Thwarts **(4, 5, 6-23)**
Transom **(6-41)**
Transom reinforcement **(6-43; 5-34)**
Walkways **(4, 5-25)**
Wood screws, centerboard trunk **(6-45)**

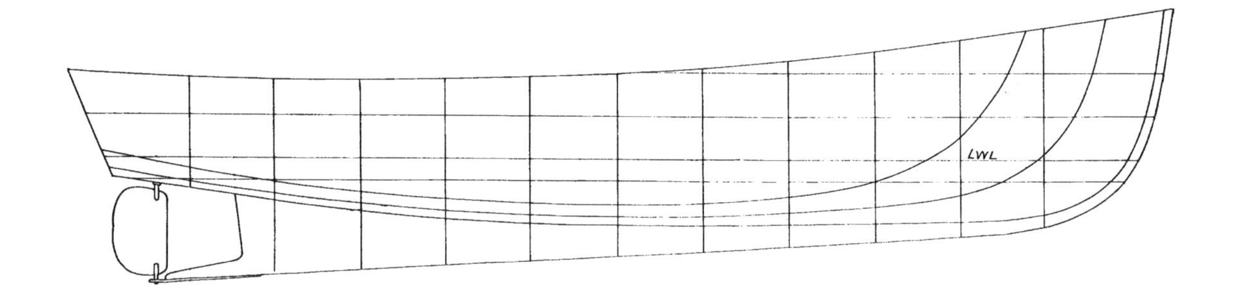

Chapter Nine

Chamberlain Launch

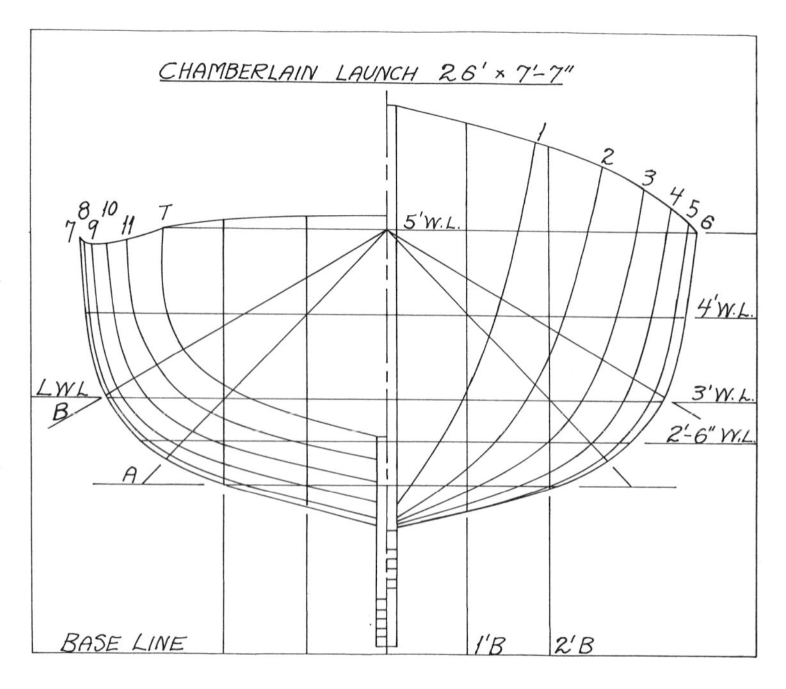

Forty years have passed since I measured the classic 26-foot power launch *Philippa* and recorded her lines and offsets in the *Maine Coast Fisherman*. Her kind had largely passed from use, and at that time she was more an item of historic interest than a craft of current use.

Since then we have come, or are about to come, full circle in the design and powering of small working launches.

Forty years ago, low-speed, low-powered boats were definitely on the shelf, and bigger engines and much higher working speeds were the wave of the future, requiring hulls to match. Fuel consumption was of little or no consideration then, for gasoline was plentiful and cheap and the supply seemed

LINES AND OFFSETS CHAMBERLAIN LAUNCH 26' × 7'-7"

		STEM	No. 1	No. 2	No. 3	No. 4	No. 5	No. 6	No. 7	No. 8	No. 9	No. 10	No. 11	TRAN.
HALF-BREADTHS	SHEER	0-1-4	1-9-6	2-7-6	3-1-0	3-5-6	3-8-3	3-9-4	3-9-1	3-8-4	3-7-3	3-5-1	3-2-0	2-8-6
	5' W.L.	0-1-4	1-7-3	2-5-5	3-0-6	3-5-1	3-8-1	3-9-4	—	—	—	—	—	2-8-6
	4' W.L.	0-1-4	1-4-0	2-2-0	2-9-7	3-2-7	3-6-2	3-7-7	3-8-2	3-7-7	3-6-5	3-4-1	3-1-0	2-8-2
	3' W.L.-LWL	0-1-4	0-11-2	1-8-4	2-4-5	2-10-6	3-2-5	3-4-5	3-5-1	3-4-4	3-2-6	2-11-0	2-5-2	1-8-0
	2'-6" W.L.	0-1-4	0-7-6	1-4-0	1-11-4	2-6-1	2-10-1	2-11-7	3-0-2	2-11-1	2-8-0	2-1-5	1-2-0	—
	KEEL & STEM RAB.	0-1-4	0-1-4	0-1-4	0-1-4	0-1-4	0-1-4	0-1-4	0-1-4	0-1-4	0-1-4	0-1-4	0-1-4	0-1-4
HEIGHTS	SHEER	6-5-4	6-0-2	5-9-1	5-6-0	5-3-3	5-1-3	5-0-0	4-10-6	4-10-1	4-9-6	4-9-7	4-10-2	5-0-0
	RABBET	6-5-4	1-9-0	1-7-0	1-6-4	1-6-1	1-5-6	1-5-6	1-6-1	1-7-2	1-9-2	2-0-2	2-3-4	2-6-6
	KEEL	—	1-5-4	1-2-6	1-1-3	1-0-0	0-9-3	0-9-1	0-7-6	0-6-3	0-5-0	0-3-5	0-2-1	0-0-7
	12" BUTTOCK	—	3-1-2	2-2-1	1-11-1	1-9-2	1-8-3	1-8-1	1-8-6	1-9-4	1-11-5	2-2-3	2-5-4	2-9-1
	24" BUTTOCK	—	—	3-6-1	2-6-2	2-1-7	2-0-0	1-11-4	1-11-7	2-0-6	2-2-6	2-5-4	2-9-1	3-1-6
	DIAGONAL A	0-2-0	1-9-2	2-6-7	3-1-6	3-6-4	3-9-1	3-10-0	3-10-0	3-9-1	3-7-3	3-4-4	3-0-5	2-8-1
	DIAGONAL B	0-1-5	1-7-3	2-5-0	3-0-3	3-5-7	3-9-3	3-11-0	3-11-5	3-11-0	3-9-4	3-6-6	3-2-7	2-10-3

DIAGONAL A UP 5' ABOVE BASE LINE, OUT 2'-6" ON 2'-6" W.L. DIAGONAL B UP 5' ABOVE B.L. OUT 3'-6" ON 3' W.L. LINES TO OUTSIDE PLANKING. OFFSETS MEASURED IN FEET-INCHES-EIGHTHS.

inexhaustible; in fact, nobody then ever bothered to think about it. How different today!

And so it is that boat designs like the *Philippa* are beginning to come off the shelf. Their low and moderate speeds, energy-efficient hulls, good seakeeping qualities and superior working characteristics are increasingly in demand.

Such a hull as the *Philippa*'s would do very well with a small modern diesel, both as a working launch or as a pleasure boat.

The lines and offsets presented here represent the basic hull; numerous adaptations can be made for a variety of uses and to suit individual needs and tastes. The lines have been completely redrawn and the offsets carefully rechecked. So often, hull lines are printed without offsets, although lines without offsets are not of much help to the amateur builder. To scale them from a print of lines reduced to a small size is generally beyond his ability and experience, or at least he does not have the confidence to attempt the job himself.

Following is the text of my original

Richard Dion's old open work launch sits on the ways at Dion's in Salem, Massachusetts, ready for launching after her rebuild into a cruiser in 1955. Built in 1927 by William Chamberlain, her light construction withstood heavy use over the first quarter-century of her working life (John Gardner photograph).

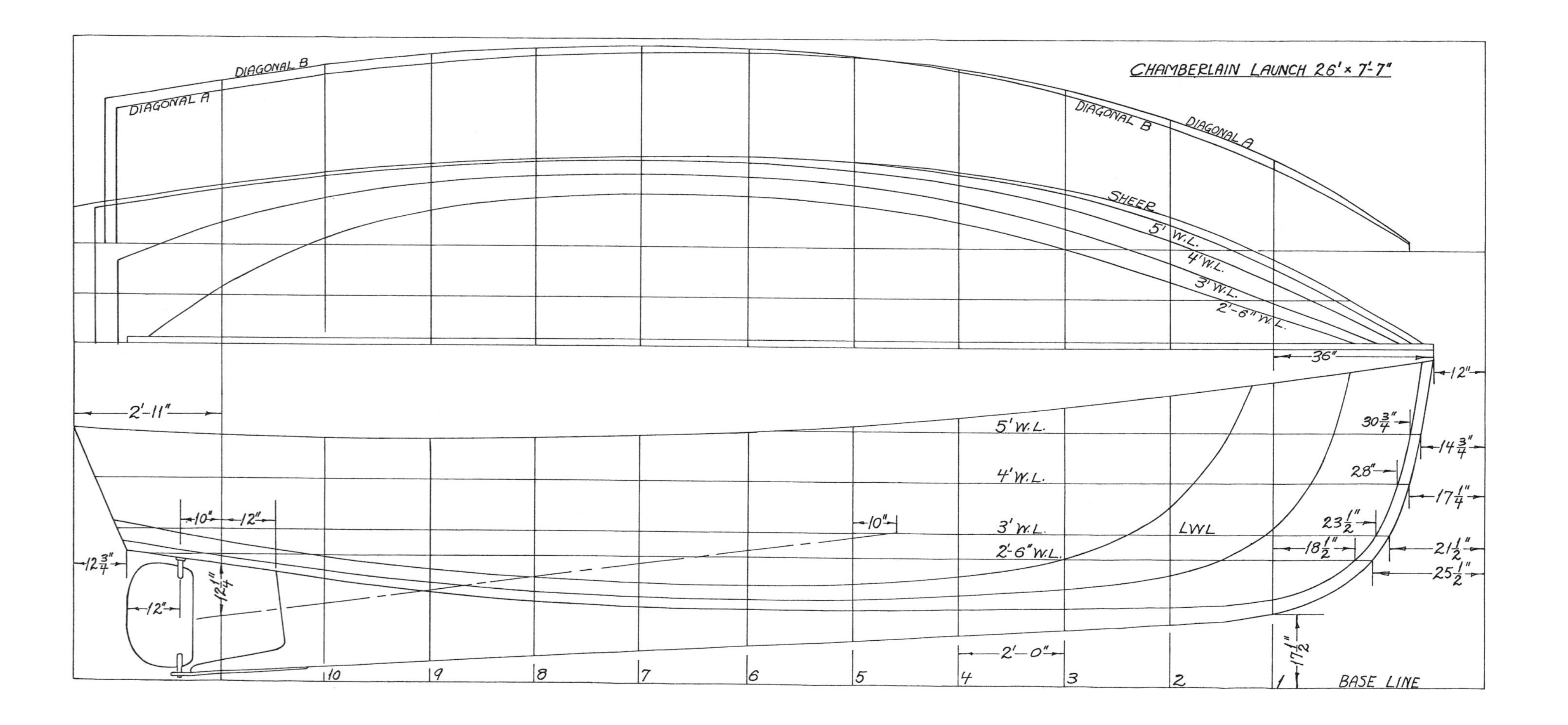
CHAMBERLAIN LAUNCH 26' × 7'-7"
DIAGONAL A
DIAGONAL B
SHEER
5' W.L.
4' W.L.
3' W.L.
2'-6" W.L.
LWL
36"
12"
2'-11"
30 3/4"
14 3/4"
28"
17 1/4"
23 1/2"
18 1/2"
21 1/2"
25 1/2"
10"
12"
12 3/4"
12 1/4"
2'-0"
17 1/2"
BASE LINE
1
2
3
4
5
6
7
8
9
10

This stern detail shows a hull more suited for low or moderate speeds. The buttocks run up to the transom, which is narrower than the hull in midsection, unlike modern high-speed hulls where the buttocks run straight aft to the transom or slightly down and the stern is widened out fuller and flattened down (John Gardner photograph).

account of the *Philippa* as printed in the *Maine Coast Fisherman* for January 1956, now long out of print. Reading it over, I find that it covers the subject adequately, and I see no reason for revising or amending it.

The *Philippa* is a boat that has proved herself in use. This 26-foot launch has a hull that is able, dry, stiff and easily driven at moderate speeds. For 20 years she was an open boat. And it was not until quite late in life that she was converted by the addition of a small house, and such conveniences as a toilet, berths and an arrangement for cooking. As the *Maybe*, she was adapted for cruising. Aft, she was floored over for a cockpit. An awning was put up, and she was generally fitted out as a small pleasure craft for fishing or cruising for a couple or small family. In the summer of 1956 she was used in the waters of Ipswich Bay and the vicinity.

The boat was built in Marblehead, Massachusetts, in 1927 by William Chamberlain. Who her designer was I don't know, although it could have been Chamberlain himself. The fact is that her lines are fairly typical of some of the better power lobsterboats then in use in this locality.

Chaisson in Swampscott, for one, was building powerboats at this time on lines much like these. In particular, sterns had not then been widened out and flattened down, as they mostly are today to meet high-speed requirements.

She was built as a working launch to carry mail, food, supplies and the like from Marblehead to Cat Island, well known also as Children's Island because of the large children's hospital and sanitarium that occupied that island for many years. The run from the wharf of the Marblehead Transportation Co. near the foot of State Street is somewhat more than a mile. The hospital was open during the summer months only, but the launch ran quite late in the fall, going into storage from December until March 30, according to Lewis Church of Marblehead, who had charge of the Children's Island boats from 1928 to 1944. Thus in the spring and fall over these years she must have seen considerable rough water. Incidentally, her original name, the *Philippa*, was taken, I am informed, from the name of the man who gave her to the hospital, Philip Saltonstall.

Church recalls that the *Philippa* was first powered with an 8 h.p. two-cylinder Palmer. In 1937, according to his recollection, this engine was replaced by a 12 h.p. Palmer of the same type, which engine was in use in 1944, the last year Church was in charge. The following year, Richard Searle, who followed Church at the Island, had a Gray 4-52 installed in the *Philippa*. This engine was retained by Richard Dion when he converted her to a cabin boat.

Searle is of the opinion that she probably did not do more than 4 m.p.h. with her original low-powered engine. With the Gray she will do 7 to 8 m.p.h. nicely. Pushed beyond this her stern settles and she begins to lose her good qualities. For this hull to operate efficiently at a higher speed than 8 m.p.h. it would be necessary to change the lines of her bottom aft. A little extra speed might be squeezed out of her by putting "shingles" under the transom but at the risk of spoiling her sea qualities to some extent.

The old* Philippa, *re-named* Maybe, *swings at her mooring in Marblehead. Her conversion included the addition of a trunk cabin with windshield, bunks for two below deck and a head (John Gardner photograph).

Compare the lines of her bottom aft with those of the modern high-speed boat of this length. In the high-speed hull the after bottom is wider, fuller and flatter. The buttocks run straight to the transom or even kick down slightly at the end. Sometimes the transom is nearly as wide as the midsection. Such a hull is no good unless running at high speeds. Slowed down in a seaway such a boat can be very difficult. They require lots of horsepower. One hundred horsepower and more is not excessive for a 30-foot boat.

The *Philippa* is an entirely different craft. She is suited for low or moderate speeds, at which her performance is excellent. Twenty-five horsepower is probably quite ample for her best performance. Of course a slow-turning engine is to be preferred, and a large wheel of low pitch.

Searle has high praise for the *Philippa.* Quoting from a letter he wrote me: "I remember loading her almost to the gunwales with Bird shingles when I was working on the roofs out there. She was a mighty good, sturdy, seaworthy launch. She would take 25 passengers, but she wasn't a passenger boat. She was used for a work boat—mail, food, supplies, *etc.* She was old but always well looked after, handled, and in top shape, not a punk piece anywhere in her. She was so well built she was dry and able."

After the hospital closed, the *Philippa* passed through several hands. Richard Dion of Salem rebuilt her extensively, putting on a house and making other changes as have been mentioned and renaming her the *Maybe.* She was damaged in the hurricane of 1954, but not seriously. For a working launch her construction was light but strong enough as her long life has proved. The original thickness of her native pine plank could not have been over 13/16", and is now about 3/4". Her bent timbers of white oak were, as I recollect, about 1" x 1$^1/_8$" and perhaps 9" on centers. Her stem and keel were of 3-inch siding only. But she was well fitted, of good lumber and well fastened. I think she has never had any rot in her. Her long and trouble-free life speaks well for the endurance of light yacht construction when it is properly done.

I think she is a pretty boat, and I think she looks well with the house, although Searle said he liked her better as an open boat. Those who have used her speak highly of her various good qualities. Richard Dion praised in particular her stiffness and stability as compared with some larger boats which he happened to know. Of course her big drawback for many today is her moderate speed. To attempt to drive her much over 8 m.p.h. is to waste gas and to lose her best performance.

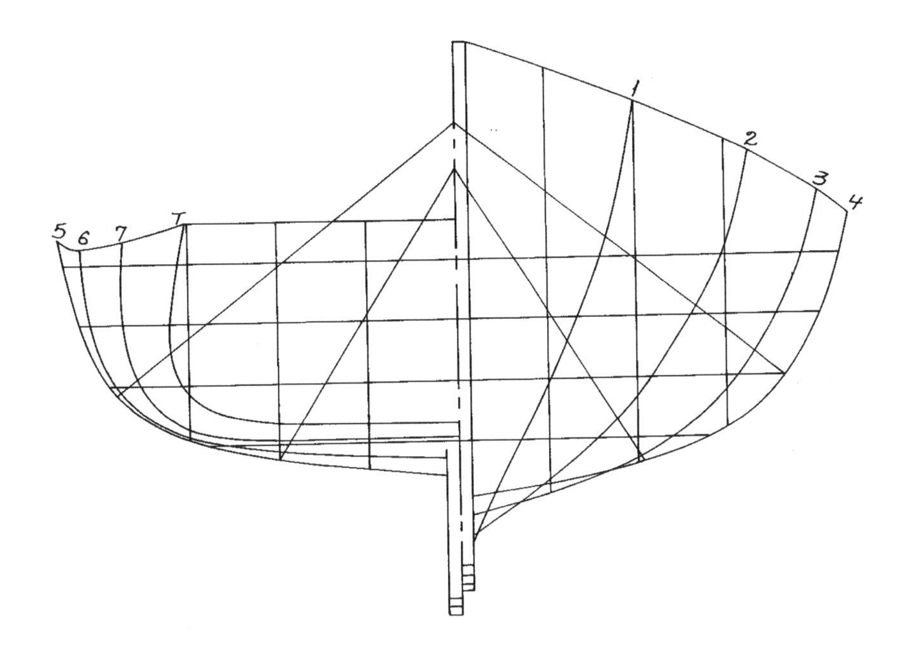

CHAPTER TEN

JONESPORTER LAUNCH

One of the handsomest, fastest, most seakindly, most hydrodynamically efficient working launches ever developed is the classic Jonesport lobsterboat. Its principal originator is generally reputed to be William Frost.

As a young man in the early years of this century, Frost moved down to the Jonesport area from Digby, Nova Scotia. There is no doubt that others had a part in perfecting this model, but it is clear that Frost stood head and shoulders above all of them. Who his teachers were, if he had any, and where his ideas came from are now a matter of hearsay, surmise and secondhand or thirdhand recollection.

It is my impression that he was a self-taught genius of sorts, one who worked from half-models he carved himself rather than from designs laid out on the drawing board. One result is that almost no lines and offsets for his boats are recorded in print. I may have overlooked a published source of working lines and offsets for the Jonesport boat, but I have to say that I do not recall anything of this kind that would be of much help to a would-be builder.

Lines do exist for the 34-foot William Frost boat that made the best showing in the model tests of some fishing launches that Professor Thomas C. Gillmer ran in the test tank at the Naval Academy in Annapolis and reported on in his paper for the Fishing Boat Section of the Food and Agriculture Organization in 1960.

The lines of this boat were used to illustrate that paper when it was published in *Fishing Boats of the World: 2*, but they appeared in such reduced size that it would not be easy to enlarge them to permit scaling offsets. This would be a job for a naval architect, or at least an experienced draftsman, and is not something that the average boat builder would feel like tackling by himself.

This 34-footer is an especially shapely example of the classic Jonesport hull that William Frost built in 1946 after he moved his shop to Portland, Maine. As a lobsterboat working out of Gloucester, Massachusetts, it caught the eye of Philip Bolger, the well-known naval architect, who was moved to take off the lines for study, subsequently furnishing them to Professor Gillmer for tank testing. (For those interested, these lines may be seen on page 342 of *Fishing Boats of the World*, Jan-Olof Traung, Editor, Fishing News Books Ltd., 1960.)

The close resemblance of these lines to those for the 32-foot West Jonesport boat printed herewith is quite remarkable. Both are essentially the same model. What differences exist are only minor and unimportant. Consequently, there is every reason to expect that the performance of the two boats would not differ significantly either. This argues well for the West Jonesport boat, considering that Frost's 34-footer tested superior for Gillmer in 1960 and likewise for DeSaix in 1981.

Indications of a revival of interest in efficient workboat hulls, plus Pierre DeSaix's report of tank-test results for lobsterboat models (see *National Fisherman* for December 1981, p. 57), recalled to mind the West Jonesport boat I had measured and reported on 20 years earlier in the *Maine Coast Fisherman*. This boat, then recently acquired by the late Captain Allison Ames, formerly of Camden and Lincolnville, Maine, was hauled out awaiting repairs at the Frederick J. Dion yard in Salem, Massachusetts, where I was employed at the time.

Captain Ames, a good friend of mine, was quite willing to have me measure the boat. There was no rush and I had plenty of time to do a careful and thorough job. The lines produced for publication in the October 1961 issue of the *Maine Coast Fisherman* were accurate. However, they were not accompanied by offsets, though the reason for the omission now escapes me.

The offsets printed here are taken from a recently completed enlargement of my original draft of the lines, redrawn to a scale of 1½" to the foot. From a drawing of this size it is possible to derive offsets accurate enough to build from. Perhaps there are offsets for other Jonesport hulls as readily available elsewhere. But, if so, I do not know where.

One striking example of the scarcity of such information is Royal Lowell's book, *Boatbuilding Down East: How to Build the Maine Lobsterboat*, which covers the construction process competently and thoroughly. Yet it contains no lines and offsets, even though the late Royal Lowell was born on Beals Island next door to Jonesport and is a grandson of William Frost, one of the two persons to whom the book is dedicated.

How anyone can be expected to build boats without lines and offsets is something I fail to understand. Nevertheless, to give this book its just due, it is by far the best source to date on the construction, that is to say, the carpentry, of Maine lobsterboats. The book is available from International Marine Publishing Company, Camden, Maine.

Selections follow from the article in the October 1961 issue of the *National Fisherman* which accompanied my lines for the 32-foot West Jonesport boat.

"One fine boat," says Lewis Church, the boat's former owner, who should know. Yes, if anyone would know, Lewis Church would. Church, a native of Jonesport, Maine, where in July 1944 this sharp 32-footer was launched, had spent much of his life in and around such fast power launches. As a lobsterman for many years and even longer as a skipper for the Marblehead Transportation Company, Lewis Church had seen and tried them all.

The boat was built for Church by Frank Smith & Sons of West Jonesport. Church says the model was worked out according to his own ideas for speed and general ability, which were based on his observations and experience with the older power launches of the Jonesport fishermen. Smith, the builder, also added touches of his own. The construction is fairly typical for the locality and the model and might be said to be substantial for a fast launch, as we shall explain in detail.

Originally the gas tank was located in the center of the boat, along with the engine,

OFFSETS WEST JONESPORT LOBSTER BOAT. LOA 32'-4". BEAM 9'. DRAUGHT 2'-7".

	STATIONS	STEM KEEL	NO. 1	NO. 2	NO. 3	NO. 4	NO. 5	NO. 6	NO. 7	TRAN.
HALF BREADTHS	SHEER	0-1-6	2-0-0	3-3-3	4-0-6	4-4-7	4-5-2	4-2-1	3-8-5	3-0-3
	4' W.L.	0-1-6	1-7-1	2-11-4	3-10-5	4-3-5	4-4-4	4-2-0	3-8-7	3-1-1
	3'-4" W.L.	0-1-6	1-4-1	2-7-4	3-7-0	4-0-6	4-2-3	4-1-0	3-8-5	3-2-2
	2'-8" W.L.	0-1-6	1-0-3	2-2-1	3-1-5	3-8-1	3-10-4	3-10-0	3-6-5	3-1-1
	2' W.L.	0-1-6	0-8-3	1-6-6	2-5-0	2-10-1	2-10-0	2-8-2	—	—
	KEEL & STEM RAB.	0-1-6	0-1-6	0-1-6	0-1-6	0-1-6	0-1-6	0-1-6	0-1-6	0-1-6
HEIGHTS	SHEER	6-4-5	5-8-4	5-1-5	4-8-3	4-5-1	4-3-4	4-2-0	4-3-0	4-5-1
	RABBET	—	0-10-5	0-11-5	1-2-1	1-4-5	1-7-2	1-9-6	2-0-2	2-2-4
	BOTTOM KEEL	—	0-7-4	0-6-2	0-5-1	0-4-2	0-3-2	0-2-1	0-1-0	—
	BUTTOCK 1	—	2-7-2	1-6-5	1-4-6	1-6-1	1-8-3	1-10-1	2-0-2	2-2-5
	BUTTOCK 2	—	—	2-5-5	1-9-2	1-8-5	1-9-7	1-11-0	2-0-4	2-2-5
	BUTTOCK 3	—	—	4-1-7	2-6-1	2-1-0	2-0-5	2-0-7	2-2-1	2-6-3
	DIAGONAL A	0-3-1	2-3-4	3-3-5	3-9-4	3-10-2	3-9-0	3-7-7	3-6-2	3-3-2
	DIAGONAL B	0-2-0	2-1-2	3-4-4	4-2-3	4-7-6	4-9-4	4-9-0	4-6-2	4-0-7

DIAGONAL A UP 5'-0"- OUT 3'2" ON BASE LINE. DIAG. B UP 5'-6"- OUT 3'-8" ON W.L. 2'-8".

placing the major weight amidships. Church says she ran quite level at top speed, which he thinks may have been 15 or 16 knots. When new, she was powered with a 110 h.p. Buick automobile engine, but the engine probably furnished considerably less horsepower than that in a marine installation. Church believes that she could have handled more power and with it would have shown an even better turn of speed.

In the eight years from 1944 until 1952, when Church sold the boat to Joe Walker, a Marblehead lobsterman, he used her a lot and in all kinds of weather.

"Never put her bow under once," he reports. Certainly she's a splendid sea boat for her size. If the model were enlarged to 40 feet, Church thinks it would give a good account of itself.

After Walker got the boat in 1952, he made some minor alterations such as moving the gas tank aft to give more working room in the cockpit. Subsequently he changed over to diesel power in the interest of fuel economy, as many lobstermen are now doing.

In 1960 the boat got jammed against a wall and was squeezed and crushed to some extent. While the lower hull was not damaged, considerable repairs were needed to restore the upper structure, and Walker took this opportunity to dispose of the boat in order to get a larger one built in line with the changing requirements of the local lobster fishery. In 1961 when I measured the boat the owner was Captain A. G. Ames of Manchester, Massachusetts.

Before going on to a discussion of the scantlings and the construction of this launch, attention should be directed to some features of the shape that deserve particular notice.

When I first viewed the hull from astern as she sat raised on the cradle, I was struck by the reverse in her run and the apparent hollow in the underside of the transom. It is very noticeable, but in actuality less than it appears.

A long straightedge applied to the horn timber and along the rabbet line forward of the sternpost shows barely 3/8 inch of reverse curve in 12 feet, the maximum hollow coming just about 6 feet from the stern. And the hollow across the underside of the stern at the transom and the first frame forward of it is less than 1/2 inch in depth.

In either case, this is not very much, but Church considers it important and even critical. Whether performance would be improved with a little more or a little less hollow would need testing to make a positive determination.

Curves throughout are moderate and easy. No excessive flare, no hard turns, no lumps or bunches anywhere. While the boat is quite sharp, with its forward floors rising

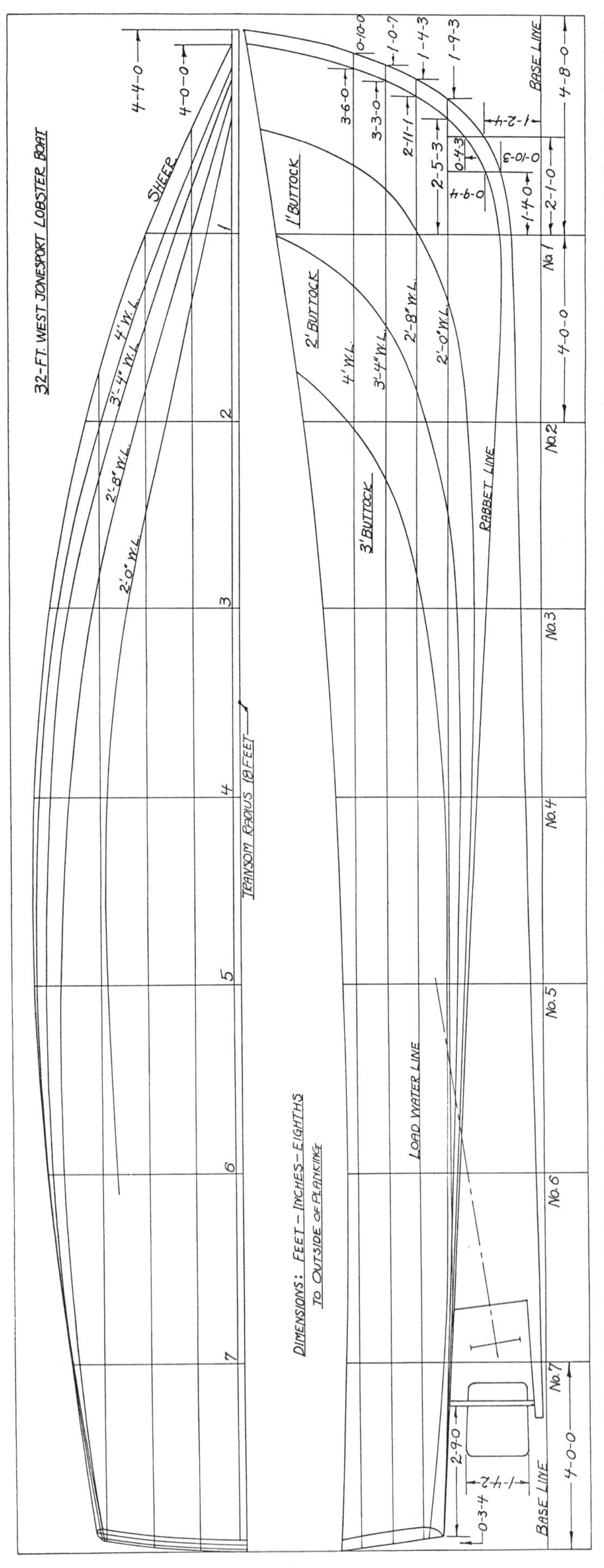

to easy bilges, the entrance along the load line, as shown, is straight rather than hollow and pinched. All portions of the hull blend together in exceptional harmony: a powerful bow, a handsome sheer—altogether an exceptionally pleasing hull.

Possibly the actual load line is lower than is shown. The LWL is drawn in the lines as it was painted on the boat, and lobstermen, as a rule, tend to paint the load line high. A lower load line would show a very slight reverse, or hollow, in the entrance.

As previously stated, the construction is substantial; not heavy as workboats come, yet heavier than most light, fast launches. After 16 years and hard use lobstering, the condition of the hull was exceptionally good. There was not a sign of rot anywhere, or any structural failures aside from those incurred in accidents, which are not to be taken into account in assessing the effect of ordinary wear and aging of materials and normal stresses and strains.

The planking of Maine white cedar is a full inch thick, which is thicker than is often found in launches of this size. Sometimes their planking finishes as light as 13/16". The extra thickness is a good thing here, in my opinion, as cedar is light and not as strong as pine, Douglas fir or mahogany.

The keel and deadwood are oak, sided 3½", probably as light as one would dare go in a boat of this size. There is no back rabbet.

The ribs, or timbers, are a bit unusual, according to what is generally seen here, being 2¼" x 1" oak, bent on the flat and spaced 10" on centers. They run across the top of the keel and help greatly to tie the boat together.

Floors, sided 1¾", are placed mostly in every third frame bay. While they are set next to the

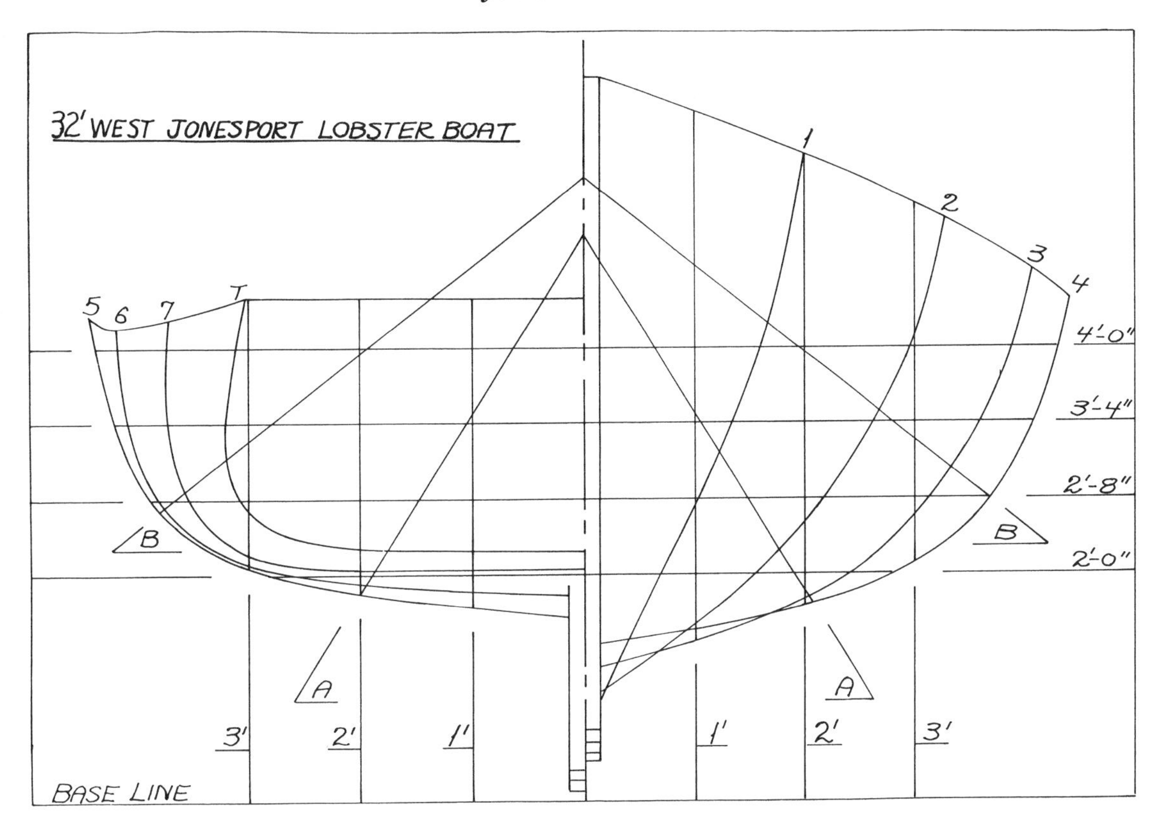

frames, they are not bolted to them and might just as well have been set in the middle of the bays. In fact it would have been better if they had been.

The ribs, while comparatively flat, are not as thin as the flat ribs commonly seen in Nova Scotia and are much stiffer in consequence, and better for it. They are thin enough, however, to allow the points of the galvanized chisel-point boat nails, then available for plank fastenings, to come through for clinching. Because of the extra width of these ribs, they probably add a few more pounds of hull weight than would narrower but thicker frames of nearly square cross section, such as are more commonly used in New England.

But narrow, lighter, frames that are thicker in cross section crack much easier, both in bending and afterwards, as they age. One of the most common ailments of round-bilged launches of any age is cracked frames along the bilge, especially on the hard turns aft. Flatter, wider frames avoid this, and of course in this boat we are considering, the bilges turn more easily than in some of these other more extreme types.

One point that should be explained, perhaps, is that the cockpit was not built watertight, and hence the cockpit floor does not slope aft for drainage through the stern, as is usual in working lobsterboats.

While I have called this a lobsterboat on the lines drawing, because it was used principally for lobstering, it is actually of a size and type that is now virtually obsolete in the New England lobster fishery.

What I had to say about the 32-foot Jonesporter in the *National Fisherman* in January 1982 brought several letters in response which contributed additional information on the characteristics and performance of this boat, as well as its origin. One was from Ralph L. Church of Marblehead, son of Lewis, the original owner.

"Your article in the January 1982 issue of the *National Fisherman* brought back some fond memories of experiences with my father, Lewis Church, in his Jonesporter. I spent many days on the *Whistler*, first working with my father and later working as a helper for Joe Walker after he bought the boat in 1952.

"When the *Whistler* arrived in Marblehead, she was probably one of the fastest work boats around and about the only Jonesporter in the area. Most lobstermen were fishing inshore as they had always done. Dad had the idea that fishing offshore, as they had been doing in Maine for years, would open up virgin territory. In the mid-to-late 1940s he did very well in these areas. After the war, more large boats were built, and many more fishermen began fishing farther offshore.

"During those times there were numerous trap wars and territorial claims by fishermen from other towns. Dad had many stories about encounters with other fishermen during this period. The speed of his boat caught many by surprise, and got us out of some tight spots in those days.

"With little provocation, Dad would drop our lobster gear to prove the boat's speed to any comer. For several years he could beat every lobster boat that came along. Her speed and clean lines got the attention of local yachtsmen, who had several like her built in the 1950s powered with large engines. It was frequently commented how very little wake was left by *Whistler* as she slipped through the water, which speaks well for the efficiency of the hull.

"... The remarkable Jonesport model showed that there could be agility, speed and seaworthiness built into a hull that also serves as a good work platform.

"Although I lobster only part-time now, I still have admiration for the Jonesporter style. Fred Lenfesty built me a 28-foot boat in 1960. She had a good turn of speed and was used for pleasure boating for several years. She is now lobstering in Marblehead and has stood up well."

Another letter commenting on the Jonesporter and discussing various theories as to the origin of the design was received from Elmer Harris, a Canadian long associated with Community Boating in Boston, Massachusetts, whose knowledge of our native watercraft is encyclopedic.

Harris begins by saying, "I had had for quite some years the belief that the Jonesport boat was based on the Nova Scotia Cape Islander, developed at Clarks Harbor, the point of land in Nova Scotia nearest to New England, and the Jonesport boat became the basis for the beautiful, more conservative, variegated Maine coast lobster boat. Also, more recently—maybe two or three years ago—a comment accompanying a picture of a Jonesport boat in an issue of the *National Fisherman* confirmed that belief. Can derive from your article some support for this belief, but also for the opposing theory that the Jonesport boat was introduced by a Nova Scotian, William Frost, who came from Digby, Nova Scotia, which is about one hundred miles north of Clarks Harbor."

At the root, these two theories about the origin of the Jonesport boat many not be as far apart as first might appear. Certainly Will Frost had an important part in the development of the Jonesport boat, possibly more than any other single individual, but no one has ever claimed, to my knowledge, that he was the sole originator and that others did not have a part in it as well.

Frost must have known about the Cape Islander almost as soon as the first one was built at Clarks Harbor in 1907 by Ephraim M. Atkinson. Changes were coming fast then to accommodate the new engines. That same year (1907) a contributor to *Yachting* noted that Casco Bay's Hampton fishing fleet had almost entirely converted from sail to power, with consequent changes in hull design.

Naturally, there were bound to be similarities in powerboats that evolved at about the same time, in neighboring waters, for much the same use and from similar sailing craft. But the point is that the Jonesporter was no copy or imitation of the Cape Islander, but went beyond the Cape Islander in all respects. It was a better sea boat and safer. It was faster, better looking and better built.

Today the old-style Cape Islander is obsolete and has all but disappeared, while modern Nova Scotian lobsterboats are very little different from the hulls of present-day descendants of the Jonesporter.

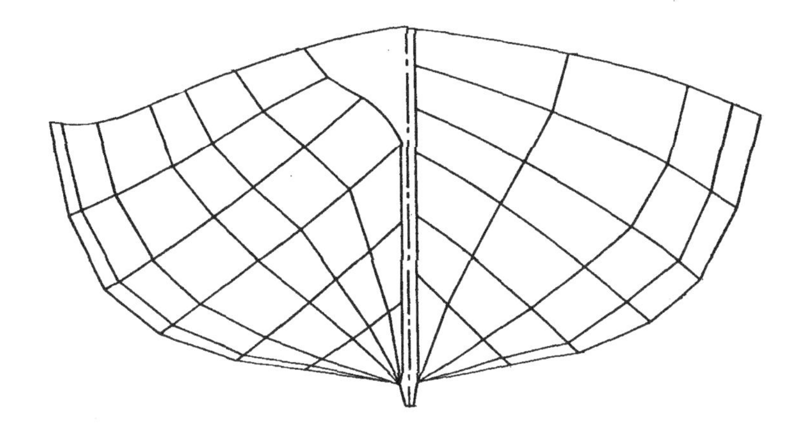

CHAPTER ELEVEN

FOUR-OARED GIGS

AMERICAN STAR AND *GENERAL LAFAYETTE*

As its contribution to the celebration of the bicentennial of the American Revolution, Mystic Seaport Museum dedicated the four-oared gig *General Lafayette*, built at Mystic Seaport and launched in June of the previous year. The *General Lafayette* is a reproduction of the winning race boat *American Star*, which was presented to General Lafayette on his last visit to the United States in 1825. The boat returned to France with him to the Chateau de la Grange, the ancestral home of Lafayette's wife where Lafayette spent the last years of his life.

There the boat had lain undisturbed and forgotten, its existence unknown to the boating world, until its chance discovery by Kenneth Durant who was researching the Whitehall boat at the time as the possible prototype of the Adirondack guide-boat.

Upon his retirement as an editor and journalist, Durant had set out to ascertain the origin of that remarkable craft. It seemed logical at the beginning to assume a connection with the Whitehall boat, and in the course of his research Durant's attention was directed to an attractive and copiously illustrated volume, *France and New England* by Allan Forbes and Paul F. Cadman, and published by the State Street Trust Company of Boston of which Forbes was president. Although banking may have been Forbes' business, his passion was New England history. His extensive collections, much of the material of a maritime nature, and exhibited and stored at the bank, had all but turned it into a museum.

During the first world war, Forbes' protégé Paul F. Cadman had served as an ambulance driver stationed in France,

The* American Star *in April 1968, photographed by Count René de Chambrun at the Chateau de la Grange, Bernay-en-Brie, France.

giving him the opportunity to travel about widely taking photographs later published in *France and New England.* From these it is apparent that he gained admittance to the Chateau de la Grange, and one of his many photographs of both the interior and exterior of the chateau is one showing a long, slim boat suspended high against a wall on two protruding brackets. The cutline printed underneath reads as follows: "Rowboat at la Grange in which the 'Whitehallers' of New York City rowed Lafayette in the year 1825 just before embarking for France." After the boat came to France it was kept for many years in a shed which was later torn down. The boat was removed to a barn, where it is now shown to any visitors who make inquiry about it. There is a painted inscription on it almost effaced by time and weather that reads "*American Star*—victorious Dec. 9, 1824." The boat was used by the 'Whitehallers' in a winning race held while Lafayette was in New York.

As soon as this photograph caught Durant's eye, he realized the importance of his find. Here was a well-preserved example of American boat building skill far older than anything previously known to exist. And besides, there was the connection with Lafayette. At a time when popular concern for our traditional small craft was on the rise, this discovery could not but help to stir wide interest among the boating public. Obviously there was much to be learned from it technically, and it opened to view a

Arrival of the* British Queen *at the Battery in New York, by J. Pringle, 1839. Note the four-oared racing gig in the center foreground (photograph courtesy of the I. N. Phelps Stokes Collection, Prints Division, The New York Public Library, Astor, Lenox and Tilden Foundations).

bygone era when rowing was an intensely popular competitive sport.

Durant lost no time in informing Howard Chapelle, at that time senior maritime historian at the Smithsonian, who promptly left for Paris to confer with his counterparts at the Musée de la Marine and to arrange to view the boat at la Grange. He was denied admittance. Next to try was Dr. Robert Bruce Inverarity, Director of the Adirondack Museum, but also without success, and an attempt by the United Press International with which Durant had connections fared no better. A wall had been set up that no one seemed able to breach. The explanation given was that access to the chateau was restricted for the time being to André Maurois, prominent French novelist, essayist and biographer who was engaged in writing a biography of General Lafayette's wife. Madame de Lafayette had inherited la Grange from her mother the Duchess d'Ayen de Noailles who had been guillotined during the French Revolution. Maurois served as an excuse, but probably the real reason was that Count René de Chambrun, descendant of Lafayette into whose possession la Grange had passed, had been annoyed and disturbed by the prying curiosity of uninvited visitors. At that time 20 years ago Count de Chambrun was an extremely busy man working a full week and long hours both as the active head of Baccarat, world-renowned manufacturers of fine crystal, and in his international law practice. It is to be noted that de Chambrun is also an American citizen by virtue of his descent from General Lafayette. After working Saturday mornings from 6 a.m. to 11 a.m., de Chambrun was accustomed to drive to la Grange, where he would spend the remainder of the weekend cataloguing and organizing the Lafayette collection, his own private property, yet protected and heavily restricted by a foundation chartered for the preservation of national cultural treasures. It was de Chambrun's intention, it turned out, eventually to open the collection for viewing to small select groups admitted by appointment arranged well in advance. At that time, de Chambrun was engaged in listing and bringing some order to the great mass of Lafayette's private papers and correspondence. As for the boat, things stood at an impasse for some time until Harold Hochschild, founder and patron of the Adirondack Museum, as well as head of a large multinational mining corporation, arranged to meet with de Chambrun. It appears he was cordially received, was shown the boat and came away with a photograph of it taken by de Chambrun himself.

Henceforth the wall was down, or at least some sections of it. An American student on a visit to Paris was granted permission by Madame de Chambrun to visit the chateau and to sketch the boat but not to photograph it. This aversion to photography lends some evidence to a story circulating at the time that when Henry Luce, publisher of *Life* magazine, and Claire Booth Luce visited la Grange they had offered a very substantial sum of money to be allowed to photograph the interior of the chateau in color, and were firmly if not indignantly turned down by Madame de Chambrun.

Finally in the summer of 1971 when Helen Grey, director of publications at Mystic Seaport Museum, was traveling in Europe, she included la Grange in her itinerary. Count de Chambrun proved to be friendly and cooperative. She was permitted to photograph the boat, and de Chambrun provided her with a full-length view of *American Star* made from two photographs pieced together that he had taken in 1968.

It was at this juncture in the early summer of 1972 that I decided to bring together what was known about *American Star* and its historical background for publication in the fall issue of the *Log of Mystic Seaport*. There is little in that account that I would change today either as to its content or the manner in which it is presented, and it is for this reason that that account, very lightly edited, is reproduced herewith.

Although she is the oldest surviving American race boat, and the victor in an international match race which was one of the most colorful events in the annals of early American sport—one which had the whole city of New York agog nearly 175 years ago—few Americans are now aware that there ever was such a boat, or that this craft is still preserved and cherished in a fifteenth-century Norman castle tucked away in the French countryside some 30 miles southeast of Paris.

From the Eno Collection, Prints Division, the New York Public Library, Astor, Lenox and Tilden Foundations.

The boat is the *American Star*, built in Brooklyn, New York, by John and William Chambers shortly before 1820[1], and the castle is the Chateau de la Grange, Bernay-en-Brie, on the family estate to which General Lafayette returned in 1799 to spend the latter years of his life after long imprisonment and exile abroad as a French patriot and liberal.

In 1824-25 the aging Lafayette visited the United States for the last time. He came as a guest of the nation to make what became a triumphal tour. Americans adored him almost as much, perhaps, as they had Washington. They had not forgotten his youthful sacrifices on their behalf, or his heroic part in their liberation from British rule. He was feted from one end of the country to the other and showered with honors and gifts. In fact, so numerous and varied were the latter that for the purpose of conveying them back to France the American government provided a new frigate which President Adams named the *Brandywine*, after the battle in which the young Lafayette had distinguished himself as a soldier.

None of the many gifts bestowed upon Lafayette during his visit better expressed the affection and esteem of the American people than this boat, which in winning the previous year against the British challenger had become a symbol of victory and national pride not only for thousands in the burgeoning new waterfront metropolis of New York, but all along the North Atlantic seaboard as well. Peverelly, in his annals of beginnings in American sports, *American Pastimes*, published in 1868, compares the triumph of the *American Star* with that of the yacht *America*, observing that the former did as much for rowing as the latter did for yachting later on.[2]

In the late afternoon on Saturday, July 9, 1825, a balloon ascension in General Lafayette's honor took place from Castle Garden at the Battery on the lower tip of Manhattan. Spectators assembled to witness the event numbered almost 5,000, some of whom had been waiting since early morning. Balloons were new then, the last word in excitement and reckless daring. No sound came from the waiting throng as the General cut the riband releasing the intrepid aeronaut in his frail wicker basket. As he rose in the air, silken flags streamed from either hand, one the Stars and Stripes, the other banner bearing the name of Lafayette in gold surmounted by tricolor streamers. And as the balloon gained

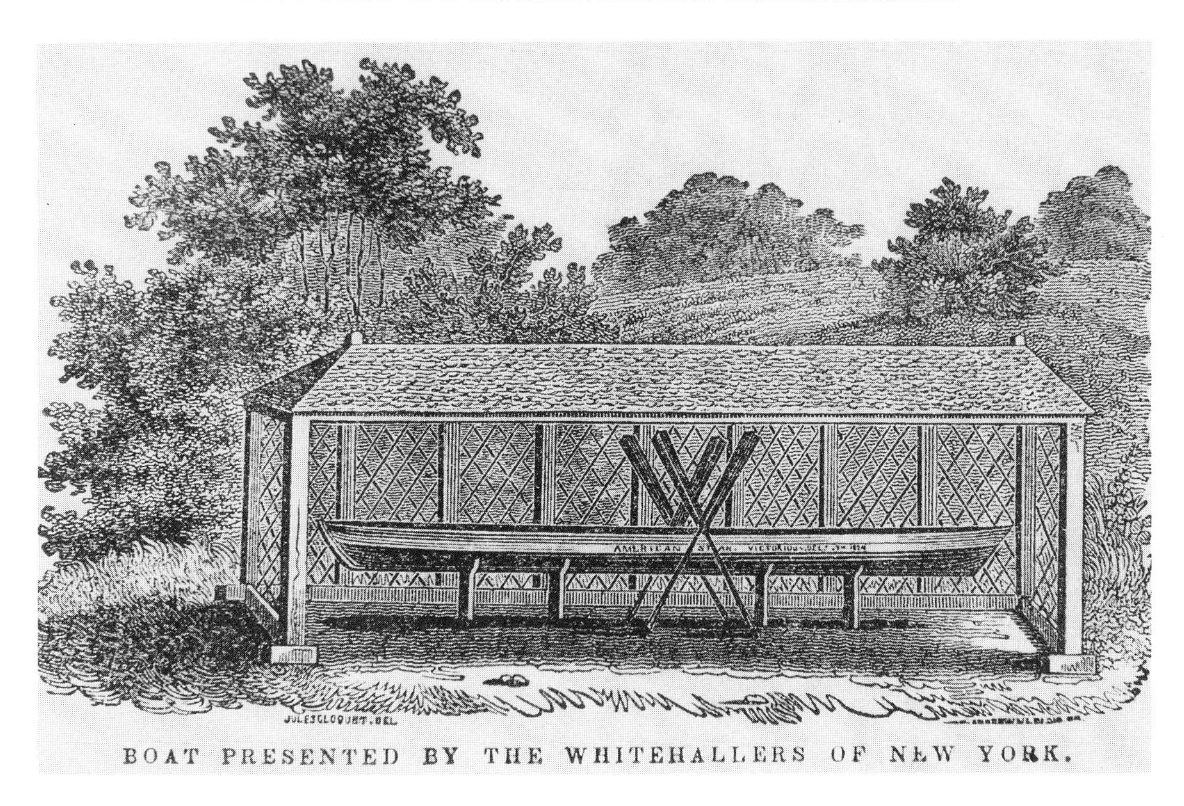

***General Lafayette's exhibit pavilion at la Grange especially constructed for the* American Star, *from Jules Cloquet's* Recollections of the Private Life of General Lafayette, 1836, Volume II.**

altitude and began to drift across the city, a shower of handbills fluttered down bearing verses in French and English eulogizing the honored guest.[3]

That evening the General was to dine with Colonel Varick at Paulus Hook, the site today of Jersey City. Moored below the Battery at the foot of Whitehall stairs, waiting to ferry him across, was the famous race boat *American Star*, appointed for the occasion with cushions, crimson carpet and silver-mounted oars. Her crew, the pick of the Whitehall boatmen, were the same who six months earlier had rowed the *American Star* to victory over the British challenger. Having learned of the General's dinner engagement some time earlier, they had asked to ferry him across, and their request had been readily granted. In the meantime it had been decided by the owners of the *American Star* to offer the boat to the General as a gift, and on the return from the Jersey shore the presentation was made. The coxswain, John Magnus, acting as spokesman, expressed the hope that the General would take the boat back to France with him "where it may occasionally remind you of your grateful friends you have left behind, the ingenuity of the mechanics of a country which you assisted to liberate, and also our great naval motto, 'Free Trade And Sailors' Rights.'"

In his reply Lafayette said he was moved by "the proud feelings of an American patriot ... with the grateful feelings of a friend, I now receive your kind present; no keepsake could be more welcome; the more gratifying, indeed, when offered from the hands of the five victors. It shall be most carefully and fondly preserved. I beg you gentlemen, will accept, and transmit to your companions, the congratulations, the thanks, and the good wishes of a veteran, heartily devoted to the great naval motto–'Free Trade And Sailors' Rights.'"[4]

The General proved as good as his word. On his return to la Grange, the *American Star* was placed on exhibit in a building specially designed and constructed for that purpose. Jules Cloquet's quaint account in his *Recollections of the Private Life of General Lafayette*, published two years after the General's death, provides a description.

"The other building is an elegant pavilion situated under the windows of the

Above is the 32-foot pilot gig* Active *under construction in Tom Chudleigh's shop in 1973 at St. Marys, England. Harold Kimber is at work in the background and* Treffry, *built in 1843, is alongside. Lightly built, and fast, these slim, 6-oared boats were used for piloting, salvage, smuggling and life saving at the Isles of Scilly off England's Cornish coast; the first was built in 1790 by the Peters family of St. Mawes, near Falmouth. Despite rough use in treacherous waters some lasted well over a century (F. E. Gibson photograph).

library, and covered with a tiled roof supported by slender woodwork. Its pillars are joined together by a wooden grating, and its form, which is extremely lengthy, explains the object of its construction. Lafayette caused it to be built for the preservation of a canoe presented to him by the Whitehallers of the port of New York.

"The canoe had been victorious in a species of sham-fight with the boat belonging to the British frigate, *Hussar*—a contest which sensibly piqued the pride of both nations. The canoe is extremely narrow and lengthy, made for rapid sailing, painted blue inside, and brown with a yellow band outside. On the benches are the names of the sailors who gained the day, and on each side is the inscription, '*American Star*, Victorious, Dec. 9th, 1824.'"[5]

The 80 years, more or less, which follow are blank. Apparently during that time Cloquet's "elegant pavilion" fell to ruins or was torn down, and the boat removed to what has been described as a barn or carriage house. This was where Paul F. Cadman of the American Ambulance Corps found it set on wall brackets close to the ceiling when he visited la Grange during World War I. The original color had become a faded gray, yet the name and victory inscription were still visible, and it was quite apparent that the boat remained structurally sound and undamaged—in fact, amazingly well-preserved for its one hundred years.[6]

Quite apart from the sentiment attached to the *American Star* as a memento of Lafayette, is the technical significance of this boat, for it is the only known surviving example of lightly built American small craft of its period. There is much to be learned from it. It should be possible now to resolve a number of questions concerning the early Whitehall boat, which so far have remained unanswered. When this boat was presented to Lafayette, it was offered as an example of the "ingenuity of American mechanics," and so it remains, a unique artifact from an era of American craftsmanship long since passed away.

American maritime history has so far dealt almost exclusively with large vessels. "But what of the humble boat," W. P. Stephens asks, "without which the ship would have been useless in those early days when piers were unknown, and the era of steam was far in the future? ... Situated on an island, New Amsterdam was dependent on close communication with the neighboring shores ... without the common rowboat in the early days of its settlement, New Amsterdam would have been completely isolated ..." Two hundred years later during the Whitehall era, rowboats were just as indispensable, while the uses to which they were put had greatly expanded.[7]

In 1824, at the time of Lafayette's second visit, the population of New York City—America's largest and busiest port and commercial city—had reached approximately 150,000, or double what it had been less than two decades previous. For such prodigious growth, shipping and shipbuilding were mainly responsible. A myriad of boats and vessels both small and large plied the busy waters surrounding Manhattan. Harbor communication and transportation were largely dependent on rowboats, the evolution of a special type of which was well advanced by 1824. These craft operated along the entire waterfront, yet they gathered in greatest numbers in a basin under the Battery wall at the foot of Whitehall Street, whence this celebrated Whitehall boat acquired its name.[8]

Long after the more distant ferries had been taken over by steam, the rowing Whitehalls retained their monopoly of harbor traffic. In fact the Whitehall boats, normally 19 feet long and pulled by two pairs of sculls, continued to work around the piers, running lines for vessels and picking up whatever jobs came their way until nearly the close of the century. But in 1824 these boats were in their prime. Without them, or craft like them, the pulsating activity of that busy port must have moved at a much slower pace, if it moved at all.

Incoming vessels, as soon as they had passed the Narrows and had been cleared by the health officers, were surrounded by boats on all manner of business—the owners' agents, friends of the passengers, competing reporters eager to learn from the Master what vessels he had spoken to and foreign news otherwise unavailable before the invention of wireless. Less savory were the waterfront sharks, crimps and keepers of saltwater boarding houses who came aboard with their offers and enticements for Jack Tar, soon to be paid and pleasure bent ashore. Passengers and ships' officers in a hurry to set foot on land engaged these swift harbor taxis. Straining oars frequently overhauled vessels outward bound, perhaps already standing in the stream with sails loosed, to deliver final orders, belated passengers, the latest edition damp from the presses; or a crimp's boat might pull alongside bringing a reluctant addition to the crew, groggy and sick, to be boarded.

Junkmen combed the waterfront in Whitehalls looking for salvage, and thugs and harbor thieves disguised as junkmen were chased by police in Whitehalls. Steamboat excursions, popular at the time, usually hired a Whitehall to follow behind to pick up passengers who fell overboard.[9]

For all, or at least most of, the uses to which these harbor boats were put, speed was essential, competition brisk and urgent. The nineteenth century in America was a pushing, driving era. Survival was a matter of getting ahead and staying ahead. From the start the pace was fast and kept getting faster. It was only natural that workboats should become race boats, that a new sport of competitive rowing should develop and,

Carl Brownstein of the Rights O'Man Boat Works in Shelton, Washington, designed and supervised the construction of this scaled-down version of the* Treffry. *One of four boats built for Evergreen State College, Olympia, Washington, the 22-foot four-oared gig* Carl Brownstein *is shown here on launch day in 1978 (courtesy Rights O'Man Boat Works).

John McGiffert's four-oared gig* Royal Oak *at Kingston on the Thames River, England. Retired from racing in 1834, the 31' 4" x 4' 0" hull was built about 1800 in County Down, Northern Ireland, for the Bailie family from oak, pitch pine and elm. She was entered in The Great River Race on the Thames in September 1992, the first competitive outing for the boat in 156 years. Lightly built with seven strakes of plank,* Royal Oak *is typical of a type wealthy families raced in regattas on Strangford Lough in northeast Ireland prior to 1816. The oldest racing gig of this type,* Royal Oak *will be on exhibit in the new River and Rowing Museum at Henley. Lionel Willis of the National Maritime Museum is at the bow and flanking the boat are John and Nora McGiffert (photograph courtesy of John McGiffert).

attaining wide popularity for a time, should be eclipsed by other competitive sports.

Although rowing races for match stakes occasionally seem to have taken place in this country as far back as the French and Indian Wars,[10] rowing as sport emerged rather gradually at first from the occupational competition of the Whitehall boatmen, and did not attain widespread, organized form until after 1830.[11]

The first recorded race of any consequence was rowed in the summer of 1811 between the *Invincible*, built by John and William Chambers of Brooklyn, formerly of London[12] where the brothers had learned their trade, and the *Knickerbocker* from the Water Street boat shop of John Baptis, leading New York City boat builder of his day.[13] The race was rowed in rough water and against a strong wind down the North River from Harsimus, New Jersey, to the Battery flag staff, a distance of two-and-one-half miles, with the *Knickerbocker* ahead all the way, easily winning by a long lead. This race caused much excitement in the city, and so stirred popular interest that the *Knickerbocker*, painted white with green gunwales and a gilt strip, was placed on exhibit in Scudder's Museum, and a public benefit arranged for Baptis. Later transferred to P. T. Barnum's American Museum at 220 Broadway, the *Knickerbocker* was unfortunately destroyed by fire when that museum burned in 1865.[14] No measurements, detailed description, or drawing of the *Knickerbocker* have come down to us. But an incidental reference in the *New York Gazette* in 1824 mentions her as a gig, and we know from contemporary accounts of the race that the crews of both boats consisted of four oarsmen and a "setter,"[15] as the coxswain was then called.

W. P. Stephens is in error when he states in his account of the race that the boats were pulled by four pairs of sculls.[16] More credible is Peverelly's description of these craft as "four-oared boats," and in fact there is every reason to believe that these 1811 race boats were very little different from the four-oared gig *American Star* which followed in less than ten years.

It was in the late fall of 1820 that arrangements were made for a match race between the *American Star*, owned by "a company of gentlemen of that place," and the *New York*, newly built by John Baptis of Manhattan for the Knickerbocker Club. The forthcoming event aroused great local interest. Rivalry was strong. A purse of $1,600, a lot of money at the time, was collected, and an "elegant flag" made up as a trophy for the winning boat; in addition, side bets variously estimated from $3,000 to $10,000 were placed. The course agreed upon ran from the upper dock of the Williamsburg Ferry down the East River to a stake boat anchored abreast of Castle Williams on Governors Island, a distance of about three miles.

At two o'clock Saturday afternoon on November 11, with snow in the air and a strong wind blowing from the northeast, the race started. The New York newspapers reported an "immense concourse" of spectators lining the pier heads and vessels lying at the wharves along the East River as far as the Battery. According to the *New York Evening Post* it was a "handsome performance and much skill was exhibited by the rowers on both sides... the *New York* won with apparent ease ... about fifty yards ahead." The *Commercial Advertiser* in its account reprinted from the *Gazette* cuts the lead of the winning boat to 20 yards, and reports the time as 12 minutes.

After the race the *New York*, with the coxswain seated in it and displaying the victory flag, was lifted from the water at the foot of Whitehall stairs, and placed on the shoulders of the winners' friends. This jubilant procession headed by a band then proceeded from Whitehall through Broadway to the Fulton Hotel on Fulton Street, where the refreshments and entertainment were waiting. "The party spent the evening in great glee," and the winning boat, suspended in front of the second story of the hotel, remained on exhibit for all the city to see for several days.[17]

According to the report in the *Evening Post*, both boats were 27 feet long, which recent measurement has verified for the *American Star*. Both were four-oared gigs and apparently not dissimilar in design, but possibly the *New York* was the better sea boat. This could have been the difference between Baptis' boats and the Chambers brothers' boats. When the *Invincible* lost to the *Knickerbocker*, heavy seas were running and strong winds were encountered, just as happened when the *American Star* lost to the *New York*. According to Peverelly, the *Knickerbocker* was a better sea boat, winning with greatest ease because of this. In smoother water the results might well have been different in both races.

There can be no doubt that the *American Star* was a fast boat: her subsequent victory over the famous British gig established that. The lines of the *American Star* show a hull slimmed and lengthened for speed, with flat sheer and scant freeboard–the evolution from workboat to race boat well advanced.[18] What the Baptis boats were like we shall never know–nothing remains of any of them. The ashes of the *Knickerbocker* were scattered long ago, and the *New York*, sold South shortly after her triumph to be raced in Mobile with a slave crew, must have long since rotted away.[19]

If the *American Star* had not been highly esteemed as a race boat we may be sure that she would never have been picked to uphold the honor of the Whitehallers and their city against the British challenger in December 1824. Although the *New York* was no longer at hand, and the *Knickerbocker* could not be removed from Scudder's Museum,[20] there must have been other boats in the city which could have been used. The *General Lafayette*, newly built by Baptis for the *New York Gazette* office as a press boat, and winner over the *Mapes* from Staten Island the previous March,[21] was a fast boat, but obviously too small to be considered at only 22 feet overall with a crew of two, each pulling a single long oar.

Layers of paint and grime concealed* American Star's *planking. The six 1/4" strakes were probably from northern white cedar stock (John Gardner photograph).

In the late autumn of 1824 the British frigate *Hussar* dropped anchor in New York harbor, reporting the rescue of a New York vessel from freebooters off Cuba. Captain George Harris was welcomed in the City's best clubs where he was wined and dined in style. As the after-dinner port passed, according to one account, Captain Harris let it be known that he had a fast race boat on board which had won on the Thames and in the West Indies.[22] Another version has it that the pilot who brought the *Hussar* into the harbor spotted a rakish race boat stowed on the foredeck, and in conversation with the captain ventured that there were boats in the city that could give her a contest.

In any case on December 4 the *Gazette* printed a communication from Harris stating that inasmuch as the boatmen of the City had learned that he possessed a "fast row-boat," he was prepared to row his boat "on the first fair day after Tuesday next ... against any one that can be produced, not excepting the one I have seen at the Museum."

The fat was in the fire. Local partisans immediately put forward the *American Star*, and no time was lost in concluding arrangements for a race with a purse of $1,000 put up by Captain Harris. Two days later the *Gazette* reported "great excitement" in the city, but "no unmanly feeling," adding that "debasing conduct" was not anticipated from either side, and it was mutually agreed that the winner should "treat the losing party to an entertainment."[23]

Yet tension must have continued to mount, for on the 8th of December the *Gazette* printed a statement from Captain Harris regretting "that his public offer to try a New York boat should have been construed as a challenge ... his only object was to gratify those ... fond of aquatic sport."[24]

The race was rowed on the morning of December 9 in rough water with a strong wind blowing—weather "Captain Harris deemed unfavorable to the best speed of his boat ... yet [he was] determined not to disappoint public expectation."[25] The city waited tense and expectant, with 50,000 spectators lining the wharves and the Battery, by far the largest crowd up to that time, and for some time after, to view an

American Star*'s diminutive, high-tucked transom stern is only 13 1/2" wide at the top and the timbers, spaced 18" apart, are 1/4" sided x 5/8" molded (John Gardner photograph).*

Preparation of this drawing of* American Star *was sponsored by the Society of the Cincinnati of the State of Connecticut.

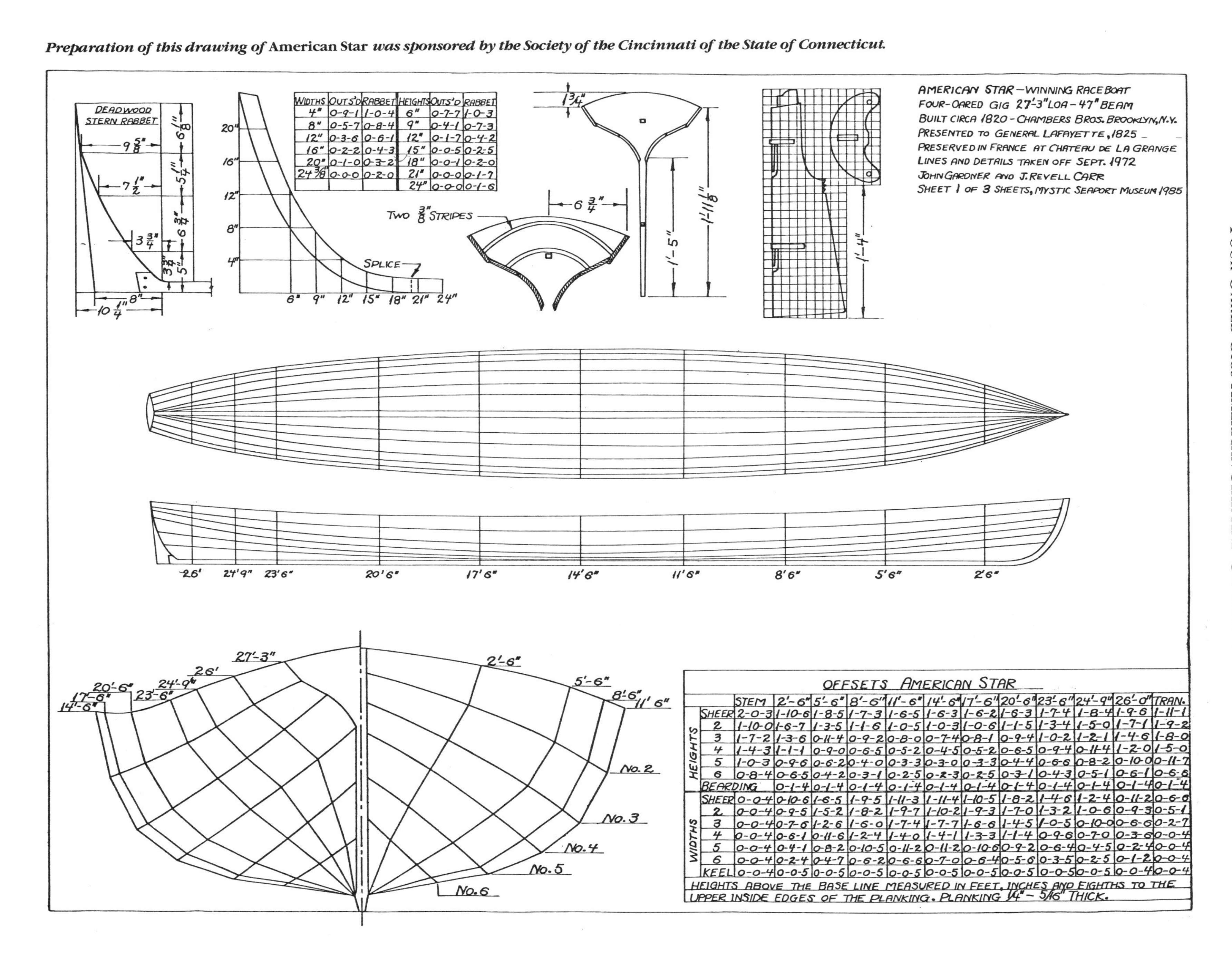

WIDTHS	OUTS'D	RABBET	HEIGHTS	OUTS'D	RABBET
4"	0-9-1	1-0-4	6"	0-7-7	1-0-3
8"	0-5-7	0-8-4	9"	0-4-1	0-7-3
12"	0-3-6	0-6-1	12"	0-1-7	0-4-2
16"	0-2-2	0-4-3	15"	0-0-5	0-2-5
20"	0-1-0	0-3-2	18"	0-0-1	0-2-0
24 3/8"	0-0-0	0-2-0	21"	0-0-0	0-1-7
			24"	0-0-0	0-1-6

OFFSETS AMERICAN STAR

		STEM	2'-6"	5'-6"	8'-6"	11'-6"	14'-6"	17'-6"	20'-6"	23'-6"	24'-9"	26'-0"	TRAN.
HEIGHTS	SHEER	2-0-3	1-10-6	1-8-5	1-7-3	1-6-5	1-6-3	1-6-2	1-6-3	1-7-4	1-8-4	1-9-6	1-11-1
	2	1-10-0	1-6-7	1-3-5	1-1-6	1-0-5	1-0-3	1-0-6	1-1-5	1-3-4	1-5-0	1-7-1	1-9-2
	3	1-7-2	1-3-6	0-11-4	0-9-2	0-8-0	0-7-4	0-8-1	0-9-4	1-0-2	1-2-1	1-4-6	1-8-0
	4	1-4-3	1-1-1	0-9-0	0-6-5	0-5-2	0-4-5	0-5-2	0-6-5	0-9-4	0-11-4	1-2-0	1-5-0
	5	1-0-3	0-9-6	0-6-2	0-4-0	0-3-3	0-3-0	0-3-3	0-4-4	0-6-6	0-8-2	0-10-0	0-11-7
	6	0-8-4	0-6-5	0-4-2	0-3-1	0-2-5	0-2-3	0-2-5	0-3-1	0-4-3	0-5-1	0-6-1	0-6-6
	BEARDING		0-1-4	0-1-4	0-1-4	0-1-4	0-1-4	0-1-4	0-1-4	0-1-4	0-1-4	0-1-4	0-1-4
WIDTHS	SHEER	0-0-4	0-10-6	1-6-5	1-9-5	1-11-3	1-11-4	1-10-5	1-8-2	1-4-6	1-2-4	0-11-2	0-6-6
	2	0-0-4	0-9-5	1-5-2	1-8-2	1-9-7	1-10-2	1-9-3	1-7-0	1-3-2	1-0-6	0-9-3	0-5-1
	3	0-0-4	0-7-6	1-2-6	1-6-0	1-7-4	1-7-7	1-6-6	1-4-5	1-0-5	0-10-0	0-6-6	0-2-7
	4	0-0-4	0-6-1	0-11-6	1-2-4	1-4-0	1-4-1	1-3-3	1-1-4	0-9-6	0-7-0	0-3-6	0-0-4
	5	0-0-4	0-4-1	0-8-2	0-10-5	0-11-2	0-11-2	0-10-6	0-9-2	0-6-4	0-4-5	0-2-4	0-0-4
	6	0-0-4	0-2-4	0-4-7	0-6-2	0-6-6	0-7-0	0-6-4	0-5-6	0-3-5	0-2-5	0-1-2	0-0-4
	KEEL	0-0-4	0-0-5	0-0-5	0-0-5	0-0-5	0-0-5	0-0-5	0-0-5	0-0-5	0-0-5	0-0-4	0-0-4

HEIGHTS ABOVE THE BASE LINE MEASURED IN FEET, INCHES AND EIGHTHS TO THE UPPER INSIDE EDGES OF THE PLANKING. PLANKING 1/4" – 5/16" THICK.

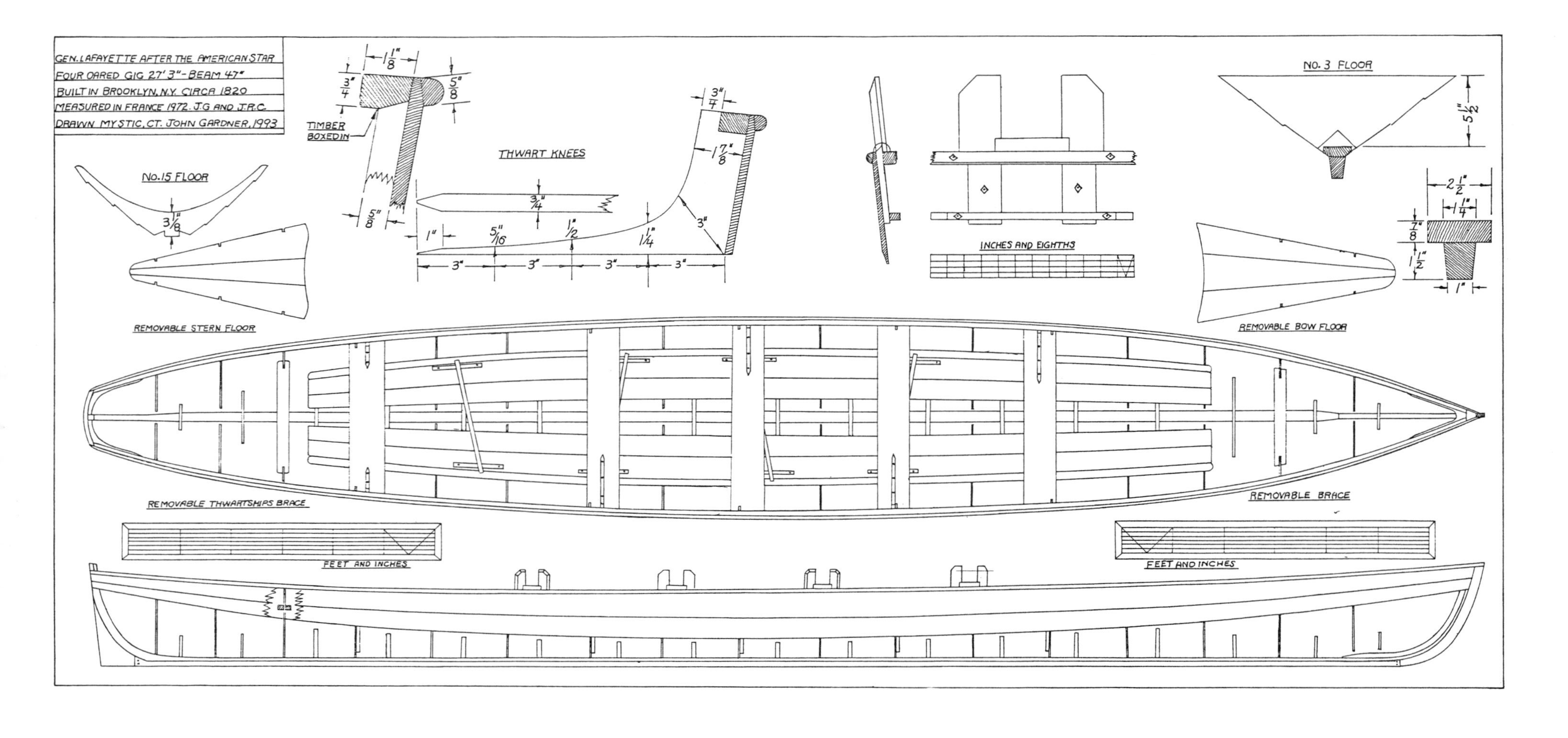
GEN. LAFAYETTE AFTER THE AMERICAN STAR
FOUR OARED GIG 27' 3" - BEAM 47"
BUILT IN BROOKLYN, N.Y. CIRCA 1820
MEASURED IN FRANCE 1972. J.G AND J.R.C.
DRAWN MYSTIC, CT. JOHN GARDNER, 1993
TIMBER BOXED IN
THWART KNEES
No.15 FLOOR
NO. 3 FLOOR
INCHES AND EIGHTHS
REMOVABLE STERN FLOOR
REMOVABLE BOW FLOOR
REMOVABLE THWARTSHIPS BRACE
REMOVABLE BRACE
FEET AND INCHES
FEET AND INCHES

American Star*'s fixed wooden tholes supported heavy 13-foot oars and the upright braces, 2" x 3/4", between the lower and upper risers were added when extra bracing was required for the thwarts (John Gardner photograph).*

Captain presented a purse to the winning crew containing a gold half eagle for each of them.[29] Never did a contest of this sort end more happily and with more good feeling on either side, an accurate reflection of the state of the political and economic climate.

Commenting on the two boats before their race, the *New York Gazette* called both "fine specimens of art," adding that both were built very much alike, although the *Star* was somewhat longer. The British boat, this account continued, had been built some two years previously by a man in his eighties who had built hundreds, but considered this his masterpiece.[30] As the Chambers brothers, builders of the *American Star*, had learned their trade in London, it may account in part for the similarity of the two boats.

It was generally agreed at that time that the two boats, as boats, were quite evenly matched, and that it was the difference in the styles of rowing which decided the day. The British crew pulled with a long, sweeping stroke, "by throwing back their whole bodies as if they were towing a man-of-war," as one account put it, while "the Whitehallers sat straight, moving with quick, short strokes, and drawing back with a jerk."[31] The *New York Evening Post* gave the number of strokes per minute for the *Star* as 46 against 39 per minute for the *Dart*. This was many years before the introduction of such refinements of rowing technique and equipment as sliding seats, outrigged oarlocks and spoon oars.

That the *American Star* is still in existence, and in sound condition, is something of a minor miracle. No other race boat of that early period has survived. No lines, specifications, or detailed dimensions for such boats are known to exist. Leaving sentiment aside, the importance of the *American Star* not only for small-craft history, but for the more general history of technology as well, can hardly be overestimated.[32]

American sporting event. Flying a small American flag from her bow, the *American Star* glistened from polishing. Her crew, dressed for the occasion, wore white guernsey shirts, blue handkerchiefs on their heads and blue pants. The British boat, *Dart*,[26] whose sobriquet was *Sudden Death*, carried her national colors with her crew dressed in their naval uniforms. Captain Harris acted as coxswain.

The New York boat pulled out to the frigate to meet her rival, and the race started at the sound of one of the frigate's guns. The course lay to a stake boat moored off Hoboken Point and back to finish at the Battery, a distance of four miles. The *American Star* sprang ahead at the start and held the lead all the way, finishing some 300 to 400 yards ahead of the *Dart* which came in all but waterlogged.[27] Time: 22 minutes.[28]

That evening both crews appeared on the stage of the old Park Theatre, where they were cheered by the audience and treated to a "perfect ovation." In the days which followed, victors and vanquished strove to outdo each other in the exchange of compliments and amenities. It was suggested that the *American Star* be presented to Captain Harris so that on his return England might get an idea of the advanced state of boat building in America, and he was actually offered the boat, but just as gracefully he declined. On his part, the

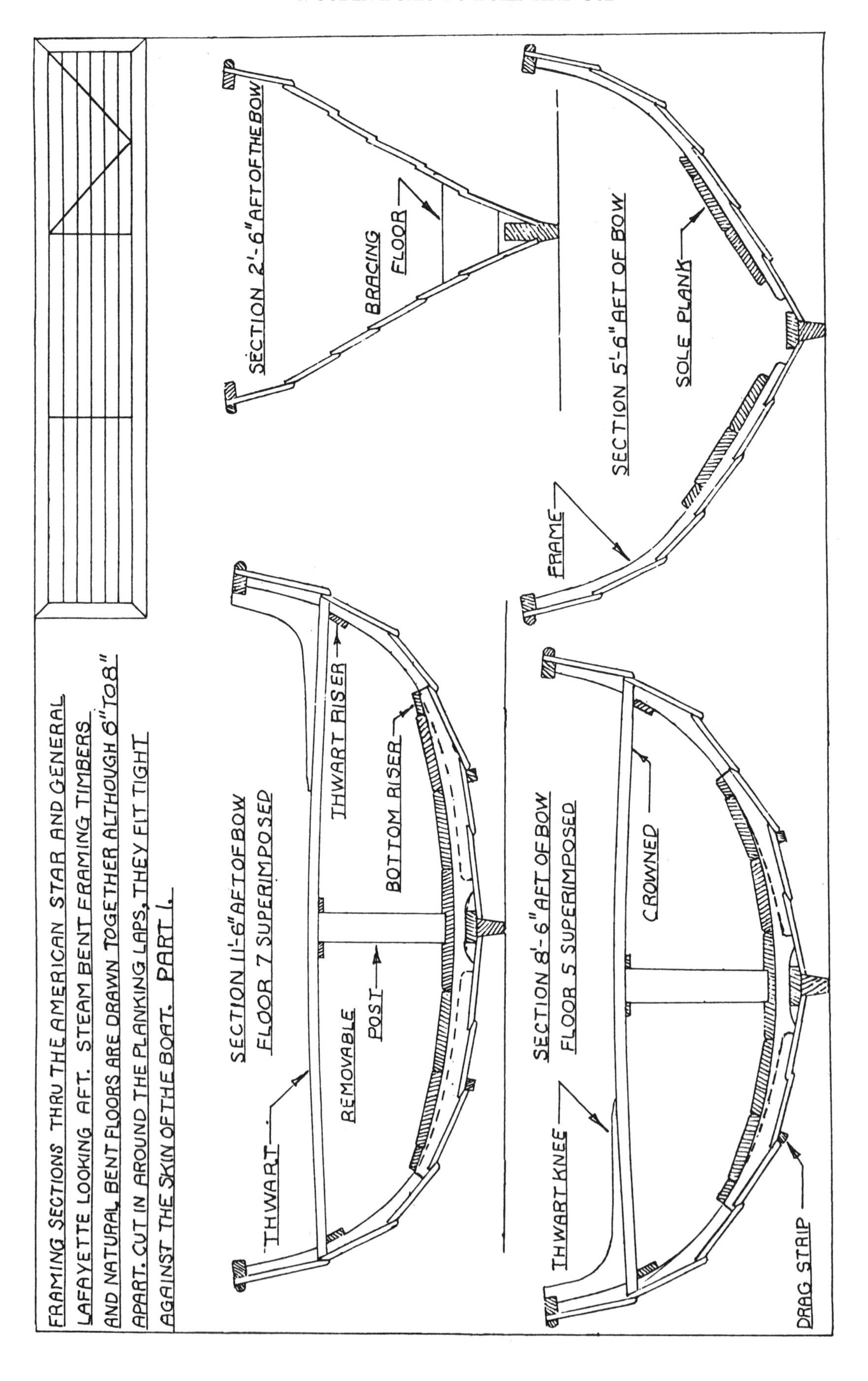
FRAMING SECTIONS THRU THE AMERICAN STAR AND GENERAL LAFAYETTE LOOKING AFT. STEAM BENT FRAMING TIMBERS AND NATURAL, BENT FLOORS ARE DRAWN TOGETHER ALTHOUGH 6" TO 8" APART. CUT IN AROUND THE PLANKING LAPS, THEY FIT TIGHT AGAINST THE SKIN OF THE BOAT. PART I.
SECTION 2'-6" AFT OF THE BOW
BRACING
FLOOR
SECTION 5'-6" AFT OF BOW
SOLE PLANK
FRAME
SECTION 11'-6" AFT OF BOW
FLOOR 7 SUPERIMPOSED
THWART RISER
BOTTOM RISER
REMOVABLE
POST
THWART
SECTION 8'-6" AFT OF BOW
FLOOR 5 SUPERIMPOSED
CROWNED
THWART KNEE
DRAG STRIP

That was how things stood in 1972, the year the first meeting of the World Congress of Maritime Museums convened in early autumn in London. Because of interest in *American Star*, arrangements were made that following the Congress two of the American delegates, J. Revell Carr, at that time curator of Mystic Seaport Museum, and John Gardner, Associate Curator of Small Craft Studies, would cross over to France to measure and take off the lines of *American Star* at la Grange. The Chateau de la Grange, a beautiful old Norman Castle, is pleasantly situated in the countryside about 30 miles southeast of Paris, where we found it convenient to stay, driving out to la Grange in the morning and returning at the end of the day. The three days allotted to the job proved to be ample. We found the *American Star* stored in an otherwise vacant room in a building which at one time had probably been a carriage house. The boat rested upright on low supports at a convenient height for working. Because of the length of the boat, photography within the confines of this room was restricted, and the light could have been better. A flashlight helped in viewing the interior, what appeared to be several coats of faded paint with encrustations of ancient grime made it difficult to see how it was put together and fastened, and in our probing we were greatly restricted as to how far we dared go, for having been entrusted with a national treasure it was obligatory that we leave it exactly as we had found it. Removing anything, or taking anything apart was not to be considered, not even scraping off a bit of grime-encrusted paint to discover the species of wood underneath. The floor boards were loosely and haphazardly piled in the bottom of the boat. Had they once been fastened down? In the condition we found them it was difficult to tell.

But taking off the lines presented no difficulties. We had worked out in advance a method for doing it, and had brought the necessary equipment with us.

When the boat had been leveled both athwartships and fore-and-aft, a fore-and-aft centerline was stretched above it to serve as a datum line. Downward from this datum line extended the imaginary vertical center plane of the boat from which all horizontal distances were squared and measured.

From an imaginary horizontal plane generated by the datum line above the boat and perpendicular or square with the vertical center plane, all vertical distances were squared and measured.

Measuring with a tape from the outermost edge of the stem at the sheer, a number of athwartships sections were laid off and marked on small pieces of masking tape attached along the gunwales.

To record the shape of the hull at each of these sections or stations, a straightedge laid across the gunwales was located on these section marks, and the vertical distance from the datum line to the straightedge was measured, giving the height of the sheer. This done, the vertical distance was plumbed down from the straightedge to the inside top edge of each plank in turn and these measurements were recorded together with the squared distance of each out from the vertical center plane, as measured along the surface of the straightedge. At the same time the depth of the bottom was ascertained, the location of the upper and lower risers, and the beam at the sheer. With this information taken at each station, the lines can be laid down, faired and minor discrepancies eliminated, as was later done, in producing the table of offsets, which appears on Sheet **1** of the drawings.

First and foremost, the *American Star* is a race boat, an early nineteenth-century equivalent of a modern racing shell, built to move through the water fast—or what was fast for a rowing machine in those days—and also built to look fast, for looks were also important in generating excitement in the crowds that thronged the Manhattan waterfront in the event of a race and placed the not-inconsiderable wagers and side bets that enlivened the occasion.

There is no gainsaying that the *American Star* is a comely craft in every way, a handsome hull with its clean curves and long sweeping lines, the graceful upsweep at the ends of its low "cut down" sheer and its diminutive, high-tucked, ornamental transom stern. Stripped of the last excess ounce of weight and timber, trimmed to the bone, as it were, to convey the impression of lightness and speed, it was

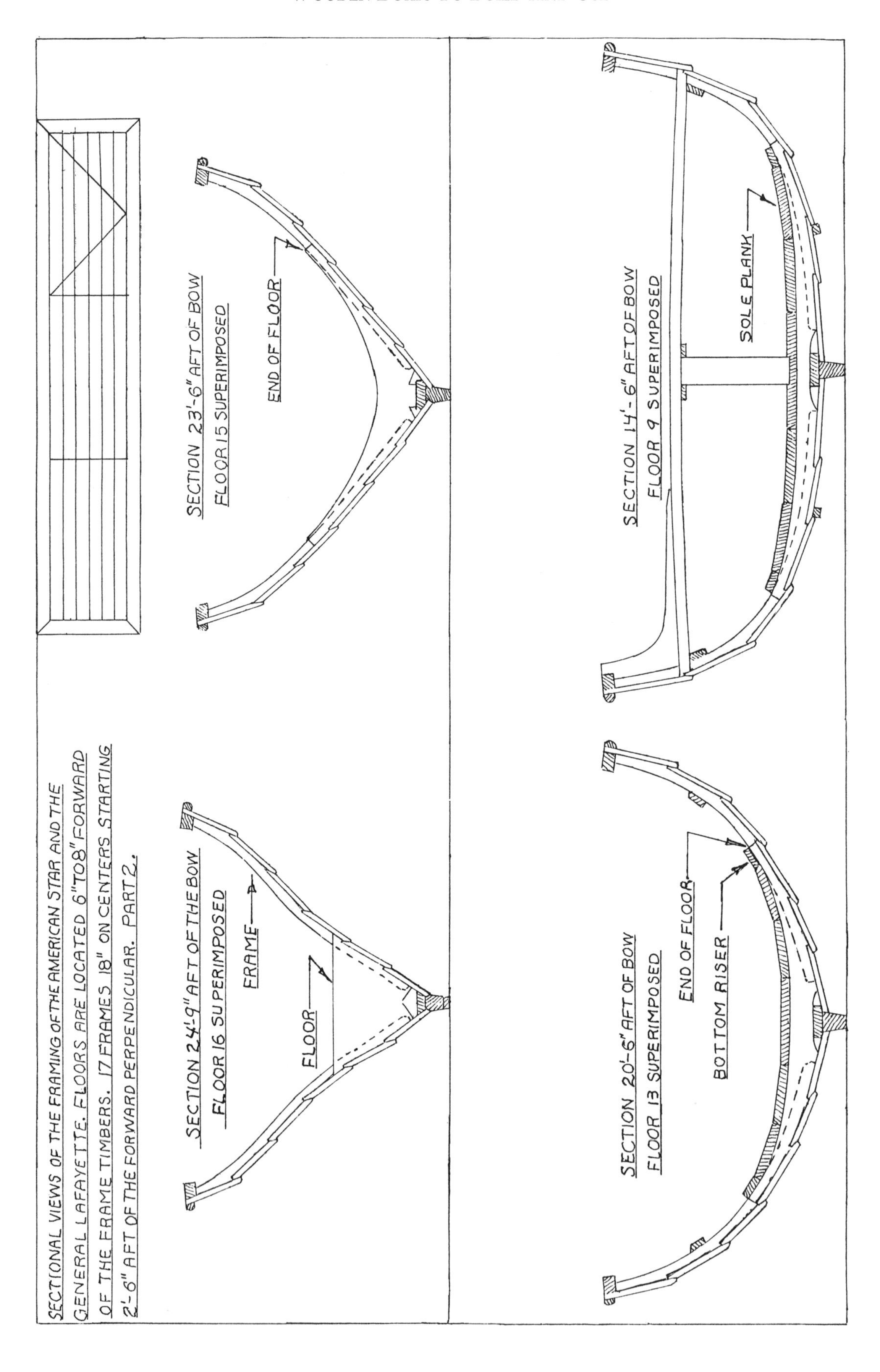
SECTIONAL VIEWS OF THE FRAMING OF THE AMERICAN STAR AND THE GENERAL LAFAYETTE. FLOORS ARE LOCATED 6" TO 8" FORWARD OF THE FRAME TIMBERS. 17 FRAMES 18" ON CENTERS STARTING 2'-6" AFT OF THE FORWARD PERPENDICULAR. PART 2.
SECTION 24'-9" AFT OF THE BOW
FLOOR 16 SUPERIMPOSED
FLOOR
FRAME
SECTION 23'-6" AFT OF BOW
FLOOR 15 SUPERIMPOSED
END OF FLOOR
SECTION 20'-6" AFT OF BOW
FLOOR 13 SUPERIMPOSED
END OF FLOOR
BOTTOM RISER
SECTION 14'-6" AFT OF BOW
FLOOR 9 SUPERIMPOSED
SOLE PLANK

purposefully so contrived, I believe, to impress and astound viewers accustomed to the heavy, roughly built working craft of that time.

A boat as lightly built as this simply had to be the fastest thing afloat, or so it would have seemed. What other reason than conveying this impression could there have been for making timbers sided barely 1/4" and molded a mere 5/8" before cutting the jogs for the laps so absurdly frail? The rivets through the laps almost cut them in two. Spaced 18" apart, how much stiffness could they have imparted to the sides of this boat? Not much. And the combined weight of all 17 pairs of these white oak frames totals barely $3^1/2$ lb. To have increased their siding by another 1/8" to allow for better nailing for the seat risers, as was done on the *General Lafayette*, would have added less than 2 more pounds, a mere nothing in a boat 27 feet long.

More than any other element of her construction, by far, it was the *American Star*'s six strakes of 1/4" clinker planking that gave it the structural strength that held the hull together. A good portion of this structural strength was supplied by the stout, well-riveted planking laps. Plank thickness was quite easily determined from the ends of the plank at the stern, as well as from several other indications, but there was no way of discovering the kind of planking material that was concealed under the several layers of paint and grime. My best guess is that it was northern white cedar, the preferred planking wood in the northeastern United States for small craft, the lightest planking wood available and yet quite tough and strong, and it was plentifully abundant at that time in long, clear boards from old-growth timber. White cedar has one drawback, its tendency to soak water and take on weight. Because of the *American Star*'s light weight and fragile construction, there would have been every reason for removing her from the water at the conclusion of a race, and for storing her under cover on land. Nor would there have been any need for her crew to train in her. Considering the primitive state of rowing technique at the time, they might well have built up their stamina, and practiced their short, rapid stroke in a heavier boat.

Whether any provision had been taken to seal her planking against the absorption of moisture, we of course do not know, as was done with the *General Lafayette*. Her 1/4" northern white cedar planking received repeated applications until the point of complete saturation, both inside and out, of boiling hot linseed oil thinned with 10 percent turpentine. By this means the planking was rendered impervious to penetration by moisture, as well as stabilized against swelling and shrinking. The effectiveness of this treatment was demonstrated at Mystic Seaport one summer when the *General Lafayette*, stored on horses in an open shed, was launched and rowed from time to time throughout the summer. During that time her planking did not shrink and open up, remaining perfectly tight and leak-free the entire season.

The *American Star*'s six strakes of plank are each made up of two sections of unequal length spliced together in the central portion of the hull. The splice in the sheer plank 16'5" from the bow overlaps the splice in the second plank 12'8" from the bow by 3'9". The splice in the third strake 15'9" from the bow overlaps the splice in the fourth strake 11'9" from the bow by 4'. The splice in the fifth strake 15'6" from the bow overlaps the splice in the garboard 12'10"from the bow by 2'8".

The two sections of each plank were joined by scarphed splices $3^1/2$" long. The forward end of the scarphed splices located on the inside, was fastened by four tacks driven through from the inside and clenched on the outside, while the after end of the splice which came on the outside was fastened by a row of rivets 3/4" and 1" apart. This, it might be mentioned, was the way it was done before there was waterproof glue. On the *General Lafayette* the plank splices were glued, which is easier, quicker and stronger.

Plank laps were 9/16" to 5/8" in width. In fitting the laps a minimum of wood was removed leaving the ends of the laps 1/8" thick, or slightly thicker, on the inside. Rivets were spaced $3^1/2$" to 4" apart, in addition to the lap rivets through the timbers. These laps, so made and fastened, greatly strengthened the hull structure as a unified whole.

On September 4, 1974 construction of the* General Lafayette *was well underway in the White Boat Shop at Mystic Seaport. This stern section of the strongback and ladder-frame shows the rabbet, stern knee and bracing supporting the stern piece which is notched aft to receive the transom and forward to receive the keel (Kenneth E. Mahler photograph, M.S.M.)

It is generally considered not to be good boat building practice to have the plank splices bunched together in the central portion of the boat, instead of breaking them up by locating some of them in the ends of the boat, yet the arrangement found in the *American Star* seems to have proved adequately strong, largely, no doubt, because of the generous splice overlap of from three to four feet. No sign of strain or failure here was detected. Nevertheless, the planking did show some damage. In several places thin pieces had been tacked over partial breaks or cracks in the first broad plank and the garboard. Other damage sustained by the boat included several broken frames, and a piece of the half round missing from the sheer on the port side. The fastenings of the sheer plank to the inwale in this section were broken, and the lap of the sheer plank with the plank below had pulled apart for a short distance. In addition, several posts or upright braces between the lower and upper risers had been put in more or less haphazardly and without regard for appearance. Apparently, these were not original but had been added later when extra bracing for the upper riser was needed to support the thwarts.

When and how this damage occurred we can never know, yet it raises some interesting speculations. The damage to the sheer with the broken fastenings and the missing section of half round trim might well have happened accidentally at la Grange in moving the boat, as evidently was done a number of times, but it seems unlikely that the crude patches on the inside of damaged planking, presumably to stop leaks, and the additional supports for the upper, or thwart risers would have been made unless the boat was being used. Besides, it seems highly unlikely that a boat given to General Lafayette to take back to France as a representative of the "ingenuity" of American mechanics would have been permitted to leave American shores with such crude repairs. Could it have been that the boat had been launched and rowed on the Seine or the Marne and

The hull of the* General Lafayette *in mold (16) with the bow section of the port garboard temporarily fitted (Kenneth E. Mahler photograph, M.S.M.).

their tributaries after the General's death? The contemporary account in the American press of the occasion when General Lafayette was transported in the boat by the Whitehallers of New York to a dinner engagement at Paulus Hook mentions "silver mounted oars" and "velvet cushions." With due allowance for journalistic hype, the condition of the boat on this occasion must have been far more shipshape and comely than these clumsy patches and reinforcements would suggest. Nor is it likely that it would have been any the worse for wear when it was packed up and shipped off to France.

Was the *American Star* built to last? In my judgment the *American Star* was too lightly constructed to endure more than a few hard races, and I venture that if records had survived of early competitive rowing in Manhattan waters, we would find that the life of the race boats of that time was short, that they soon gave out and had to be replaced.

This assumption was taken into consideration when the *General Lafayette* was built. While both the *American Star* and the *General Lafayette* have the same overall dimensions and are closely identical in most respects, there are some differences, particularly in the size of the internal framing which was purposely made heavier and stronger in the *General Lafayette* to withstand continued hard use over a long life of recreational rowing.

One of the differences between the *American Star* and the *General Lafayette* is in their floors, the timbers in the bottom of the boat extending from side to side across the keel, strengthening the bottom, holding the two sides together and providing a supporting foundation for the longitudinal floor or bottom boards.

In the *American Star* these supporting floors, cut from natural bends of oak, were sided 3/4", and in the central portion of the boat they were cut away leaving only 1" of wood or less above the inner keel. In the *General Lafayette* the siding was increased to 7/8" and the molded depth above the inner keel was also increased enough to leave $1^{1}/_{4}$" of wood connecting the two sides of the floor.

The greatest difference between the *American Star* and the *General Lafayette* is in their floor boards. In the *American Star*, judging from all appearances, they were laid in loose on top of the floors and were not fastened down. If they had been, it would have been done with iron nails. Wood screws and bronze nails were still far in the future and copper nails are too soft. Also it would have been done before the boat left the United States, in all probability when the boat was built. What conceivable reason might there have been for removing these nails, assuming there were nails, after the boat came into General Lafayette's possession, not an easy thing to do, and to have done it so carefully and thoroughly? It just doesn't make sense.

The effect of loose floor boards in a boat of the *American Star*'s long, narrow proportions, cut down sides and lack of other internal bracing is to make for an almost snake-like flexibility in response to the movement of the sea. The force of the waves is evaded somewhat, resistance is lowered, speed is increased, but at the expense of imposing severe strains on hull structure that consists of little more than its planking. Very little additional fore-and-aft reinforcement is supplied by the long slight keel and equally slight gunwales and risers. A boat so built may be able for a time to withstand moderate seas without becoming strained and developing leaks, but in the rough water of the open sea it would soon begin to work itself apart.

This weakness is overcome to some extent in the *General Lafayette* by screwing down full-length floor boards to the floors which are fastened underneath by screws through the planking from the outside. The result is a solid platform resting on a truss-like structure which ties all the floors together in a single structural unit, reinforcing and strengthening the entire hull.

The intricate design of the *American Star*'s fixed wooden tholes is ornamental, and no doubt this arrangement served quite well before the introduction of metal crutches or oarlocks, but the fixed tholes have one disadvantage, the oar can get wedged between the uprights and the leverage exerted by the heavy 13-foot oar is enough to damage the sheer strake. This happened with the *General Lafayette* on one occasion when an oar got jammed resulting in a short split in the sheer plank. Not so picturesque but more practical are removable thole pins, such as were used on early lifesaving boats. When the pin broke, as they were intended to when the oar jammed, they were readily replaced from a bag of spare pins carried for such an

In late February 1975, the hull was off the molds and right side up, and the frames installed. The narrow transom is tucked up high with the plank gains giving a smooth edge to the transom. Note the rugged bracing for the strongback, the cross spalls and the braces fastened to the overhead timbers to keep the shape of the hull (Kenneth E. Mahler Photograph, M.S.M.).

eventuality. The Cornish pilot gigs built for working in rough water were fitted with thole pins, the forward one of iron, the after one of wood, which broke instead of the oar or topsides, and was easily replaced if by chance the oar caught a crab.

We have assumed that the oars found with the *American Star* were original, although there is no indication that they had ever been "silver mounted" as reported in the contemporary New York press. The 13'-3" spruce oars were perfectly straight without any curvature or spoon in the blade, for this refinement was yet to come with the introduction of the feathering stroke.

At its inner end the oar handle is $1^{7}/_{16}$" in diameter enlarging to $1^{13}/_{16}$", 5" from the inner end, and swelling into the loom $2^{1}/_{2}$" in diameter, 11" from the inner end. Four feet from the end the diameter is $2^{5}/_{16}$", and 8' from the end $1^{3}/_{4}$". The juncture of the loom and the blade 9' from the end is oval in shape, $2^{7}/_{16}$" in width by $1^{3}/_{16}$" thick. At 11' the blade is $4^{1}/_{16}$" wide by 3/4" thick, at 13', $5^{1}/_{8}$" by 1/2". Ten-inch leathers start 3' from the inner end.

To support the *American Star*'s thwarts, upright braces or posts 2" wide by 3/4" were nailed in place between the upper and lower risers, under and midway of the ends of the thwarts. Stretchers of $1^{3}/_{4}$" by $1^{1}/_{8}$" oak for the rowers to brace against were put in loose against these uprights. In the *General Lafayette*, as shown in the drawing, stretchers fit into notches cut into cleats screwed down to the floor boards.

The exposed corners of the edges of interior members, gunwales, risers, thwarts, floors, spreaders, posts and others, are not left square and sharp but are rounded off in a small ornamental bead of 1/4" or less. On some flat surfaces this beading is repeated at one or more intervals for the decorative effect.

The diamond-shaped roves or burrs used for the rivets on the *General Lafayette* were made on the job, as we assumed they had probably had been made for the *American Star*. They were laid out on sheet copper by scratching a series of equidistant parallel lines, and superimposing over them at an angle of 60 degrees an identical series of lines. In this way the diamond-shaped burrs were outlined en masse. The holes for the nails were punched before the burrs were cut apart with snips. To brace the ends of the boat there are two oak spreaders $2^{1}/_{2}$" wide by 3/4" thick. As can be seen in the drawing, the spreader in the bow is fitted around the second pair of frames, and the one in the stern around the fifteenth pair. The ends of the spreaders are slotted for a snug fit around the frames, and they should also bear equally against the inside of the planking and the inside of the frame.

Construction Details

Interior measurements within plus or minus 1/8" taken from the *General Lafayette*. Vertical distances measured from a line 1/2" above the sheer at the stem and 3" above the sheer at the stern. Fore-and-aft distances measured from the forward perpendicular dropped from the outermost portion of the stem at the sheer.

First floor: Forward Perpendicular (F.P.) to first floor 2'$2^{1}/_{4}$"; line to top of first floor $17^{3}/_{8}$"; line to the bottom of first floor $21^{7}/_{8}$"; floor sided 5/8".

First frame: F.P. to first frame 2'6"; line to sheer 3"; beam at the sheer $20^{5}/_{8}$"; gunwale 1" wide.

Second floor: F.P. to second floor 3'4"; line to top of second floor 18"; line to bottom of second floor $23^{1}/_{8}$"; floor sided 5/8".

Second frame: F.P. to second frame 4' 0 $^{3}/_{8}$"; line to sheer $4^{1}/_{2}$"; beam at the sheer $29^{5}/_{8}$"; line to inner keel $23^{1}/_{4}$".

Third floor: F.P. to third floor 4'$10^{1}/_{2}$"; line to top of third floor 18"; line to bottom of third floor and inner keel $23^{1}/_{4}$"; beam at the sheer 34"; floor sided 5/8".

Third frame: F.P. to third frame 5'$6^{3}/_{8}$"; line to sheer $5^{1}/_{2}$"; line to inner keel $23^{1}/_{8}$"; beam at sheer 36"; gunwale $1^{1}/_{4}$" wide; lower riser extends 2" forward of third frame; lower riser's upper edge $10^{3}/_{8}$" below the sheer; floor boards same length as lower riser; width of upper floor board $4^{1}/_{8}$", lower $4^{1}/_{8}$", middle 6".

Fourth floor: F.P. to floor 6'5"; line to top of floor at its center 21"; line to inner keel $23^{1}/_{8}$"; top of lower riser from the sheer 11"; width between the tops of the lower risers 2'$2^{1}/_{8}$"; floor sided 7/8".

Fourth frame: F.P. to top of frame 7'$0^{1}/_{4}$"; line to sheer 6"; line to top of upper riser $12^{1}/_{4}$"; line to top of lower riser 1'$5^{1}/_{4}$"; line to

Ben Philbrick and John Gardner fastening the oak floors jogged to fit the plank laps. Note the limbers either side of the keel and the diamond-shaped copper rivet roves spaced 3 1/2" to 4" along the plank laps and on the 1/4" thick x 3/4" molded, steam-bent white oak frames that are also jogged over the plank laps (Kenneth E. Mahler photograph, M.S.M.).

inner keel 23 1/8"; beam at sheer 3'4 1/4"; end of upper riser extends 2" forward of fourth frame; width of floor boards, upper 4 1/2", lower 4 3/8", center 6 7/8"; horizontal distance between tops of lower risers 2'3".

Fifth floor: F.P. to fifth floor 7'11"; line to sheer 6 3/8"; line to top of upper riser 12 3/4"; line to top of lower riser 12'5 7/8"; line to top of center of floor 21 1/8"; line to inner keel 23 1/8"; beam at the sheer 3'6"; beam between bottom risers 2'4 1/2"; width of upper floor board 4 5/8", lower 4 1/2", center 7 3/8".

Fifth frame: F.P. to top of frame 8'6 1/4"; line to sheer 6 1/2"; line to inner keel 23 1/8"; beam at sheer 3'7 1/4"; F. P. to forward edge of thwart 8'5"; width of thwart 7 1/2"; line to bottom of thwart 12 3/4"; bottom of thwart to inner keel 10 3/8"; gunwales at the fifth frame 1 1/4" wide; forward edge of the thwart 3'4 1/4" wide.

Sixth floor: F.P. to top of sixth floor 9' 5 1/4"; line to sheer 6 7/8"; line to top of floor at its center 21 5/8"; line to inner keel 23 1/8"; sheer to top of lower riser 12"; sheer to top of upper riser 6 1/2"; beam at the sheer 3'8 1/2"; width of inner keel 2 1/2"; floors are 7/8" thick; width of upper floor board 5", of lower 4 3/4", middle 7 1/4"; width between the tops of the lower risers 30 1/2"; width between upper risers 39 3/4".

Sixth frame: F.P. to sixth frame 10'0"; oarlock opening 3 1/2", forward side 3/4" from frame.

Seventh floor: F.P. to top of floor 10'11"; line to sheer 7 1/8"; line to the top of floor at the center 21 7/8"; line to inner keel 23 1/8"; top of floor to inner keel 1 1/4"; beam at the sheer is 3'10"; beam at top of upper risers 3'8"; beam at top of bottom risers 2'8"; width of upper floorboards 5", lower floorboards 5", middle

$8\frac{1}{4}$"; depth of sheer to upper riser $6\frac{1}{2}$"; depth sheer to lower riser $12\frac{1}{4}$"; width of gunwale $1\frac{1}{4}$"; F. P. to forward edge of second thwart 11'2"; width $7\frac{1}{2}$"; depth from the line to the bottom of the thwart $12\frac{3}{4}$".

Seventh frame: F.P. to seventh frame 11'$6\frac{1}{2}$".

Eighth floor: F.P. to eighth floor 12'5"; line to sheer $7\frac{1}{2}$"; line to top of floor at the center $21\frac{7}{8}$"; line to inner keel $23\frac{1}{8}$"; sheer to upper riser $6\frac{1}{4}$"; sheer to lower riser $12\frac{1}{4}$"; beam at sheer to outside of plank 3'$11\frac{1}{2}$"; beam outside edge top risers 3'6"; beam outside edge lower risers 2'$8\frac{7}{8}$"; width upper floor boards $5\frac{1}{4}$", lower floorboards $5\frac{1}{4}$", middle $8\frac{3}{8}$".

Eighth frame: F.P. to frame 13'$0\frac{1}{4}$"; F.P. to forward edge of oarlock 12'$10\frac{1}{4}$".

Ninth floor: F.P. to floor 13'$10\frac{3}{4}$"; line to sheer $7\frac{3}{4}$"; line to top of center of floor $21\frac{7}{8}$"; line to inner keel $23\frac{1}{8}$"; sheer to top of upper riser $6\frac{3}{8}$"; sheer to top of lower riser 12"; beam outside plank at sheer 3'$11\frac{1}{2}$"; beam outside top of upper riser 3'$6\frac{3}{8}$"; beam outside top of lower riser 2'$9\frac{1}{8}$"; width of floor boards, upper $5\frac{3}{8}$", lower $5\frac{1}{4}$", middle $8\frac{3}{8}$"; third thwart forward edge from the F.P. 14'$9\frac{1}{2}$"; thwart $7\frac{1}{2}$" wide, down from line $13\frac{1}{4}$".

Ninth frame: F.P. top frame 14'7"; beam 3'11" outside plank.

Tenth floor: F.P. to floor 15'$4\frac{1}{4}$"; line to sheer $7\frac{3}{4}$"; line to top of floor at its center $21\frac{7}{8}$"; line to inner keel $23\frac{1}{8}$"; floor boards, top $5\frac{1}{2}$", bottom $5\frac{1}{4}$", middle $8\frac{1}{4}$"; beam outside upper riser 3'6"; beam outside lower risers 2'9"; top riser to sheer $6\frac{1}{4}$"; bottom riser to sheer $11\frac{3}{4}$".

Tenth frame: F.P. to frame 16'$1\frac{1}{2}$"; oarlock port side forward part $5\frac{1}{2}$" forward of frame; beam outside plank 3'$10\frac{1}{4}$".

Eleventh floor: F.P. to floor 16'$1\frac{1}{2}$"; line to sheer $7\frac{7}{8}$"; top of floor at center to line $21\frac{7}{8}$"; inner keel to line $23\frac{1}{4}$"; beam at the sheer outside plank 3'$9\frac{5}{8}$"; outside distance between the upper risers 3'$4\frac{5}{8}$"; outside distance between the lower risers 2'$8\frac{1}{2}$"; fourth thwart 17'$0\frac{1}{2}$" from F.P. and $7\frac{1}{2}$" wide; fourth thwart 1'$1\frac{3}{8}$" below line; beam at the middle of fourth thwart 3'$9\frac{3}{8}$"; floor boards top $5\frac{3}{8}$", bottom $5\frac{1}{4}$", middle 8".

On June 7, 1975 the* General Lafayette *was launched on the Mystic River estuary before a cheering crowd of about 450 small-craft enthusiasts and amateur boat builders attending the Sixth Annual Small Craft Workshop at Mystic Seaport (Mary Anne Stets photograph, M.S.M.).

Eleventh frame: F.P. to eleventh frame 17'7$^{5}/_{8}$".

Twelfth floor: F.P. to twelfth floor 18'5$^{3}/_{4}$"; line to sheer 7$^{7}/_{8}$"; line to top of floor at the center 22"; line to top of inner keel 23$^{1}/_{4}$".

Twelfth frame: F.P. to twelfth frame 19'1$^{1}/_{2}$"; line to sheer 7$^{7}/_{8}$"; line to inner keel 23$^{1}/_{4}$"; sheer to top of upper riser 6$^{1}/_{4}$"; sheer to top of the lower riser 10$^{3}/_{4}$"; beam outside of plank at the twelfth frame 3'6$^{7}/_{8}$"; beam outside of upper risers 3'1$^{7}/_{8}$"; beam outside of lower risers 2'6$^{3}/_{8}$"; floor boards, upper 5", lower 5", middle 7$^{1}/_{4}$"; inside forward part of starboard oarlock 11" to after edge of the fourth thwart.

Thirteenth floor: F.P. to floor 19'11$^{1}/_{2}$"; line to sheer 7$^{7}/_{8}$"; line to top of floor at its center 21$^{1}/_{2}$"; line to inner keel 23$^{1}/_{4}$"; sheer to top of upper riser 6$^{1}/_{2}$"; sheer to top of lower riser 10$^{3}/_{4}$"; beam outside upper risers 3'0$^{1}/_{4}$"; beam outside bottom risers 2'4$^{3}/_{4}$"; floor boards, upper 4$^{3}/_{4}$", lower 4$^{3}/_{4}$", middle 6$^{7}/_{8}$".

Thirteenth frame: F.P. to frame 20'7"; line to sheer 7$^{3}/_{4}$"; line to inner keel 23$^{1}/_{4}$"; sheer to top of upper riser 6$^{1}/_{2}$"; sheer to top of lower riser 10$^{1}/_{2}$"; beam at the sheer to outside of plank 3'4$^{1}/_{4}$"; beam outside upper risers 2'10$^{5}/_{8}$"; floor boards upper 4$^{1}/_{2}$", lower 4$^{1}/_{2}$", middle 6$^{1}/_{2}$"; width of gunwale 1$^{1}/_{4}$".

Fourteenth floor: F. P. to floor 21'6"; F.P. to the forward edge of the fifth thwart 21'6"; line to center of floor 21$^{1}/_{4}$"; line to top of inner keel 23$^{3}/_{8}$"; line to sheer 7$^{5}/_{8}$"; beam at the sheer 38$^{5}/_{8}$"; beam outside upper risers 2'8$^{1}/_{2}$"; beam outside lower risers 2'1$^{3}/_{8}$"; width of floor boards, upper 4$^{1}/_{4}$", lower 4$^{1}/_{8}$", middle 5$^{7}/_{8}$".

Fourteenth frame: F.P. to fourteenth frame 22'1$^{1}/_{4}$"; fifth thwart 8" wide.

Fifteenth floor: F.P. to floor 22'11"; line to sheer 7$^{1}/_{4}$"; line to top center of floor 20$^{1}/_{4}$"; line to inner keel 23$^{3}/_{8}$"; beam at sheer 35$^{1}/_{4}$"; beam lower risers 20$^{3}/_{4}$"; floor boards, upper 3$^{5}/_{8}$", lower 3$^{1}/_{2}$", middle 3$^{7}/_{8}$"; bottom riser ends 2" aft of floor; top riser ends at the fifteenth frame; fifteenth floor sided 3/4".

Fifteenth frame: F.P. to frame 23"; line to sheer 6$^{7}/_{8}$"; line to inner keel 23$^{1}/_{2}$"; beam at sheer 33$^{1}/_{4}$".

Sixteenth floor: F.P. to floor 24'4"; line to sheer 6$^{3}/_{8}$"; line to top of floor 17$^{7}/_{8}$"; line to inner keel 23$^{3}/_{8}$"; beam at the sheer 30$^{1}/_{2}$"; floor sided 1/2".

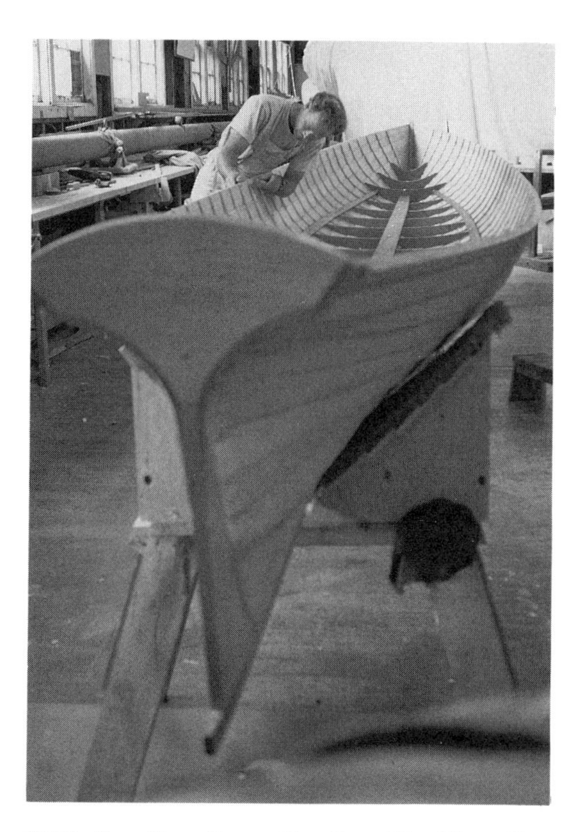

With funding from the Camilla Chandler Family Foundation, the Port Townsend Wooden Boat Foundation in Washington State undertook to build a copy of the* General Lafayette *in the winter of 1995. Builders Ed Louchard, Alex Spear and a high school apprentice, Marc Logan, used Port Orford cedar for the planking, mahogany for the keel and keelson and black locust for the stem, forefoot, stern post, stern knee and transom. They used eight, 5/16" thick, strakes of plank with the garboards thickened to 3/8" and a synthetic seam compound in the laps. The boat, launched at the twentieth anniversary of the Port Townsend Wooden Boat Festival in September 1996, will be used by school students and the community for education (Ed Louchard photograph).

Sixteenth frame: F.P. to frame 24'10$^{1}/_{4}$"; line to sheer 6"; line to inner keel 23$^{1}/_{2}$"; beam at the sheer 28$^{1}/_{4}$".

Seventeenth floor: F.P. to floor 25'6$^{3}/_{8}$"; line to sheer 5$^{1}/_{4}$"; line to top of floor 18"; line to inner keel 23$^{1}/_{2}$"; beam at sheer 25"; floor sided 1/2".

Seventeenth frame: F.P. to frame 26'0$^{1}/_{2}$"; line to sheer 4$^{1}/_{2}$"; line to inner keel 23$^{1}/_{2}$"; beam at the sheer 22$^{1}/_{2}$"; F.P. to outside of the transom 27'4$^{1}/_{2}$"; beam at the transom to the outside of the plank 14".

Footnotes

1. *New York Evening Post*, November 13, 1820; *New York Gazette*, December 10, 1824, and Robert F. Kelley, *American Rowing*, 1932, p. 16.
2. Charles A. Peverelly, *American Pastimes*, 1868.
3. *New York American*, July 11, 1825.
4. *Ibid.*
5. Jules Cloquet, *Recollections of the Private Life of General Lafayette*, 1836, Vol. II, p. 37.
6. Allan Forbes and Paul F. Cadman, *France and New England*, 1925, Vol. I, p. 60.
7. W. P. Stephens, "Old-Time Boat Building in New York," typescript, G. W. Blunt White Library, Mystic, Connecticut.
8. Rodman Gilder, *The Battery*, 1936.
9. Stephens, *op. cit.*
10. *New York Mercury*, April 26, 1756; *Boston Weekly News Letter*, May 6, 1756.
11. John Gardner, "Early Days of Rowing Sport," *The Log of Mystic Seaport*, Winter 1971.
12. *New York Commercial Advertiser*, June 22, 1807, July 2, 1810, November 13, 1820; *New York Evening Post*, November 13, 1820.
13. *New York Gazette*, March 13, 1824.
14. Peverelly, *op .cit.*
15. *New York Gazette*, December 6, 1824. Editorial note.
16. Stephens, *op. cit.*
17. *New York Evening Post*, November 11 and 13, 1820; *New York Commercial Advertiser*, November 13, 1820.
18. Approximate dimensions of the *American Star* from measurements by René de Chambrun in 1967 and John Beeuwkes, 1968. LOA 27'. Beam amidships and inside planking, 4'. Depth amidships, 17". Four oars, length, 12'9".
19. *New York Gazette*, December 10, 1824. Kelley, *op. cit.*, p. 7, "The *New York* was sold to a wealthy resident of Mobile, and went South for a career in the most obscure of our early racing periods, the period of slave racing ... unprecedented since the days of the Roman galleys!" See also James P. Baughman, "A Southern Spa: Ante-Bellum Lake Pontchartrain," *Louisiana History Journal*, Louisiana Historical Association, Winter 1962, Vol. III, No. 1.
20. *New York Gazette*, December 4, 1824. Editorial note.
21. *New York Gazette*, March 12 and 13, 1824.
22. Kelley, *op. cit.*
23. *New York Gazette*, December 6, 1824.
24. *New York Gazette*, December 8, 1824.
25. *New York Gazette*, December 10, 1824; Kelley, *op. cit.*
26. Cadwallader D. Colden, *Memoir, Celebration of the Completion of the New York Canals*, 1825, p. 235.
27. *New York Gazette*, December 10, 1824.
28. Kelley, *op. cit.*
29. *New York Gazette*, December 13, 1824.
30. *New York Gazette*, December 8, 1824.
31. *New York Evening Post*, December 9, 1824; *New York Gazette*, December 10, 1824, and Kelley, *op. cit.*
32. Kenneth Durant, *American Star*, typescript and notes, G. W. Blunt White Library, Mystic, Connecticut. Much of the basic research, including collection of illustrations and investigation of newspaper sources, was done by Kenneth Durant.

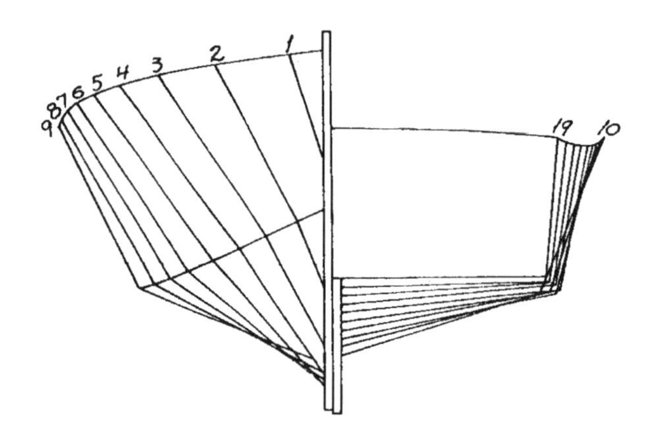

Chapter Twelve

37-Foot V-Bottom Fishing Launch

The 37-foot V-bottom powerboat detailed in the account that follows is much more than a relic dredged up from the all but forgotten annals of work-boat history. It has current possibilities well worth considering. The sea has not changed and the unforgiving demands it places upon a boat have not changed. What was a good boat then is a good boat now. And this is a good boat, an exceptionally good boat proven so by scientific testing and three generations of productive fishing by the Zdanowicz family as a summertime lobster-boat and as a small dragger in the not-so-gentle winter waters off Salem, Massachusetts.

The contemporary possibilities of M-2's proven hull are well worth taking note of and bringing to the attention of those who might want a hull with its many positive features, not only as a workboat, but one easily converted to recreational use. Here is a safe, stable, dry, comfortable, seakindly hull of exceptional carrying capacity waiting to make someone a good boat.

On the somewhat remote chance that the prospective owner might want to build the boat himself, one of the original strong points of this hull was the simplicity and ease of its construction in wood. Understood that lumber is much more expensive and harder to get than it was 35 years ago, yet savings in labor might still be enough to offset the increased cost of materials.

M-2 is the code designation given by Professor Thomas C. Gillmer, the well-known naval architect and yacht designer, to this 34'-37' V-bottomed lobsterboat I designed in 1958 and built the following year at Fred Dion's yard in Salem, Massachusetts, for Joseph F. Zdanowicz, a Salem lobster fisherman. In 1958, Gillmer

Above is the tank model of M-2, a Maine-coast lobsterboat of V-bottom type (LOA 37', LWL 35' 5", beam 11', draft 3', displacement 5.3 tons, prismatic coefficient 0.589) being tested by Professor Thomas C. Gillmer at the United States Naval Academy towing tank in Annapolis, Maryland. Although this model was hard-chined, the chine did not show on the waterline, especially at the bow where the planking was faired in (photograph courtesy of John Gardner).

performed a series of model tests of small, light-displacement fishing launches at the towing tank at the United States Naval Academy in Annapolis, Maryland.

A preliminary analysis of the results of these tests was published in the *Maine Coast Fisherman* in July 1959. Gillmer's final report, "Model Tests of Some Fishing Launches," followed. It was submitted as a paper at the second FAO World Fishing Boat Congress, which was held in Rome that same year. His findings were subsequently published in *Fishing Boats of the World, Vol. 2.*

In his final report, Gillmer narrowed the field to four representative models; however, only two will be discussed here: M-1 and M-2. M-1 is a classic, Maine round-hulled lobsterboat designed by William Frost. Several examples of this design were built in Portland, Maine, by Frost after he moved from the Jonesport area.

It is a handsome example of the Jonesport model, which Will Frost developed to a state of near perfection. The lines for the M-1 test model were supplied by Philip C. Bolger, the Gloucester, Massachusetts, naval architect. He took them off one of the Frost boats built for a Gloucester lobster fisherman in 1946.

Gillmer describes M-2 as "a fairly recent design modification among Atlantic Coast fishing boats. This boat has a V-bottomed, hard-chined hull form. The lines show a powerful hull with large freeboard, large flare and a raking stem."

"In terms of displacement, it was one of the largest of the boats tested. Her dimensions are 37' LOA (35'5" load waterline). She has an 11' beam, with a 3' draft at heel of skeg. Her displacement is 11,800 lbs. to the load waterline."

The lines redrawn and reproduced here, it should be noted, show slightly less rake in the stem than those of the test model. The forefoot is also somewhat more rounded, which adds slightly to the length of the load waterline. It narrows the entrance angle ever so slightly, as well. Otherwise the lines and dimensions are the same as the original. These changes are so minor that no appreciable effect on hull characteristics or performance may be expected.

They follow the minor alterations made in the *Stella Z. II,* the M-2 prototype built at Salem in 1959. These changes were made at the insistence of the owner, Joseph Zdanowicz, who was motivated solely by cosmetic considerations. He felt that the stem profile as originally designed departed excessively from the stem profiles he was accustomed to, and hence was too radical in appearance.

Gillmer's tests were limited to the measurement of hull resistance: the amount

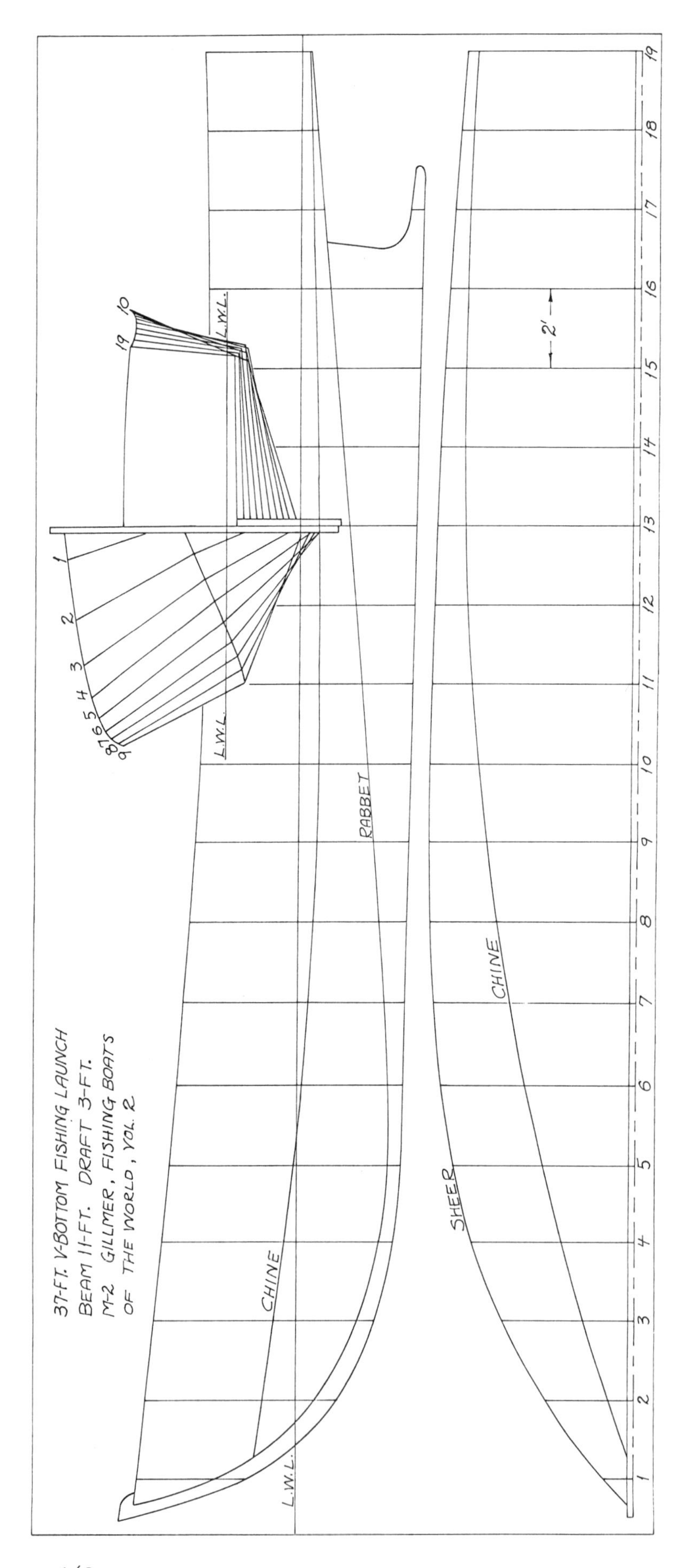

of applied force required to move the hull through the water. The results bear directly on horsepower requirements and fuel consumption, which are extremely important considerations today and for the future. There is every indication that fuel costs will continue to rise.

These test results have no bearing, however, on a number of other characteristics that are desirable and even essential for a superior fishing launch, as we shall presently see.

It will be sufficient here to summarize Gillmer's results (those who wish to examine his study in detail can read his paper in *Fishing Boats of the World, Vol. 2*). The Frost Jonesporter, M-1, proved to be the easiest-driven hull, yet its superiority over M-2 in this respect was relatively slight. It was not great enough to have any significant effect on powering costs.

Gillmer's work gave us firsthand figures to study. Previously we had to rely on the subjective judgments of fishermen/owners, who were undoubtedly influenced to some extent by regional preferences and prejudices.

In 1947, one year after the Frost-built vessel (M-1) arrived in Gloucester, Ipswich builder Justin Hull constructed a V-bottomed lobsterboat for a Beverly, Massachusetts, fisherman. It was approximately the same size as Frost's craft.

Hull, with a lifetime on the water as a fisherman and a party-boat operator, had pioneered V-bottomed fishing launches on the

37-FT. V-BOTTOM FISHING LAUNCH

	STATIONS	STEM	1	2	3	4	5	6	7	8	9
HEIGHTS	SHEER	+4-3-5	+4-2-2	+3-11-5	+3-9-0	+3-6-4	+3-4-2	+3-2-0	+3-0-0	+2-10-0	+2-8-0
	CHINE		+1-2-3	+0-10-2	+0-6-6	+0-3-6	+0-0-7	−0-1-4	−0-3-4	−0-5-0	−0-6-1
	RABBET		+2-1-4	−0-3-6	−1-7-2	−2-2-3	−2-5-0	−2-5-0	−2-3-6	−2-1-6	−1-11-6
	BOTTOM KEEL		+1-4-2	−1-0-6	−2-0-3	−2-6-7	−2-9-3	−2-10-0	−2-10-4	−2-11-0	−2-11-3
HALF-BREADTHS	SHEER		0-9-4	2-4-1	3-6-2	4-4-2	4-10-5	5-2-4	5-4-4	5-5-6	5-6-0
	CHINE		——	0-8-2	1-4-1	1-11-0	2-5-2	2-11-1	3-4-0	3-8-0	3-11-6
	RABBET		0-2-0	0-2-0	0-2-0	0-2-0	0-2-0	0-2-0	0-2-0	0-2-0	0-2-0

	STATIONS	10	11	12	13	14	15	16	17	18	19
HEIGHTS	SHEER	+2-6-6	+2-5-6	+2-5-1	+2-4-6	+2-4-4	+2-4-4	+2-4-4	+2-4-6	+2-5-1	+2-6-0
	CHINE	−0-6-4	−0-6-4	−0-6-2	−0-6-0	−0-5-6	−0-5-2	−0-4-6	−0-4-1	−0-3-3	−0-2-5
	RABBET	−1-9-5	−1-7-5	−1-5-3	−1-3-3	−1-1-3	−0-11-2	−0-9-1	−0-7-0	−0-5-0	−0-3-0
	BOTTOM KEEL	−3-0-1	−3-0-5	−3-1-1	−3-1-5	−3-2-1	−3-2-6	−3-3-2	−3-3-7	−3-4-3	−3-5-0
HALF-BREADTHS	SHEER	5-6-0	5-5-5	5-5-0	5-4-2	5-3-4	5-2-0	5-0-3	4-10-4	4-9-0	4-7-2
	CHINE	4-2-4	4-5-0	4-6-4	4-7-2	4-7-2	4-7-0	4-6-3	4-5-4	4-4-5	4-4-0
	RABBET	0-2-0	0-2-0	0-2-0	0-2-0	0-2-0	0-2-0	0-2-0	0-2-0	0-2-0	0-2-0

HEIGHT MEASUREMENTS FROM THE LOAD WATERLINE. THOSE ABOVE MARKED +. THOSE BELOW MARKED —. MEASUREMENTS IN FEET, INCHES, EIGHTHS. TO OUTSIDE OF PLANKING. STATIONS TWO FEET APART. CHINE KNUCKLE DIES OUT AT STATION 5. KEEL AND STEM SIDED 4".

Massachusetts North Shore. He built his first V as early as 1918. Hull did not, however, go in for yacht construction. His boats were honest but rough, and the price was cheap.

Regardless of their crude construction, these Hull V-bottoms were much in demand by North Shore fishermen in the 1930s and 1940s. One of these, a 38' fish-trap vessel built by Hull in about 1938 for Lott C. Peach of Marblehead, was well-known and respected. This boat did a comfortable 12 knots with her 90 h.p. Buda engine.

The positive qualities of the Hull boats and the favor they came to enjoy with local fishermen stimulated other builders to turn their hand to V-bottomed construction. Furthermore, changing conditions in the lobster fishery that favored larger, steadier boats with more working room, encouraged experimentation. Undoubtedly, the local builder who contributed most to the development of the North Shore's V-bottomed fishing boats following the end of World War II was Albert Cloutman of Marblehead.

Cloutman, who had previously been employed as a foreman at the James E. Graves Yacht Yard before setting up his own shop, was a fine craftsman. He did much to improve the rough, slap-dash construction of the Hull boats, yet without increasing building costs unduly.

Some of the Cloutman V-bottoms that won the high regard of North Shore fishermen included Ed Hawkes' *Lizzie H.*, Fred Bartlett's *Millie II*, John Zdanowicz's *Patty Z.*, as well as boats fished by Don Russell, Watson Curtis and Albert Cloutman's son Bob.

Fishermen seemed to prefer the fishing performance of the new V-bottoms over the old slack-bilged, round-hulled types. Frank Zdanowicz, when talking about his brother's boat, the *Patty Z.*, said, "You don't have to spread your legs all the time to brace, and you don't come in pooped."

The wider, hard-chined hulls proved more than adequately seaworthy. Don Russell found this out on one memorable occasion when he was forced to repair a disabled rudder in a howling blizzard six miles off Nahant.

The Cloutman boats were a great improvement over the Hull vessels. This was true not so much from the standpoint of their lines, but principally in their construction. They did, however, show certain defects that indicated the need for further experimentation and improvement.

This soon became obvious to me. In the winter of 1956-1957, I made a detailed survey of V-bottomed lobsterboats that were fishing at that time from the North Shore towns of Salem, Beverly and Marblehead.

Most of these boats were hauled out for

Bessie Ann, *the 32-foot V-bottom lobsterboat built by Justin Hull of Ipswich, Massachusetts, for Amerigo Angelina of Beverly, Massachusetts, in 1947, hauled out at Fred Dion's yard in Salem in December 1956 (John Gardner photograph).*

After studying the hull lines and discussing the boats with their owners, it became clear to me that there were two principal defects that called for correction. They were wet when pushing into head seas. And, more importantly, they required an excessive amount of power to drive them.

At the time, most were powered with converted automobile engines. John Zdanowicz had a 265 h.p. Buick Roadmaster in his 31-footer. Amerigo Angelina's 32' Hull-built vessel was powered with a 250 h.p. Buick V-8. Naturally, gas consumption was horrendous.

It was pretty obvious what needed to be done. The bows of these V-bottoms were stubby, with little flare. Their entrance angles were quite wide. Also, the chine knuckles that broke out of the water forward and ran exposed all the way to the stem were the cause of much of the loose water that came aboard.

the winter months, thus enabling me to measure them and take off their lines, which up to that time had not been recorded. The results of this survey were published in the *Maine Coast Fisherman* in six articles that ran from January to June in 1959.

These were the particular features that I set out to eliminate in making the half-model from which the lines of M-2 were developed. Thus, in M-2, the chine knuckles do not rise above the surface, but die out dead flat between 8' and 9' aft of the stem.

The 38-foot fish-trap boat built about 1938 for Lott C. Peach of Marblehead by Justin Hull of Ipswich, moored in Marblehead harbor in December 1956. Although of rough construction, this was a fine workboat hull doing about 12 knots with a 90 h.p. Buda engine. The model is quite similar to the 32-foot lobsterboat Bessie Ann *(John Gardner photograph).*

Above is* Patty Z.*, a 31-foot V-bottom lobsterboat designed and built by Albert Cloutman and launched in 1949 for John S. Zdanowicz of Beverly and Salem, Massachusetts. Here she is moored near Pickering Coal Wharf in Salem in December 1956 (John Gardner photograph).

When the boat is afloat, it does not appear to have any chine at all. The entrance has been sharpened, and the bow flare markedly increased. The result is a dryer boat that is much easier to drive. Gillmer's tests point this out.

While the older 31- and 32-footers required well over 200 h.p., the much larger and heavier 37' *Stella Z. II* found a 100 h.p. diesel quite adequate for both summer lobstering and the much more exacting demands of winter trawling and netting.

In spite of their splendid sea qualities and low power requirements, boats like the Frost M-1 were obsolete as working fishing vessels 25 years ago. They lacked the room for modern fishing requirements and were too difficult and expensive to build.

Ease and economy of construction is where M-2 shines, however. No special boat building skills, tools or equipment are necessary. No steam bending is required. No backing, hollowing or rounding of planks is called for. The M-2 has no curved frames. A minimum of beveling has to be performed.

Relatively inexpensive lumber from small local sawmills will do nicely. A wide choice of local timber species can be used, allowing one to take advantage of price breaks. Galvanized nails, even galvanized common nails, will suffice instead of screws.

In short, here is a proven hull that a fisherman can build himself, possibly at a significant savings over fiberglass.

After the full-size laydown of the lines, the first operation in the construction of the 37-foot V-bottomed fishing boat is the assembly of the foundation or "backbone." This consists of the stem, keel, deadwood, split shaft log, horn timber and stern post.

In my judgment, putting these pieces together is the most difficult and exacting operation that the non-professional or first-time builder will face in the construction of the 37-foot boat. There are several reasons for this, among them the size and weight of the timber required.

There is also a great deal of shaping and fitting, and much of it will have to be done with hand tools. This is particularly true if the builder lacks both access to a large band saw and help in handling these heavy pieces at the saw. For acceptable results, the scarphs and splices where these timbers join must fit exactly.

Holes for bolts and drifts must be just right, both as to size and location. This is critical in fastening the split shaft log, where tolerances are minimal. (In fact, the shaft log must be just right or the boat will leak.) Lastly, once drift pins are driven, or partly driven, they are in to stay. Errors made in pinning are not easily corrected.

The 31-foot V-bottom lobsterboat, Millie II, *built by Albert Cloutman of Marblehead for Fred Bartlett. In spite of the large stern, Bartlett reported that this boat behaved well in following seas and would "run away from anything" (John Gardner photograph).*

All this adds up to careful, precise workmanship, with little margin for carelessness or error—a job to separate the men from the boys, as the saying goes. The would-be first-time builder had best be aware of this from the start, and he must face up squarely and honestly to what is involved and what his capabilities are.

Unfortunately, some do not have what it takes. Still, on a number of occasions I have been surprised, and even amazed, by the high quality of boat building work turned out on the very first try by inexperienced non-professionals.

For the first-time builder who is not quite sure of himself, there is one safe way to go: he can secure part-time assistance from an experienced professional to help him over the hard spots. Outside the noted exceptions, the basic carpentry called for in this boat has been simplified to the point where almost any ordinary wood butcher can do it.

Simplicity of construction is an outstanding feature of this design. Framing and planking have been simplified to a point where no special skills are required. As for the interior and superstructure, no fancy joinery is required—just plain, straightforward carpentry.

What follows are directions for assembling the foundation. Because the description is intended primarily for first-time builders, we will go into considerable detail, with which the experienced boat builder need not bother.

Before the foundation, or backbone, can be assembled, its several parts must be got out to shape and fitted so they will go together properly. This requires patterns, which is to say, molds. In turn, molds necessitate a full-size layout such as is represented in the accompanying diagram.

In making this layout, a full-size lay-down of the lines is necessary. This is not the place to explain lofting procedures. How to make such a laydown is adequately dealt with elsewhere, in a variety of good boat building books and manuals.

When the lofting of the lines has been completed, the foundation (divided into its parts) is outlined on the laydown, as diagrammed here. Next, the shape of each part can be lifted—that is, transferred from the floor to pieces of 1/4" plywood. When cut out, these reproduce the shapes of the parts exactly.

A simple but accurate method for lifting the shape of the molds involves driving small nails or brads into the outline marked on the floor, spacing them 2" to 3" apart (or closer on sharp curves). The heads are clipped off so that the remaining stubs project slightly less than 1/8" above the lofting surface. A piece of 1/4" plywood is laid over these nails and tapped smartly all around with a hammer.

The points of the projecting stubs mark the under surface of the plywood. (They should be all the same length to mark evenly.) When lines are drawn through these points with a straightedge and flexible battens (for the curves), the shape laid out on the floor will be marked exactly on the plywood.

Next, the wood is cut to shape, producing the mold. The nail stubs are then drawn out with nippers so that the mold can be checked by placing it over its outline on the laydown. If minor deviations need correction, this is done to achieve a perfect fit.

Once the molds are got out and fitted together, the rabbet line and bearding line are transferred to the plywood from the floor, following the same procedure used to obtain the mold outlines. Small holes spaced 2" to 3" apart are then bored along these lines, holes just large enough to admit the point of a marking awl.

By placing the molds on the timbers of the backbone, pricking through the holes and drawing a line through the marks, the rabbet line and bearding line are transferred to the foundation. It should be obvious that the holes through the mold must be bored square with the surface to ensure that both sides of the foundation are marked the same.

No pains should be spared to see that the molds and the laydown of the foundation correspond exactly. These molds should fit together like a well-made jigsaw puzzle, overlying the foundation outlines precisely. If they do not, the pieces of the foundation simply won't fit together.

Oak is the timber generally specified for the foundations of fishing boats such as this. Ideally, it should be dry. Because seasoned oak of these dimensions is not easily obtainable, the builder may have to be satisfied with less. I do not recommend using fresh-cut white oak, as such wood is prone to warp and twist badly in seasoning.

Red oak "stays put" better, and sound solid growth from a red oak species like "yellow bark" will do nicely. There are many different kinds of red oak and many different grades of red oak timber. Avoid lightweight, porous wood and wood that smells "sour," the so-called "piss oak."

Sapwood should also be rejected and, obviously, so should wood with unsound knots and similar defects. Don't forget that timber containing the heart of the tree ("a boxed heart") generally checks sooner or later to the nearest outside surface. Such checks can lead to the interior of the hull, causing leaks.

When oak of suitable quality or size is unavailable, either prime-quality longleaf southern (hard) pine or similar select, old-growth Douglas fir can be substituted. Again, sapwood, bad knots, checks or like defects are unacceptable.

Because this is a large boat—and in order to gain a little more width for pinning the split shaft log—a siding of 4 1/2" is specified for the foundation throughout. Frequently stems, keels and deadwood for 30' fishing launches will be sided 4", but this is a heavy workboat that can easily tolerate another 1/2" of siding in the foundation.

If at all possible, the builder should obtain seasoned wood for the split shaft log. Its two halves should be straight and accurately sized. In fact, warped or twisted timber should be avoided throughout the foundation.

It is best if all pieces in the backbone are planed to their sided dimension at the mill. This may cost a little more, but it will be well worth the extra. Don't skimp on the timber for this foundation. False economy and "cutting corners" here can produce much frustration and grief later on.

If a large power band saw cannot be secured for cutting the scarphs and splices, this can be done quite adequately the old-fashioned way—with a sharp handsaw and then a plane. A large ripsaw is required for the task; its teeth must be properly filed to run true and must incorporate enough set to clear easily.

In this day of power tools, few carpenters and boat builders are accustomed to using a ripsaw. They have no idea how much can be accomplished when

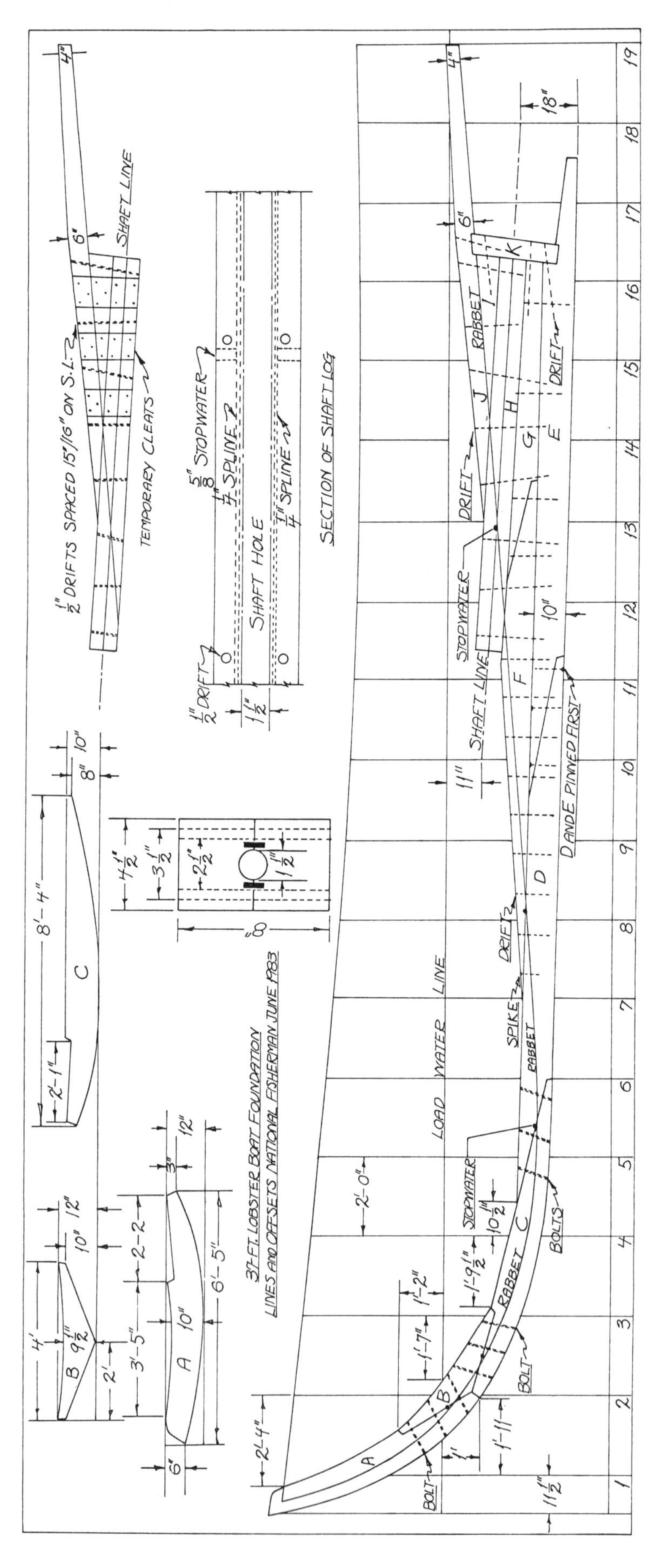

the tool is properly set and sharpened. Even if someone has never used such a saw, he will experience no great difficulty if he goes about it right.

Using the plywood mold, mark the scarph line on both sides of each backbone timber, or "stick," making sure that both sides correspond exactly. The two lines must lie in an imaginary plane that cuts through the wood precisely at right angles to the outer surfaces of the stick.

In other words, if the lines on either side do not overlie each other exactly, a cut through the stick will not be square with the outside surfaces of the foundation. As a result, the scarph joint will not pull together for a tight fit.

With the stick laid down on its side, start sawing along the line on the upper face of the timber, beginning from the outer end and keeping the saw plumb and square with the outside surface. Hold the saw at a rake of about 45 degrees and follow the line, backsawing until the cut is through the end of the stick.

Here, you will have followed the line that is squared across the timber's end and will have connected it with the line on the underside. Next, turn the stick over and saw back along the line now in view, being careful to keep the saw in the kerf previously started on the other side.

After a bit, turn the stick back over and continue sawing from the top. By sawing first on one side and then on the other—and always staying in the kerf already cut—it is easy to follow the lines and to cut the scarph square with the outside of the stick.

Finally, a few strokes with the plane are usually enough to finish the job, although to ensure a tight fit, the builder should pass a fine saw through the joint before it is drawn tight.

For fastening, 1/2" diameter, hot-dipped galvanized drift rod is used. Where bolts are required, the rod can be threaded for nuts. Both the location of the bolts and drifts, and the size of the holes are critical.

The recommended locations for the fasteners are shown on the accompanying diagram. Holes for the bolts offer no problem. They should be tight enough for a snug fit to prevent leakage, but not so tight as to cause the bolt to "cripple" when it is driven.

Incidentally, both bolts and drifts drive easiest (and with less chance of crippling or "freezing") when a heavy maul is used for the driving.

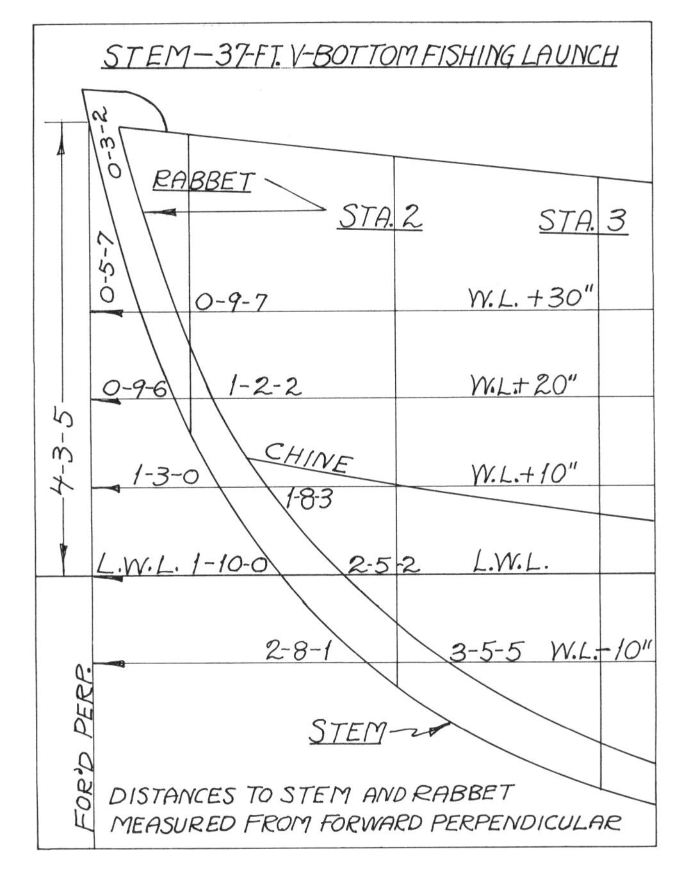

Good galvanized drift rod is well-coated with zinc; this makes it somewhat oversized, which must be taken into consideration. Moreover, the size of the optimum hole will depend, in part, on the length of the drift; on whether the wood is green or seasoned, hard or soft; and on the species of wood, whether it is oak, hard pine or Oregon fir.

Once drifts stop "going," there is no way to pull them out. All that can be done is to saw off the excess and go on to the next one.

Frequently, the difference between the diameters of the drift and the drill bit will be 1/32" or even less. Yet, I know of no way to be sure about the proper drill bit sizes except by boring some test holes in pieces of the wood used and driving some test drifts.

So-called "barefoot" augers cut faster and bore straighter. However, they are not always to be had in exactly the right size. Holes that are slightly small are easily reamed with a twist drill (of exactly the right size) that has been lengthened by brazing it to a section of drill rod.

When barefoot augers are not available for the initial boring, ordinary twist drills with extensions will do the job, but they cut much slower and go harder. I am assuming here that all boring will be done with an electric power drill. Although at one time such holes were bored by hand, those days are gone beyond recall.

Remember one thing when boring deep holes with a barefoot twist auger in a fast-turning power drill: the auger must be frequently withdrawn from the hole to clear the chips. If these are ever allowed to build up in the hole above the twist, it will be impossible to pull out the auger. It will be stuck in there for keeps, and all that can be done is to saw it off and forget it.

The following order of assembly is recommended. First, the stem assembly (parts **A**, **B** and **C**) is fitted together and bolted as a unit. This is done with the pieces lying flat on their sides on a large bench, on several wide planks supported by horses or on something similar.

Parts **D** and **E** are got out of 10" wide stock, according to the molds, and are scarphed and bolted together as shown. This partial assembly of stem and keel is then set upright on previously prepared building stocks and is shored and braced in position.

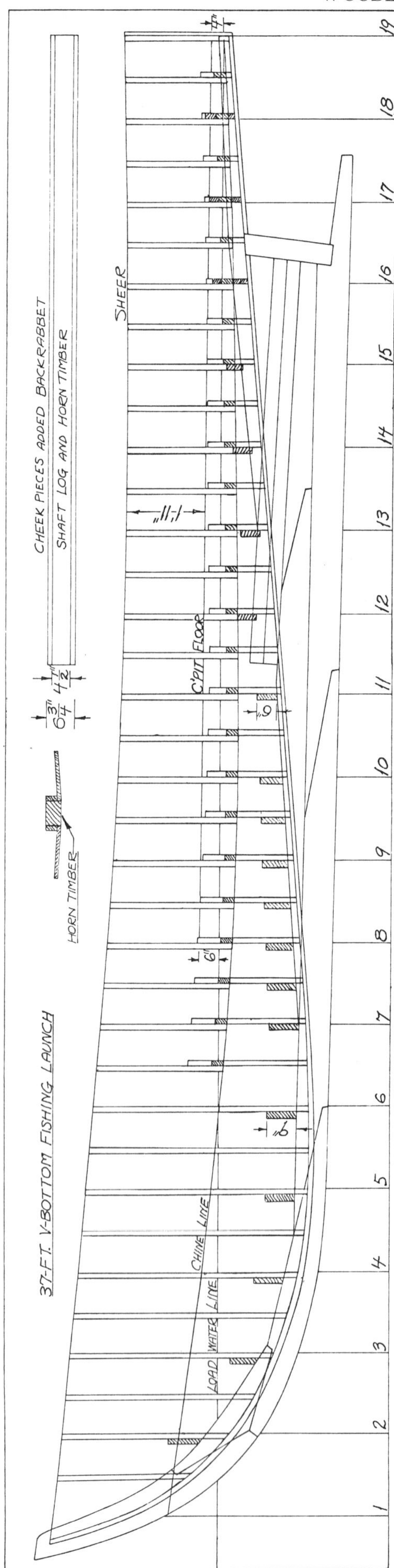

Parts **F** and **G** can now be added, and bolted and drifted in place.

The split shaft log, **H**; the deadwood wedge piece, **I**; and the horn timber, **J**, are pre-assembled as a separate unit, bored for drifts and added to the rest of the foundation.

The shaft hole through the center of the split log will be 1 1/2" in diameter, which is ample to accommodate a 1 1/4" shaft. The shaft hole will be unlined. Exactly one-half of the hole will be in the upper half of the log.

There are several ways of putting in this hole. One is to rough it out with a power saw before cleaning it out and finishing it to size with a gouge and/or round-nosed plane.

On either side of the shaft hole, grooves are cut on the table saw to take white pine splines measuring 1/4" thick by 1" wide. Each is half recessed into the upper part of the log and half into the lower. Great care must be taken to ensure that the upper and lower halves of the grooves come together exactly. This will allow the splines to fit tightly, without keeping the two halves of the split lot apart.

The splines are made of dry wood, white pine preferably, for a snug fit when moisture swells them. They are set in a thick paste of white lead and linseed oil, but if white lead can't be obtained, a non-hardening bottom seam compound thinned with linseed oil (to the consistency of thick cream) may be substituted.

The object of the splines is to make the seam between the two halves of the split log watertight. They may run the whole length of the log, but they are not needed aft of the stopwaters. The latter are located on either side, where the rabbet line crosses the shaft log seam.

The inner ends of these stopwaters bear tight against the splines, forming a seal that prevents any leakage of water from the unlined shaft hole into the interior of the boat.

The stopwaters should be driven hard enough for a snug fit against the splines, but not so hard as to break them. A small amount of paste is applied to seat them, but too much could cause undue pressure on the splines.

The stopwaters for this boat measure

5/8" in diameter and are got out of dry pine. They are initially made slightly larger in diameter than the specified dimension, planed round and then sized by driving them through a hole of the exact size bored through a dry hardwood board.

The sharp upper edges of the hole are relieved slightly and a bit of grease applied. A heavy hammer or maul is used to drive the stopwaters. This compresses them as they are sized, so that they will swell tight as the wood absorbs moisture.

The 37' V-bottom hull of* Stella Z. II *under construction in 1959. The bow is slightly less raked than the M-2 model as preferred for cosmetic reasons by the owner. Frames were spaced on 1' centers and, as shown here, the first 10 were one-piece timbers running straight from the rabbet to the sheer (John Gardner photograph).

The shaft log assembly—including the deadwood wedge and the horn timber—is pinned as a unit to the rest of the foundation assembly with drifts located on either side of the shaft hole.

Because of the limited amount of space, the location of the holes bored for these drift pins is extremely critical. They must be positioned perfectly to prevent their breaking through into the shaft hole or out through the side of the deadwood.

They will center exactly 3/4" in from the outside of the foundation, leaving 1/2" of wood between the drifts and the outside surface. To ensure that the holes do not "wander" as they're being bored, the following procedure is indicated.

Place the assembly of parts **H**, **I** and **J** together on its side. Clamp these three pieces together tightly, and mark the location of the drifts on the outside surface with pencil lines representing the centers of the holes for the drifts. Do this on both sides.

Take the assembly apart and square the lines across the edges of each of the pieces, front and back, working from the outer surface. Using a combination square as a gauge and setting its blade at 3/4", intersect with the lines previously drawn across the edges, working from the outer surface.

Set a carpenter's auger bit with a sharp worm exactly on the intersections just made, and bore a shallow hole—only a few chips' worth. This will be enough to start the barefoot auger.

In boring for the drifts, each of the four pieces is handled separately. It is laid on its side—outside up and level—on a bench or table and is clamped in place. A short straightedge is laid flat on the upper surface, exactly along the line of the hole to be bored.

Two people are required for drilling. The operator of the drill sights along the straightedge to line up the bit. His assistant, standing off at right angles, tells him how much to raise or lower the tool to keep the bit parallel with the straightedge.

The hole is bored halfway, and then—following the same procedure—it is finished from the other side of the stick. If care is taken, the two holes from opposite sides will meet exactly, or nearly so, at the center, nevertheless an extended twist drill of the right size should be passed through them. All of the holes in each of the separate pieces are bored in this way.

When the parts are assembled again, all of the holes will line up and connect perfectly. If the auger used was a little small, the holes can be reamed to exact size with an extended twist drill, as previously mentioned.

Before this is done, the assembly should be drawn tightly together and secured in position with temporary cleats, nailed as diagrammed, so that the holes cannot slip out of register.

If the shaft log assembly cleated in this manner is placed in position on the rest of the foundation and temporarily secured there, the drill can be run through the holes, marking the timbers underneath. The shaft log assembly can then be lifted off, and the holes for the drifts bored to their full depth.

In doing this, the drill may be tipped slightly in to get more wood, but only very slightly in order that the drifts will still follow when driven. Note that the drift holes are not bored parallel but are canted in and out so the drifts will pull against each other.

The last piece to go in is the stern post. The keel and horn timber are notched to receive it, and these notches are initially made slightly narrower than the post itself. A piece with a straight side (sided the same as the post but molded smaller fore-and-aft) can be wedged against the after ends of the shaft log and deadwood pieces. A sharp saw can then be passed through the joint to even it.

Several wedgings and passes may be required before the joint fits tightly. When the post is finally driven into place, the fit will be snug if the notches (top and bottom) are given a slight draft for wedging.

Of course, before the post is installed and fastened, the continuation of the shaft hole will have been located and bored. Fastening will be with drifts, as diagrammed.

The 37-foot wooden V-bottomed fishing launch and lobsterboat we have been discussing here has been planned to meet the requirements of non-professional builders. To briefly recapitulate, tank tests at the Naval Academy at Annapolis gave the hull a favorable rating for low hull resistance and easy, economical powering.

There are no curved frames in the boat, no steam bending, no backed-out planking and no troublesome internal chine stringers. The novel method of setting up the vessel should simplify and speed up the framing operation. Fastenings are all galvanized, consisting of drifts, bolts and nails. Both cut boat nails and ordinary galvanized wire nails are used, the latter extensively.

Lumber can be native timber species directly obtainable from small local sawmills at minimum prices. If possible, this lumber should be bought somewhat in advance of its use to allow some time for partial seasoning.

Species for planking can be northern white pine, white cedar, hackmatack, juniper, cypress, Douglas fir, Port Orford cedar and Alaskan yellow cedar, depending on where the boat is to be built. Oak of good quality is the first choice for the backbone and foundation, i.e., the stem, keel, deadwood, shaft log and so forth. In the South, however, prime quality longleaf hard pine is an acceptable substitute, as is Douglas fir of similar quality on the West Coast. In both cases sapwood should be avoided.

The same alternatives apply for the frames, with the addition of hackmatack (sometimes called eastern larch). As oak is both heavier and harder, frames got out of this wood can be somewhat smaller than for the other three. No fiberglass, polyester or epoxy glue is specified. For bulkheads, other interior structures—and possibly for decking—marine plywood can probably be used to advantage.

In the lines drawing printed here there are 19 numbered sectional stations, of which No. 19 represents the transom. Numbered stations are spaced 2' apart. Frames are spaced on 1' centers, being located both on the numbered stations and halfway between them (making 36 frames in all).

The forward 10 frames are each one-piece and straight, or nearly so, extending in continuous lengths from the rabbet to the sheer. Each of the remaining 26 frames consists of two pieces joined at the chine knuckle, where they are reinforced by a chine block.

At all numbered stations, the port and starboard halves of the frames are joined by bolting or nailing them to floors. These, in turn, rest on the foundation and are drift-bolted to it. Floors are omitted for frames located between the numbered stations—

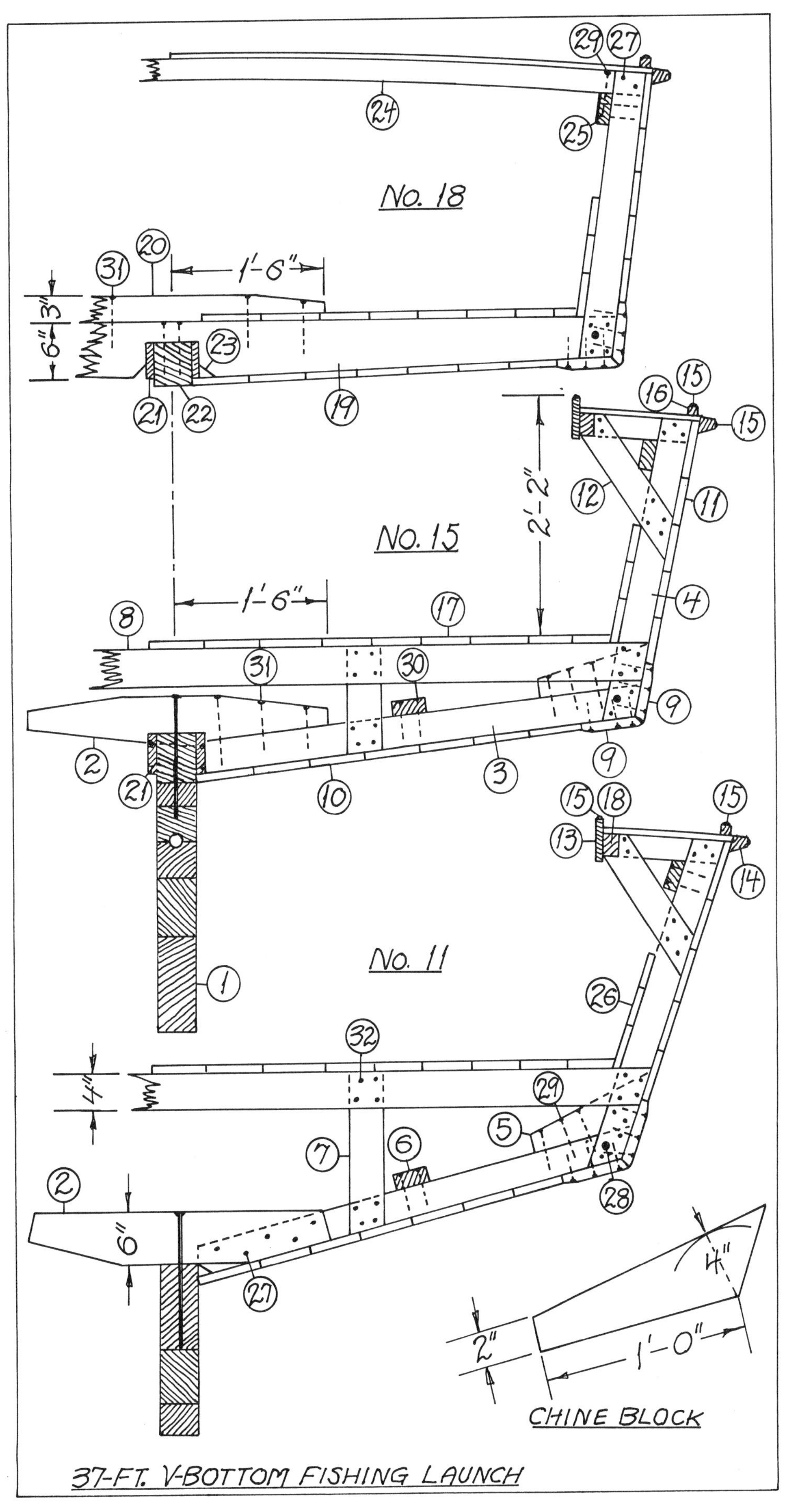

29
27
24
25
No. 18
20
31
1'-6"
3"
6"
23
21
22
19
16
15
15
2'-2"
12
11
No. 15
1'-6"
4
8
17
31
30
9
2
9
21
10
3
15
18
15
13
14
No. 11
1
26
32
4"
29
5
7
6
28
2
6"
27
4"
2"
1'-0"
CHINE BLOCK
37-FT. V-BOTTOM FISHING LAUNCH

The pre-assembled frames were positioned on the leveled foundation and secured along with the floors, and then the 7/8"-thick sheer plank was nailed to the seven frames serving as molds (John Gardner photograph).

simplifies this part of the hull construction process, thereby making it easier. No temporary molds are required as these are replaced by seven pre-assembled frames and the transom, which remain in the boat as part of the completed structure.

Of the seven frames serving as molds, the three forward ones (on numbered stations, 2, 4 and 6) have single-piece sides, which will simplify pre-beveling. Pre-beveling the other four on stations numbered 8, 11, 14 and 17 should not be too difficult.

The other 26 frames are beveled as they are fitted in place, and this should cause no trouble at all. The pieces are simply cut and tried until they fit.

As previously mentioned, there are none of

except for those under the engine beds where extras are needed for reinforcement.

Frames not attached to floors are spiked to the foundation and receive additional support from a bottom stringer running the length of the hull. It is located on top of the lower portion of the frames, on either side of the foundation.

The dimensions of the frames are generous to provide for secure and solid nailing. This may add somewhat to hull weight, but not enough to increase the working load unduly or raise fuel cost appreciably.

Where longleaf hard pine, Douglas fir or hackmatack are to be used for framing stock, I have specified a molded width of 3½" and a sided thickness of 1½", and the accompanying diagrams are so drawn.

Where oak is used, the molding could well be reduced to 3" and the siding to 1¼". Molded thickness of the floors should be 1¾" for oak and 2" for hard pine, Douglas fir and hackmatack.

The method proposed for framing

the internal chine stringers that are usually found in V-bottom construction. It is therefore unnecessary to notch the frames and fit the stringers into them, a tedious job at best. Bolted chine gussets are also dispensed with, meaning a considerable savings in both labor and materials.

This alternate method of chine construction is not only simpler and easier, but also, in my opinion, stronger.

After accurately leveling the foundation fore-and-aft (to the load waterline), the pre-assembled frames are positioned on the foundation at their proper locations and are carefully plumbed, leveled, horned and secured in place. This includes the floors, which are drift-bolted to the foundation with 1/2" diameter galvanized drift rod.

Now the sheer strake can be installed. It will serve as a fairing ribband and will remain in place as part of the permanent structure. Planed and sanded to a finished thickness of 7/8", this strake is nailed to the frames serving as molds with 2½" hot-dipped galvanized cut boat nails with their heads set about 1/8" below the surface for puttying.

Because the boat is 37' long, it is likely that the sheer strake will have to be made up from three boards. Some long ones—up to 18' or 20' in length—should be included with shorter boards so that as planking proceeds, the butts can be well staggered. (The rule is that planking butts should be separated by at least three frame bays horizontally and three planks vertically.)

Butt block locations for the sheer strake should be planned so as not to interfere with any of the frames when they are put in. Butt blocks will be 10½" long, the distance between frames. A good 1/2" will be chamfered off the inside corners at either end to provide a way for condensation to drain.

The butt blocks here are best made of oak a full 1½" thick. When the plank ends are nailed with 2½" galvanized cut boat nails (five nails to each plank end), about 1/4" of the nail will come through for clinching. Holes for such nails must be bored, of course.

After the sheer strake is on and fastened, it should be checked for fairness. If a slight hump or flat place should show anywhere, the cross spall on the offending frame can be slackened and the frame pulled in or pushed out slightly. Once the problem is corrected, the cross spall is refastened.

As already stated, the sheer strake serves as a fairing batten, or ribband, for locating and fitting the rest of the frames. Another batten must be laid on the topsides for the same purpose and is located about 7"-8" above the chine knuckle. This temporary batten will be 3" wide and 3/4"-7/8" thick. Where its sections are joined, the cleats should be nailed on the outside.

On the underside (bottom) of the frames, between the chine knuckle and the keel rabbet, there will be two such temporary 3" wide longitudinal battens or ribbands running the full length of the boat. These may be nailed in place with 3" double-headed staging nails, and although they will be removed later, they should be solidly fastened.

One of the two battens is located about 6" below the chine knuckle, the other about the same distance from the keel rabbet, give or take a little for convenience.

Fitting and installing the rest of the frames, 26 in all, is a fairly simple operation and not as lengthy as might be supposed. The first thing to do is mark the section lines, locating the position of the frames on the inside of the sheer planks and the longitudinal battens.

The imaginary cross sections on which the frames are located cut square across the hull from side to side and stand square with the plane of the load waterline, which was leveled when the foundation was set up, as previously noted.

To fit and install one side of a two-piece

Deck beams, side deck braces, carlin and clamp forward of the cabin top and deck house, prior to the installation of plywood decks (John Gardner photograph).

Save for the shutter plank,* Stella Z. II *is almost ready to leave the shop (John Gardner photograph).

When the planing has been completed, the inboard end of the frame piece is trimmed to fit against the keel, or foundation. Finally, the lower portion of the frame is clamped back in position. There should be enough left on the outboard end of the piece so that it extends slightly beyond the chine knuckle. The upper and lower parts of the side frame can then be bolted together.

The other side of the frame is put together in the same manner. If the frame is to be attached to a floor, the latter is first fitted, then drift-bolted to the keel or to the foundation. The inboard ends of the frame are then fastened to the floor. Either bolts or several 3" galvanized cut boat nails can be used for this.

frame, select a section of frame stock several inches longer than its final length. Set this on edge along the forward side of the section line for the frame, as marked on the inside of the sheer plank and on the batten below it.

At the same time, this section of frame is clamped against a straightedge laid across the boat on top of the sheer planks and along the section line marked on both sides of the boat. This will line up the upper frame piece in cross-section, enabling the builder to accurately measure the amount of beveling required.

When this frame piece has been planed off for a good fit against the inside of the sheer plank and the ribband below, it is clamped in position with 1" or so of its length extending below the apex of the chine knuckle.

A piece for the bottom part of the frame, also a few inches longer than its finished length, can now be laid along the after side of the section line as marked on the lower ribbands (or battens) and clamped to the upper portion of side frame already in place. This brings the lower section into a vertical position for beveling.

Finally the chine blocks are nailed to the bottom sections of the frame on which they rest with a 4" and a 5" galvanized wire spike.

Likewise, they are nailed onto the upper portion of the side frame with 3" galvanized cut boat nails, well staggered and bored for a snug, drawing fit. It is all to the good if the ends of the nails protrude a bit for clinching.

Lastly, and before planking starts, the protruding ends of the upper and lower portions of the side frame (and, if need be, the clinch block) are trimmed off flush and fair with a sharp handsaw. Actually, the whole framing operation is much simpler than it sounds. Doing it is easier than describing it, and once the procedure has been grasped, there is really not much to it.

Overlapping the two parts of the side frames or, rather, letting the ends run by each other—for bolting and nailing—results in more than adequately strong construction. Adding the chine block provides still more strength.

On the 37-footer, however, the chine

knuckle is further reinforced on the outside by the two chine planks, which measure 1$^1/_8$" thick—1/4" thicker than the other planking. To blend smoothly with the planking above and below, the outer edges of these chine planks are beveled to 7/8", letting the bevel run back from the edge by about 2".

Likewise, the forward ends are thinned enough to project into the rabbet cut by 7/8". Because there is no transverse curvature in the sides or bottom of this boat, no planing has to be done after the planks have been put on the boat.

There is no backing or hollowing. Planks can be finished flat and smooth at the bench and need only minimal sanding before painting.

Where the edges of the two chine planks come together at the apex of the chine knuckle, they divide the apex angle equally. A touching fit on the inside with an opening of about 1/8" on the outside is desirable for a good caulking seam. Because of the extra thickness of these planks and the solid backing inside, this seam should take and hold cotton extremely well.

Only driven cotton, closely tucked and solidly set down, will do here. Soft, synthetic caulking compounds are not recommended.

The galvanized fastenings specified for this boat call for some comment. Drift bolts and standard bolts have already been discussed in connection with the foundation. The same drift bolts made up in various lengths from 1/2" diameter galvanized rod will fasten the floors to the foundation.

One 3/8" galvanized carriage bolt measuring 3$^1/_2$" long is required at each frame knuckle where the top and bottom parts of the side frames run by each other. The total needed will be something like four dozen, all told.

Bolts of the same size could be used for fastening the frames to the floors (and they would be excellent for this), but 3$^1/_2$" galvanized cut boat nails will serve quite well here at considerably less expense. One-quarter inch of each 3$^1/_2$" nail will come through for clinching, which results in a very strong job. Boring snug lead holes for these nails before they are driven is, of course, mandatory.

Ordinary galvanized wire nails and spikes hardly need comment. Most of the interior of the 37' vessel will be put together with such nails, which will range in size up to 5" and 6" in length.

The same sort of fastener will attach the chine blocks to the lower sections of the side frames and will join the bottom frame stiffeners and cross-keel reinforcements in the after part of the boat. For the larger sizes, lead holes are often needed to prevent splitting, and here again, care must be taken not to get the holes too large.

Hot-dipped galvanized square-cut boat nails measuring 2$^1/_2$" long have already been designated as plank fastenings, with the length increased to 3" for thicker chine planks. Such nails were once universally used for boat construction, but in recent years, they have been replaced almost entirely by the more fashionable wood screws. Galvanized steel wood screws,

The keel, sided 4 $^1/_2$", and the horn timber were both notched to receive the stern post. The 3-bladed prop swung from the 1 $^1/_4$" diameter shaft was driven by a 100 h. p. diesel (John Gardner photograph).

The* Stella Z. II *hauled out at Fred Dion's Salem yard (John Gardner photograph).

however, leave much to be desired. For heavier work such as we are considering here, galvanized square-cut boat nails are at least as good and are less expensive.

Fortunately, there is still a source of supply for these nails, namely the Tremont Nail Company of Wareham, Massachusetts. Tremont and its predecessors have been continuously engaged in the nail-making business for the past 160 years. The firm produces a superior nail.

Cut boat nails hold well because when their square, wedge shape is driven into the run of the grain, it jams the wood fibers against the sides of the nail. Square-cut boat nails are much stouter for their length than ordinary wire nails.

They have large, solid heads that seat firmly in the wood and do not rust away. No matter how hard they are pounded, they won't break off, as the heads of wire nails sometimes do. Tremont's hot-dipped nails are well coated with zinc, giving good corrosion resistance.

I am told that at the suggestion of Canadian boat builders who still use the old-type fasteners extensively, Tremont has thinned the square points of its boat nails somewhat to make it easier to clinch over the ends.

Cut boat nails have been specified here to fasten the planking, both because of their holding power and their large solid heads. For joining the two-piece side frames, the 3" cut nails are much stouter than their wire counterparts and are a better substitute for bolts.

There is a considerable overlap in the possible application of these two kinds of nails, and the choice must be left to the boat builder's judgment. Wire nails are cheaper and drive more easily in the smaller sizes. When the wire nails will do, it is sensible to use them.

Except for the heavier chine strakes, there is nothing exceptionable about planking the 37' boat—nothing that departs from standard carvel planking procedure. It bears repeating that seams should be caulked with cotton set down hard, not with flabby modern synthetics.

Except for experienced plankers and professional boat builders, I don't believe it would be a waste of time to make a pine half-model of the hull to a scale of 1"=1'. This can be used to lay out the planking, using small pine battens and pins to hold them in place.

To mark the planking shapes, wisdom dictates a pencil soft enough to erase easily when corrections are made. Of these, I daresay, there should be plenty.

This hull, designed for ease and economy of construction and use in the inshore fishery in 1958, is still valid for a one- or two-man operation today. However, fishermen in the 1990s face not only strict, inhibiting fisheries regulations but all sorts of government regulations that directly affect their livelihoods and the working waterfronts they inhabit. As a footnote to changing times, on June 11, 1992, the National Park Service issued an eviction notice to Salem's last commercial fisherman, Michael Zdanowicz, who left Salem's Central Wharf for Beverly. His grandfather began operating a family lobster and fishing business in Salem in 1918.

List Of Parts

(1) Keel, sided $4\frac{1}{2}$"; oak, hard pine, Douglas fir

(2) Floors, sided $1\frac{3}{4}$", molded 6" to 9"

(3) Bottom frames, $1\frac{1}{2}$" x $3\frac{1}{2}$"; oak, hard pine, Douglas fir

(4) Side frames, $1\frac{1}{2}$" x $3\frac{1}{2}$"; oak, hard pine, Douglas fir

(5) Chine block, sided $1\frac{1}{2}$"

(6) Stringer, $1\frac{3}{4}$" x 4"; Douglas fir, oak, spruce

(7) Post, $1\frac{3}{4}$" x 4"; Douglas fir, oak, spruce

(8) Cockpit beam, $1\frac{3}{4}$" x 4"; Douglas fir, oak, spruce

(9) Chine plank, $1\frac{1}{8}$" thick

(10) Bottom plank, 7/8" thick

(11) Side plank, finish 7/8" thick

(12) Side deck brace, $1\frac{1}{2}$" x $3\frac{1}{2}$"

(13) Coaming band, 7/8" x 4"; oak

(14) Guard, $1\frac{1}{2}$" x $1\frac{3}{4}$"; oak

(15) Galvanized 3/4" half oval

(16) Toe rail, 1" x $1\frac{1}{4}$"; oak

(17) Cockpit flooring, 7/8" thick; white pine

(18) Deck Carlin, $1\frac{1}{2}$" x $2\frac{1}{2}$"; oak, Douglas fir

(19) Frame/floor, sided $1\frac{3}{4}$"

(20) Reinforcement, sided $1\frac{3}{4}$"

(21) Cheek pieces for backrabbet, $1\frac{1}{4}$" thick

(22) Horn timber, sided $4\frac{1}{2}$"

(23) Limber, cut $1\frac{1}{2}$" each way

(24) Deck beam, molded $2\frac{1}{2}$", sided $1\frac{3}{4}$"

(25) Clamp, $1\frac{3}{4}$" x $3\frac{1}{2}$"; Douglas fir, oak

(26) Sheathing, 3/4" or 7/8" thick

(27) $3\frac{1}{2}$" Galvanized cut boat nail

(28) Galvanized carriage bolt, 3/8" x $3\frac{1}{2}$"

(29) 5" Galvanized wire spike

(30) 4" Galvanized wire nail

(31) 6" Galvanized wire spike

(32) 3" Galvanized wire nail

The sharpened entrance of* Stella Z. II *and increased bow flare helped result in a dry, easily driven hull economical to run. In the water the chine knuckles did not rise above the surface and the thicker chine planks were faired so that the bow appeared smooth (John Gardner photograph).

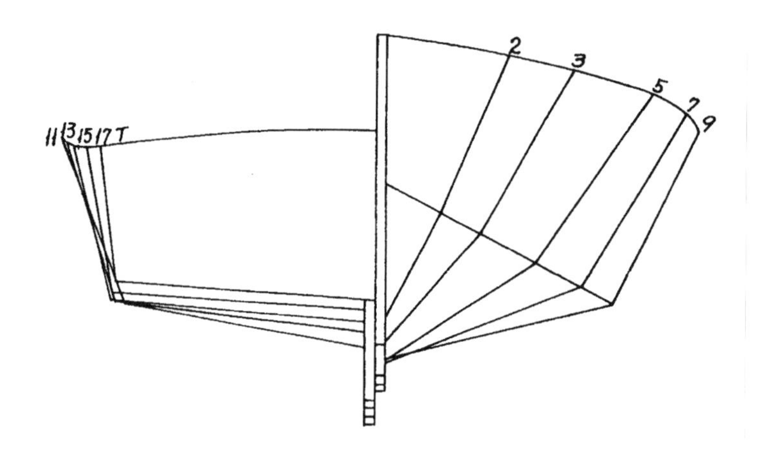

CHAPTER THIRTEEN

25-FOOT V-BOTTOM WORK LAUNCH

As regular readers of my *National Fisherman* columns hardly need to be told, there are a lot of good boats in the 25-foot range. Yet there seems to always be room for one more in the inshore fisheries, and neither interest nor demand appear to be tapering off.

Fishermen like to pick and choose, and they have their own ideas about what will work best for them. You can bet that their new boats aren't going to be exactly like the old ones.

One disadvantage of mass-produced fiberglass jobs built from the same mold is that you can't make changes or improvements in individual boats as you can with one-of-a-kind wooden or steel vessels. When the opportunity for experimentation is restricted, design freezes into rigid, familiar shapes, and both improvement and evolution slow down or come to a stop.

The objective here is not only to provide the dimensions required for building this particular boat, but to offer as well a basis for further experimentation. This will give something for fishermen and builders to start from when working out craft to suit their own ideas and special requirements.

Limiting the boat's overall length to 25 feet allows us to keep construction costs affordable at a time when they are very high and bound to go even higher. Also, 25 feet is not too large for a builder to attempt single-handedly, making it possible for a fisherman to build this boat himself.

The hull shape is one that will be easily built in either steel or wood. There are no difficult curves or twists to form. The sectional shapes are all straight lines, and much of the hull surface is flat, or nearly so. If the hull is built of wood in the conventional plank-on-frame manner, the

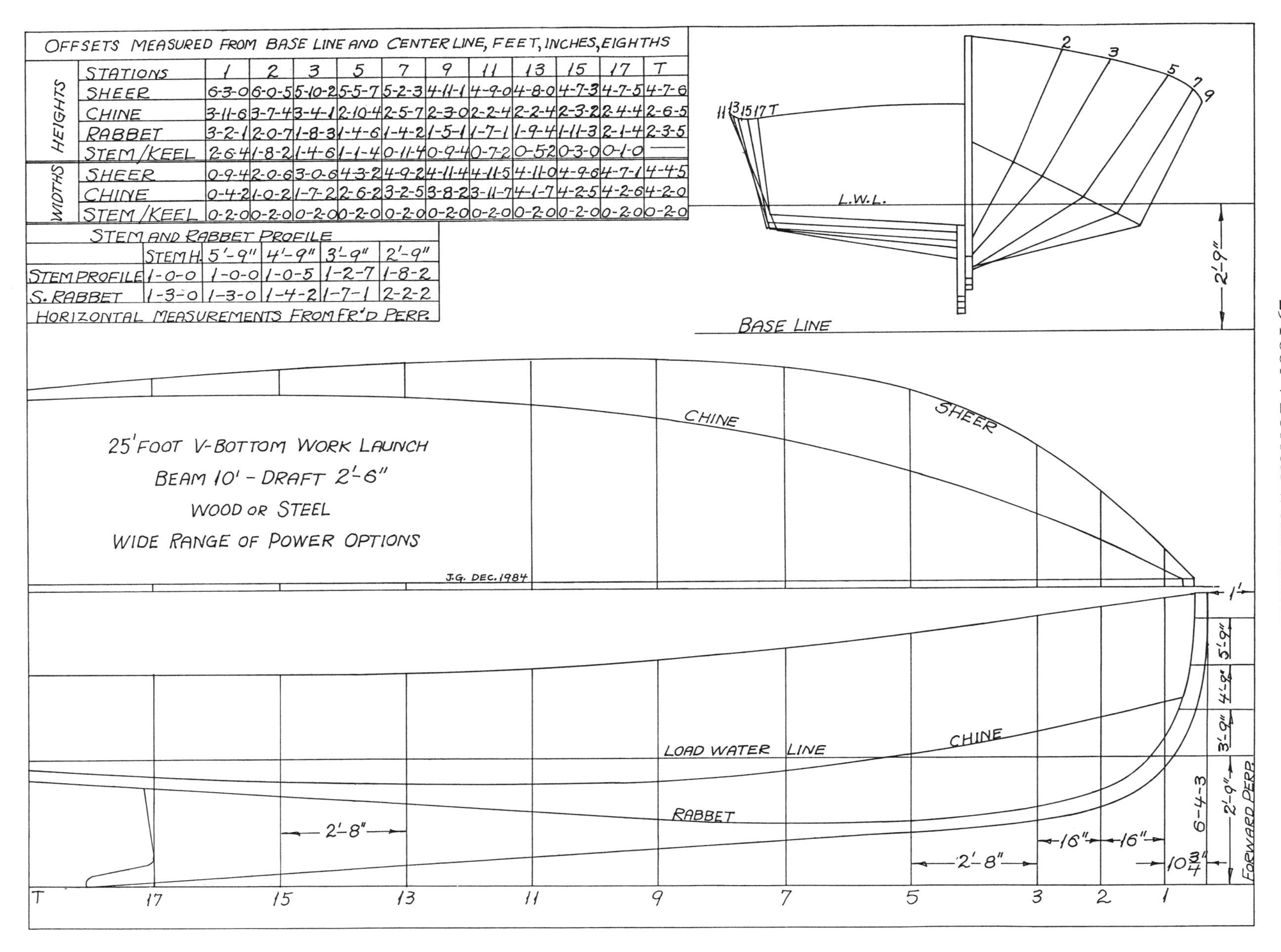

OFFSETS MEASURED FROM BASE LINE AND CENTER LINE, FEET, INCHES, EIGHTHS

	STATIONS	1	2	3	5	7	9	11	13	15	17	T
HEIGHTS	SHEER	6-3-0	6-0-5	5-10-2	5-5-7	5-2-3	4-11-1	4-9-0	4-8-0	4-7-3	4-7-5	4-7-6
	CHINE	3-11-6	3-7-4	3-4-1	2-10-4	2-5-7	2-3-0	2-2-4	2-2-4	2-3-2	2-4-4	2-6-5
	RABBET	3-2-1	2-0-7	1-8-3	1-4-6	1-4-2	1-5-1	1-7-1	1-9-4	1-11-3	2-1-4	2-3-5
	STEM/KEEL	2-6-4	1-8-2	1-4-6	1-1-4	0-11-4	0-9-4	0-7-2	0-5-2	0-3-0	0-1-0	—
WIDTHS	SHEER	0-9-4	2-0-6	3-0-6	4-3-2	4-9-2	4-11-4	4-11-5	4-11-0	4-9-6	4-7-1	4-4-5
	CHINE	0-4-2	1-0-2	1-7-2	2-6-2	3-2-5	3-8-2	3-11-7	4-1-7	4-2-5	4-2-6	4-2-0
	STEM/KEEL	0-2-0	0-2-0	0-2-0	0-2-0	0-2-0	0-2-0	0-2-0	0-2-0	0-2-0	0-2-0	0-2-0

STEM AND RABBET PROFILE

	STEM H.	5'-9"	4'-9"	3'-9"	2'-9"
STEM PROFILE	1-0-0	1-0-0	1-0-5	1-2-7	1-8-2
S. RABBET	1-3-0	1-3-0	1-4-2	1-7-1	2-2-2

HORIZONTAL MEASUREMENTS FROM FR'D PERP.

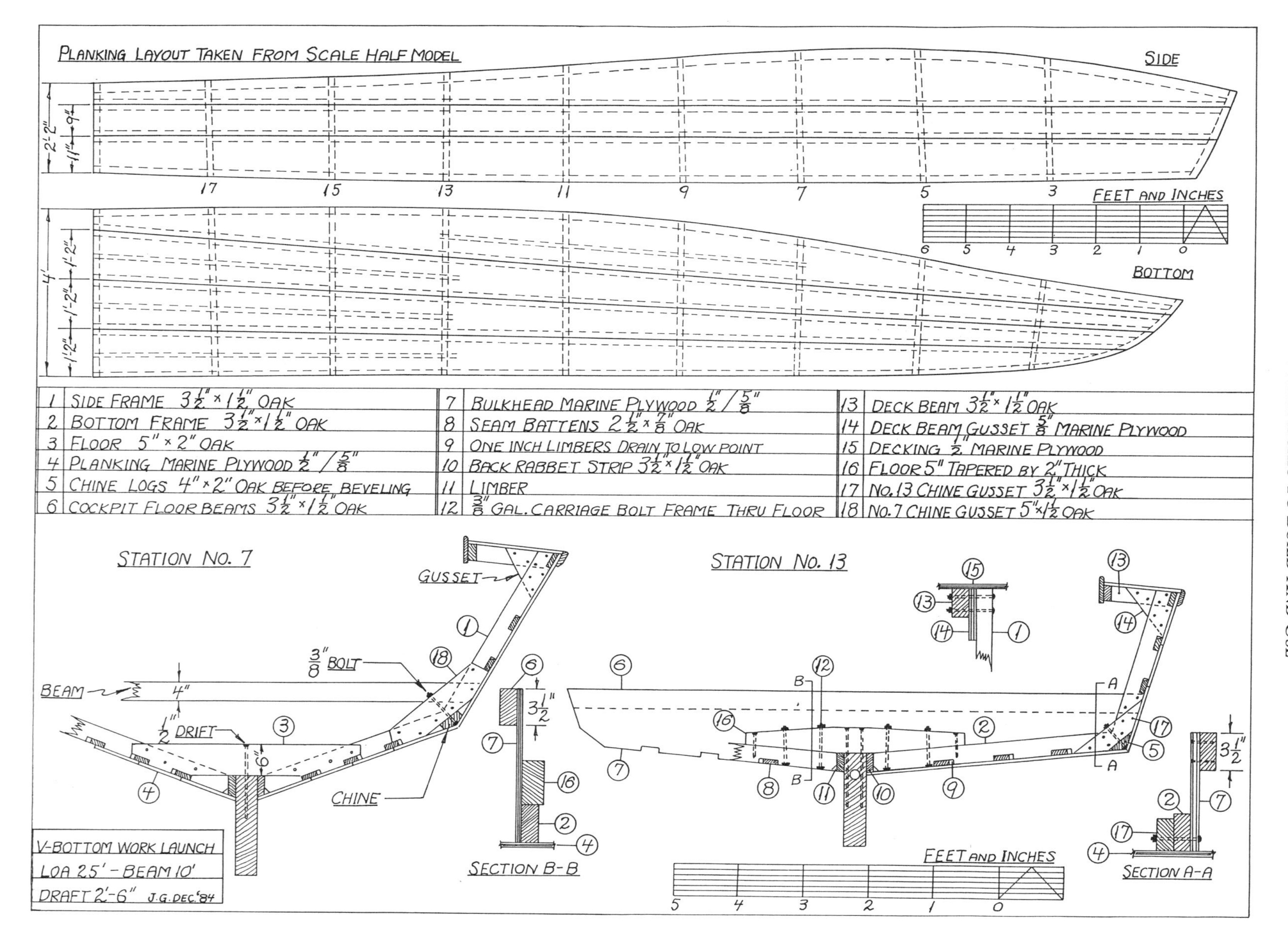
PLANKING LAYOUT TAKEN FROM SCALE HALF MODEL
SIDE
BOTTOM
FEET AND INCHES
1 SIDE FRAME 3 1/2" × 1 1/2" OAK
2 BOTTOM FRAME 3 1/2" × 1 1/2" OAK
3 FLOOR 5" × 2" OAK
4 PLANKING MARINE PLYWOOD 1/2" / 5/8"
5 CHINE LOGS 4" × 2" OAK BEFORE BEVELING
6 COCKPIT FLOOR BEAMS 3 1/2" × 1 1/2" OAK
7 BULKHEAD MARINE PLYWOOD 1/2" / 5/8"
8 SEAM BATTENS 2 1/2" × 7/8" OAK
9 ONE INCH LIMBERS DRAIN TO LOW POINT
10 BACK RABBET STRIP 3 1/2" × 1 1/2" OAK
11 LIMBER
12 3/8" GAL. CARRIAGE BOLT FRAME THRU FLOOR
13 DECK BEAM 3 1/2" × 1 1/2" OAK
14 DECK BEAM GUSSET 5/8" MARINE PLYWOOD
15 DECKING 1/2" MARINE PLYWOOD
16 FLOOR 5" TAPERED BY 2" THICK
17 No.13 CHINE GUSSET 3 1/2" × 1 1/2" OAK
18 No.7 CHINE GUSSET 5" × 1 1/2" OAK
STATION No. 7
STATION No. 13
SECTION A-A
SECTION B-B
GUSSET
CHINE
3/8" BOLT
1/2" DRIFT
BEAM
V-BOTTOM WORK LAUNCH
LOA 25' - BEAM 10'
DRAFT 2'-6" J.G. DEC '84

planks will all go on flat without backing or hollowing. No steaming would be necessary. For the most part, the only beveling required would be for caulking. And, except for smoothing the outside of the hull for painting, there would be only a minimum of planing and sanding.

The frame futtocks are all perfectly straight and can be economically cut out of straight planks. The frame bevels are the simplest possible. Furthermore, the frames also serve as molds when bolted together in accordance with the full-sized layout of the sectional shapes.

This means a considerable saving of labor and materials. There can be no doubt that with this 25-footer we have the simplest, easiest and quickest form of plank-on-frame construction there is.

However, another method of construction can also be employed: using seam battens overlaid with 1/2" thick marine plywood put on in strips. The plywood strips are glued to the battens with epoxy adhesive and secured to them with annular ("ring") nails.

This method is exceptionally strong. It is also leak-proof, requiring no caulking or recaulking. This last feature is important, as plank-on-frame caulking seems destined to become a lost art. Even now, it's difficult to get a proper caulking job.

Standard-length plywood panels can be used and cut up to advantage with very little waste because the plywood is put on in tapered strips that are straight except along the keel and chine. Also, these strips can be made up of a number of short lengths that are easily spliced on any convenient frame. These relatively narrow strips are easily scarphed and glued as they are put in place on the boat. This is so much easier than scarphing and pregluing full-width panels.

Good-quality sawn lumber suitable for plank-on-frame construction is getting increasingly difficult to find and is generally expensive. Sometimes it's not to be had at all. Marine plywood, on the other hand, is readily available almost everywhere without any shortage, at least so far. Although it's not inexpensive, if used economically it may well prove cheaper than high-priced, hard-to-obtain conventional planking lumber.

One of the features that characterizes the superior fishing launch is a good-sized cockpit that serves as a stable working platform in most weather conditions. This was something that many of the older, narrower, round-bottomed hulls did not have, although their performance was excellent in other respects. Even in moderate or relatively quiet seas, the active motion of these narrow, round hulls was uncomfortable and tiring for the crew, who had to be on their feet for the entire working day.

Although this boat is only 25 feet overall, a load waterline width of 8½' in the cockpit area guarantees exceptional room there. Also, a waterline width of more than 8', combined with a hard chine and nearly flat after sections, ensures an easy, restrained motion and steady footing in ordinary weather suitable for fishing.

But even in the inshore fisheries, boats are bound to encounter rough weather and get caught out in storms. Needless to say, a boat should be able to take it when that occurs, and bring the crew home safely. There is every reason to believe that this boat can do just that.

The design was reworked directly from a somewhat larger, but otherwise closely similar lobsterboat, the M-2. That design has proven surprisingly able in rough water and has done exceptionally well in the severe conditions of winter fishing. I do not believe the reduction in size is sufficient to greatly reduce the boat's seakeeping ability.

Another thing on the plus side for this design is its broad range of power options. Given ample horsepower, this hull, with its wide, flat after bottom, could be brought up to planing speeds. With considerably less horsepower, it would drive economically at lower speeds.

Finally, this is not an untried or untested design. Its basic features have been proven in use. Boats built with the same characteristics have given a good account of themselves.

It is obvious that the boat has a great deal in common with the 37-foot V-bottomed lobsterboat, M-2, that was described and discussed in detail in 1983 in the June, July and September issues of the

Joe Richman of Sable River, Nova Scotia, contracted Camilie D'eon of Middle West Pubnico, Nova Scotia, to build this fiberglass one-off 35-foot lobsterboat and inshore longliner in 1987 based on a Bill Oehrle design with input from John Gardner (photograph courtesy of Joe Richman).

National Fisherman. To recap, M-2 was the outgrowth of an in-depth study that I made in the late 1950s for the *Maine Coast Fisherman*, of V-bottomed lobsterboats then being developed in Salem, Beverly and Marblehead, Massachusetts.

M-2 resulted from an attempt to incorporate some of the best features of several of these boats in one hull while avoiding several undesirable features that had been observed. In 1959, the first boat built from M-2 lines, the *Stella Z. II*, was constructed for Joseph Zdanowicz, a Salem lobsterman who also fished for flounder in the winter. The *Stella Z. II* proved to be all that had been hoped for and a little more. As far as I know, this boat was still fishing until quite recently. Her design and construction are discussed in Chapter 12 of this book.

Because of extensive publicity in the columns of the *Maine Coast Fisherman*, M-2 came to the attention of Godfrey Kalat of East Haddam, Connecticut. Among the things that caught Kalat's eye were the favorable reports of M-2 tank tests carried out by Professor T. C. Gillmer at the U. S. Naval Academy Engineering Laboratories in Annapolis, Maryland. Professor Gillmer's findings were reported in a paper published in *Fishing Boats of the World, Vol.2*.

Kalat liked the lines so much that he decided to build a smaller version for his own use as a pleasure boat. His reduction, apparently a strictly proportional one, was approximately two-thirds the size of the original, giving an overall length of 25 feet.

He also made a few other, what he called "slight," alterations, one of these being to remove some or most of the twist or "warp" out of the after bottom. It is possible that the performance of the boat at planing, or close to planing, speeds may have been improved by that modification. However, the prototype M-2 was never intended to plane.

Kalat finished construction in the early months of 1961. For power he selected a 100 h.p., direct-drive Flagship engine. The first trial runs were made with a 13" x 11" propeller. Kalat figured the slippage was close to 45% with that prop. When he changed to a 14" x 9" Equipose wheel, slippage at 3,100 r.p.m. still figured at 40%, although the boat moved somewhat better and got up to 15 m.p.h. on several clocked runs.

Kalat told me at the time, as reported in the April 1962 edition of the *National Fisherman,* that he wanted a cruising speed of 12 to 14 m.p.h. at 2,200 r.p.m. A 16" x 9" wheel might have been able to do this, he believed, but a 14" propeller was the largest he could swing. His next best choice would have been 14" x 10".

Recent attempts to locate the boat have been unsuccessful. Kalat was 75 years old in 1961 when his boat was completed, so it's unlikely that he is still alive. Basically, Kalat was much pleased with his boat, reporting that it was dry and behaved nicely in a head sea.

He did, however, consider adding wedges to hold up the stern at planing speeds. Wedges attached to the bottom directly under the transom in the attempt to improve planing capability were much in vogue at the time. They were frequently tried on stock powerboats built by Chris-Craft, Owens and others. It would appear that Kalat made his reduction of the lines strictly proportional and thus got his stern too narrow.

Kalat did not provide me with the lines that he worked out for his reduced version of M-2. I made the reduction to the lines that were printed with my report commenting on Kalat's boat in April 1962. The lines that appear here are a revision of these. I have further widened the stern slightly, as well as increased the clearance for the propeller; the boat can now carry a 22" wheel. The 8' 5" width at the transom on the load waterline will be ample to prevent it from sucking down unduly at planing speeds, I am sure.

Kalat chose to plank the hull with plywood strips over seam battens, but he did not use epoxy glue. Rather, he used a so-called marine glue with little or no adhesive strength. The use of epoxy glue greatly increases structural strength and is recommended not merely for the planking, but throughout the hull.

One further suggestion: anyone considering building this boat would be well-advised to make an accurate half-model of the hull, scale 1"=1', on which to lay out the planking. This would be particularly helpful when the boat is to be planked with plywood.

Due to requests from *National Fisherman* readers, the editor asked boat designer and retired fisherman, William H. S. Oehrle of Achusnet, Massachusetts, to work these plans up for steel construction, and Oehrle's account was published in the November 1985 edition.

Working from individual personal experience, fishermen frequently adapt designs to their needs, and as I wrote to Bill Oehrle back in 1990, if man in general had never made changes as he went along, we would still be sitting on a log, paddling with a branch to get across the river. In 1989, Nova Scotia fisherman Joe Richman came to visit after reading my *National Fisherman* article on the Seagoin 21, built by Franklin Post. After some discussion I sent him to Bill Oerhle who designed a 35-foot inshore lobsterboat to satisfy the special requirements of lobstering at Sable River. Richman was pleased with the boat built by Camilie D'eon of Middle West Pubnico, Nova Scotia, and Oerhle wrote to say that, "Joe Richman's boat was an interesting job. It isn't often that a client tells his architect that he doesn't have to go faster than seven to eight knots or more than a couple of hours from his home harbor." Both Richman and D'eon made slight changes while the boat was under construction.

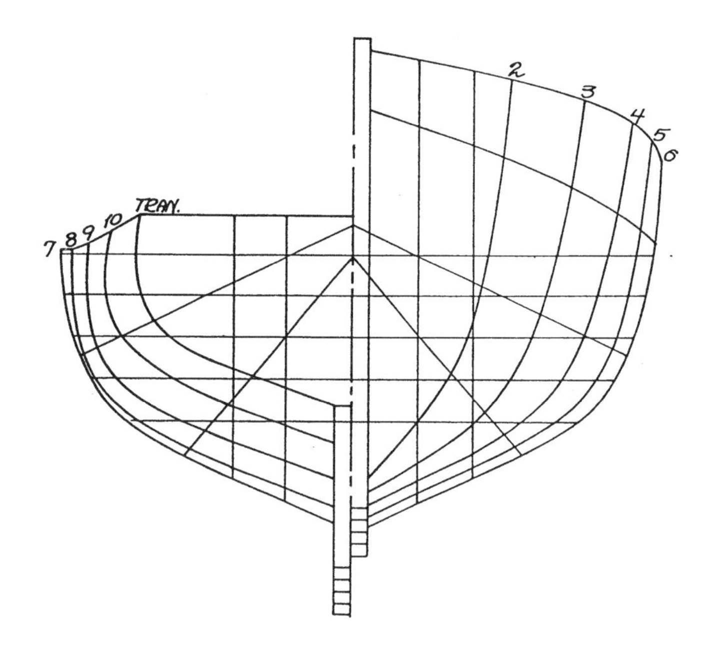

CHAPTER FOURTEEN

SEAGOIN-32 FISHING BOAT

Among the sturdiest and most distinctive small commercial fishing boats developed on the Atlantic Coast following the introduction of the gasoline engine were the "Seagoin" Block Islanders, designed and built by Franklin G. Post of Mystic, Connecticut.

The first of the Seagoin line, a 25-footer, was built by Post in 1914, the year he opened his Mystic yard. He designed it for a Block Island fisherman who used the boat successfully to catch tuna and swordfish in the open Atlantic outside Block Island for more than 20 years.

The reputation quickly acquired by this boat for sturdy performance and seakindly ability soon brought demands for others like it. In the early years of the Post yard, more than 40 of these Seagoin boats were built for local fishermen as they changed over from sail to power.

In time, the Seagoin line was expanded to include both enlarged commercial models and a series of popular adaptations for sportfishing and general pleasure use. Included were commercial fishing boats from 25'-35' in length and a series of Seagoin cruisers built on the same fishing-boat lines but lightened, streamlined and outfitted as pleasure craft and sportfishermen. The Seagoin-25 was described as "a real all-weather sportfishing boat." Other models were built in lengths of 26', 28', 31' and 35', and while they did not depart radically in design from the original commercial type, they were built a little more streamlined, with additional length and beam, and a slightly shallower and faster bottom.

In 1935 the Post yard brought out the Seagoin-32, an enlargement closely modeled on the 25-footer and intended primarily as a commercial fishing boat, although by

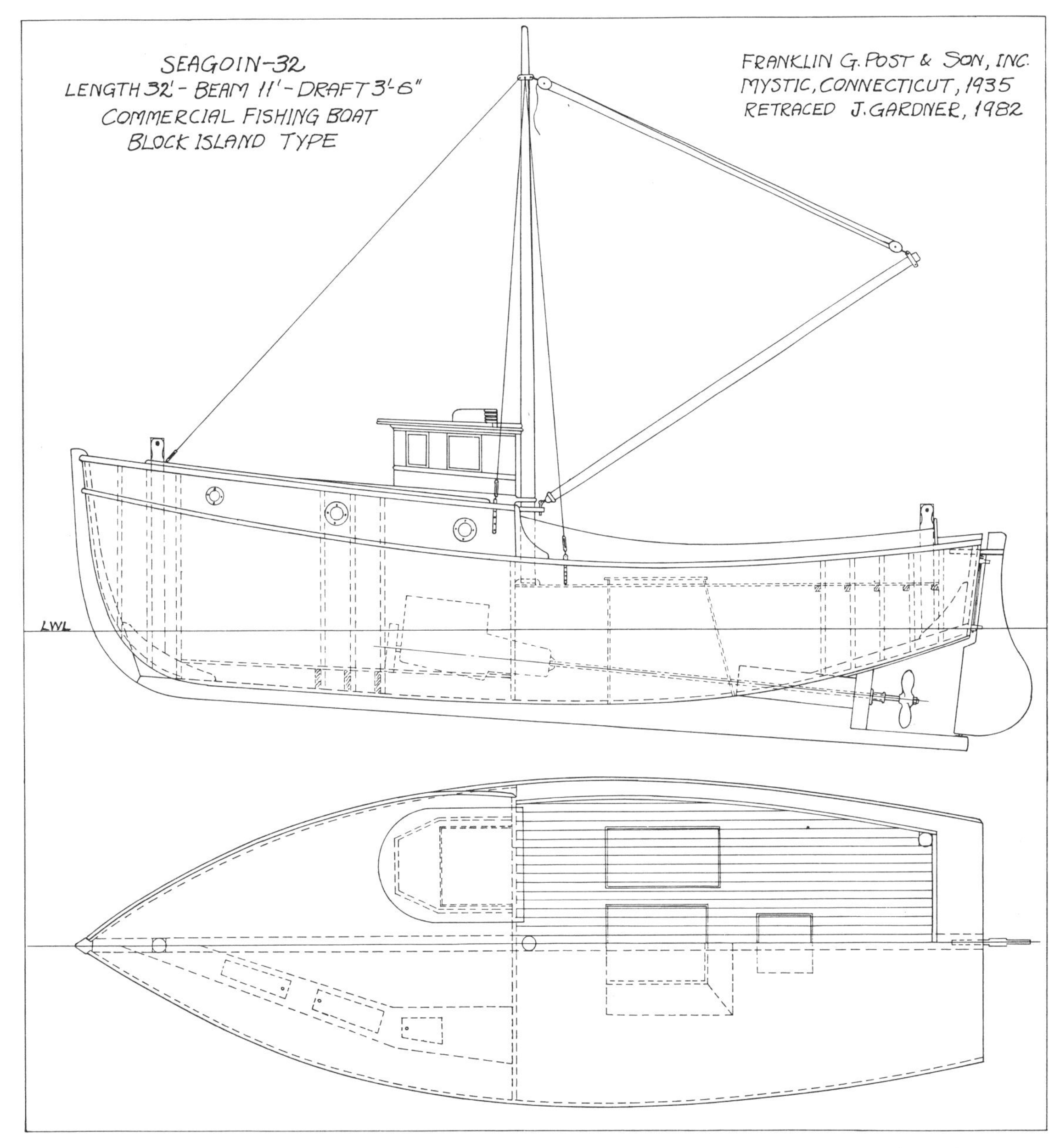

locating the motor under the cockpit floor instead of in the cabin, the boat was easily converted to pleasure use. The designer's description reads in part as follows:

"The Seagoin-32, like the 25, was originally designed as a fishing boat to be operated as such in the waters around Block Island, Point Judith and Montauk Point, where a good, rugged, easy-driving hull which will carry a heavy load and catch fish in the open Atlantic in any kind of weather is of prime importance."

When adapted for pleasure use, the Seagoin-32 slept four, with upper and lower berths, clothes locker and Shipmate stove. A 35-foot model was advertised as an "offshore fisherman-hull type with double cabins and semi-enclosed bridge, a staunch and roomy home afloat with all the seagoing qualities of the Block Island commercial fisherman."

No lines or offsets for the original Seagoin-25 or the 32' enlargement have been preserved, so far as I have been able to determine. Nor does it appear that any of these boats were built after the yard passed out of the hands of the Post family when it was sold in 1959, although a number of these boats previously built were still sound and in service in the early 1980s.

Still, enough information has survived to provide the basis for what I believe to be a reliable reproduction of the original design, the same that is given here. And although it would be too much to expect

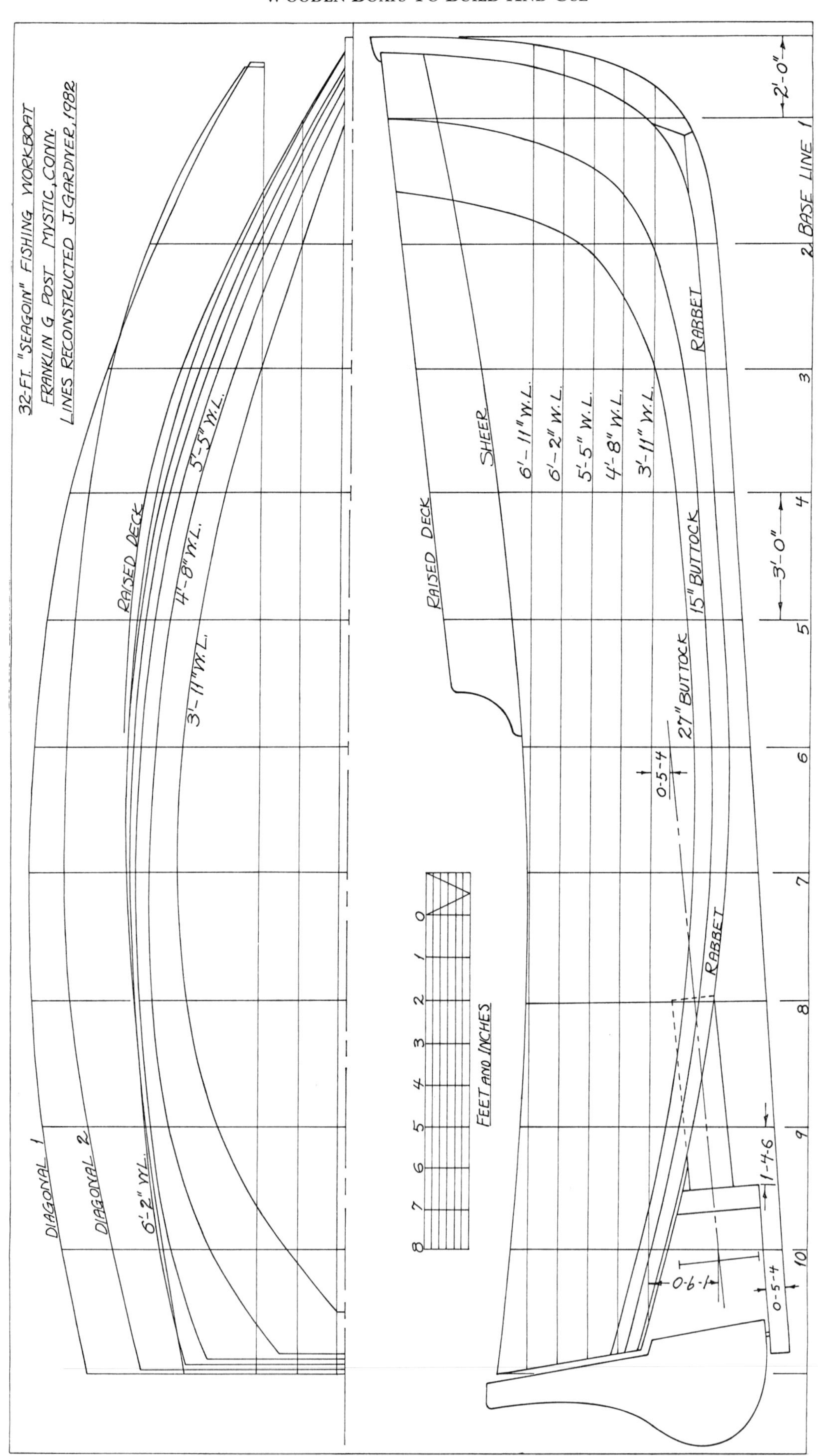
32-FT. "SEAGOIN" FISHING WORKBOAT
FRANKLIN G. POST MYSTIC, CONN.
LINES RECONSTRUCTED J. GARDNER, 1982
RAISED DECK
5'-5" W.L.
4'-8" W.L.
3'-11" W.L.
DIAGONAL 1
DIAGONAL 2
6'-2" W.L.
SHEER
6'-11" W.L.
6'-2" W.L.
5'-5" W.L.
4'-8" W.L.
3'-11" W.L.
RAISED DECK
RABBET
15" BUTTOCK
27" BUTTOCK
RABBET
BASE LINE
2'-0"
3'-0"
0-5-4
1-4-6
0-6-1
0-5-4
FEET AND INCHES

Jen, a Seagoin-32 used for lobstering. Designed with heavy scantlings in the 1930s, some Seagoin-32s were still afloat after more than half a century (Franklin G. Post & Son Boatyard Collection, M.S.M.).

these reconstructed lines to duplicate the original lines exactly, I submit that they come close, that they embody the essential features of the original 32-footer, and that the performance of a hull built from them should not differ significantly from that of the Post-built 32-footers. In a word, I am confident these reconstructed lines for the Post Seagoin-32 are true to type.

The sources drawn upon in making this reconstruction of lines were found in the Franklin G. Post & Son Boatyard Collection now housed at Mystic Seaport Museum, Mystic, Connecticut.

Two items in particular were my most important sources. One was an ink tracing on linen of a profile view of the Seagoin-32 with some construction detail which was drawn to a scale of 1/2"=1' by Ernest F. Post, and dated December 10, 1935. Apparently this drawing was intended to introduce the newly designed model to the boating public, for it is labeled, "Arrangement Plan, Proposed Deep Sea Cruising and Fishing Boat," and was reproduced in reduced size on 8 1/2" x 11" printed sheets for general distribution. This tracing was expertly drawn and is presumably dimensionally accurate. It also carries a specifications list.

The 21-inch model used in the reconstruction of the hull lines for the Seagoin-32 (Franklin G. Post & Son Boatyard Collection, M.S.M., Claire White-Peterson photograph, M.S.M.).

OFFSETS 32' POST "SEAGOIN" FISHING WORKBOAT. J. GARDNER. 1982

	STATIONS	STEM	2	3	4	5	6	7	8	9	10	TRAN.
HALF-BREADTHS	RAISED DECK	0-2-4	2-10-6	4-3-0	5-1-4	5-5-5	5-7-6	—	—	—	—	—
	SHEER	0-2-4	2-9-3	4-0-0	4-10-2	5-3-4	5-6-5	5-6-4	5-3-7	5-0-2	4-7-0	4-0-5
	RABBET	0-2-4	0-2-4	0-2-4	0-2-4	0-2-4	0-2-4	0-2-4	0-2-4	0-2-4	0-2-4	0-2-4
	3'-11" W.L.	0-2-4	1-2-6	2-3-2	3-1-5	3-9-0	4-2-1	4-3-4	4-0-5	3-3-0	1-5-6	
	4'-8" W.L.	0-2-4	1-9-2	2-11-0	3-10-2	4-5-7	4-10-1	4-11-4	4-10-5	4-5-3	3-5-0	1-8-3
	5'-5" W.L.	0-2-4	2-1-4	3-3-5	4-3-0	4-10-3	5-2-0	5-3-4	5-2-3	4-11-0	4-4-3	3-6-0
	6'-2" W.L.	0-2-4	2-4-1	3-6-6	4-5-7	5-1-0	5-4-5	5-5-5	5-3-5	5-0-5	4-8-1	4-0-3
	6'-11" W.L.	0-2-4	2-6-1	3-9-1	4-8-1	5-2-5	5-6-3	5-6-3	5-3-7	5-0-4	4-8-1	4-1-4
HEIGHTS	RAISED DECK	10-8-2	10-2-0	9-9-4	9-4-6	9-0-5	8-8-2	—	—	—	—	—
	SHEER	9-7-2	8-8-6	8-2-2	7-8-5	7-4-5	7-1-5	7-0-0	7-0-0	7-1-3	7-4-3	7-7-5
	RABBET	—	2-10-0	2-7-2	2-4-4	2-1-6	1-11-7	2-0-2	2-4-0	2-10-1	3-6-2	4-2-4
	KEEL	—	2-4-0	STRAIGHT LINE							0-7-2	—
	BUTTOCK 27"	—	5-8-4	3-10-3	3-3-6	3-0-5	2-10-2	2-10-1	3-0-5	3-6-3	4-2-2	4-10-5
	BUTTOCK 15"	—	3-10-7	3-1-6	2-9-7	2-7-0	2-4-6	2-5-0	2-8-0	3-2-1	3-10-0	4-6-4
	DIAGONAL 1	0-3-7	2-7-6	3-9-6	4-8-5	5-3-3	5-6-7	5-8-3	5-7-4	5-4-1	4-9-7	4-2-0
	DIAGONAL 2	0-5-6	2-9-6	3-9-1	4-3-4	4-7-1	4-9-4	4-10-3	4-7-7	4-2-7	3-7-1	2-10-4

LINES TO OUTSIDE OF PLANKING. DIAGONAL 1 UP 7'-6" AND OUT 6' ON THE 4'-8" W.L.
DIAGONAL 2 UP 6'-11." OUT 6' ON BASE LINE. MEASUREMENTS IN FEET, INCHES & EIGHTHS

Judging from the professional caliber of his drawings and design work, it is clear that Ernest Post was qualified as a naval architect. On graduation from college in 1924, he joined his father at the yard and seems to have assumed increased responsibility for design as the business expanded. In time it became one of the largest building and service yards between Boston and New York.

In the early 1930s, Franklin G. Post & Son Inc. was building on the average 40 boats a year, from 25-foot fishing launches to 100-foot pleasure yachts, with an operating force of between 80 and 90 men when working at capacity. The Post yard must have produced several hundred Seagoin boats in all.

The other item that provided basic information was a nicely finished model of what appeared to be the Seagoin-32 when compared with the profile drawing already mentioned. However, it was only 21" long—too small for either a 3/4" or a 1" scale, but too large for 1/2"=1'.

By using a machinist's surface gauge, it was possible to make an accurate lines take-off from this model. The lines so obtained were laid out on detail paper and reduced, using proportional dividers, to correspond to lines for a 32-foot boat drawn 1/2"=1'. A tracing on transparent vellum was made of the profile view. This was then superimposed over the profile view of the Seagoin-32 as drawn by Ernest Post in 1935. The fit was remarkably close. Only a few minor adjustments were required to make the two profile views correspond almost exactly.

The next step was to enlarge the adjusted lines obtained from the model to a scale of 1"=1'. This was done to facilitate fairing and to permit a more accurate scaling of offsets. In fact, very little fairing was found to be necessary. The lines as taken from the model proved surprisingly fair to begin with, an indication that the model had been carefully and accurately made and that it was a reliable representation of the full-size boat.

The list of scantling dimensions as given on the Post tracing follows:

Keel, white oak, 5".
Transom, 2".
Deadwood, 5".
Frames, 1 1/2" x 2".
Topsides planking, yellow pine, 1 1/4".
Bottom planking, cedar, 1 1/4".
Deck frames, white oak, 2" x 2 1/2".
Cockpit floor frames, white oak, 2" x 2 1/2".
Deck planking, fir, 1 1/4".
Cockpit floor planking, fir, 1 1/4".
Cabin floor planking, fir, 7/8".
Mast and boom, spruce.

All fastenings and deck hardware were to be hot-dipped galvanized iron. Power was

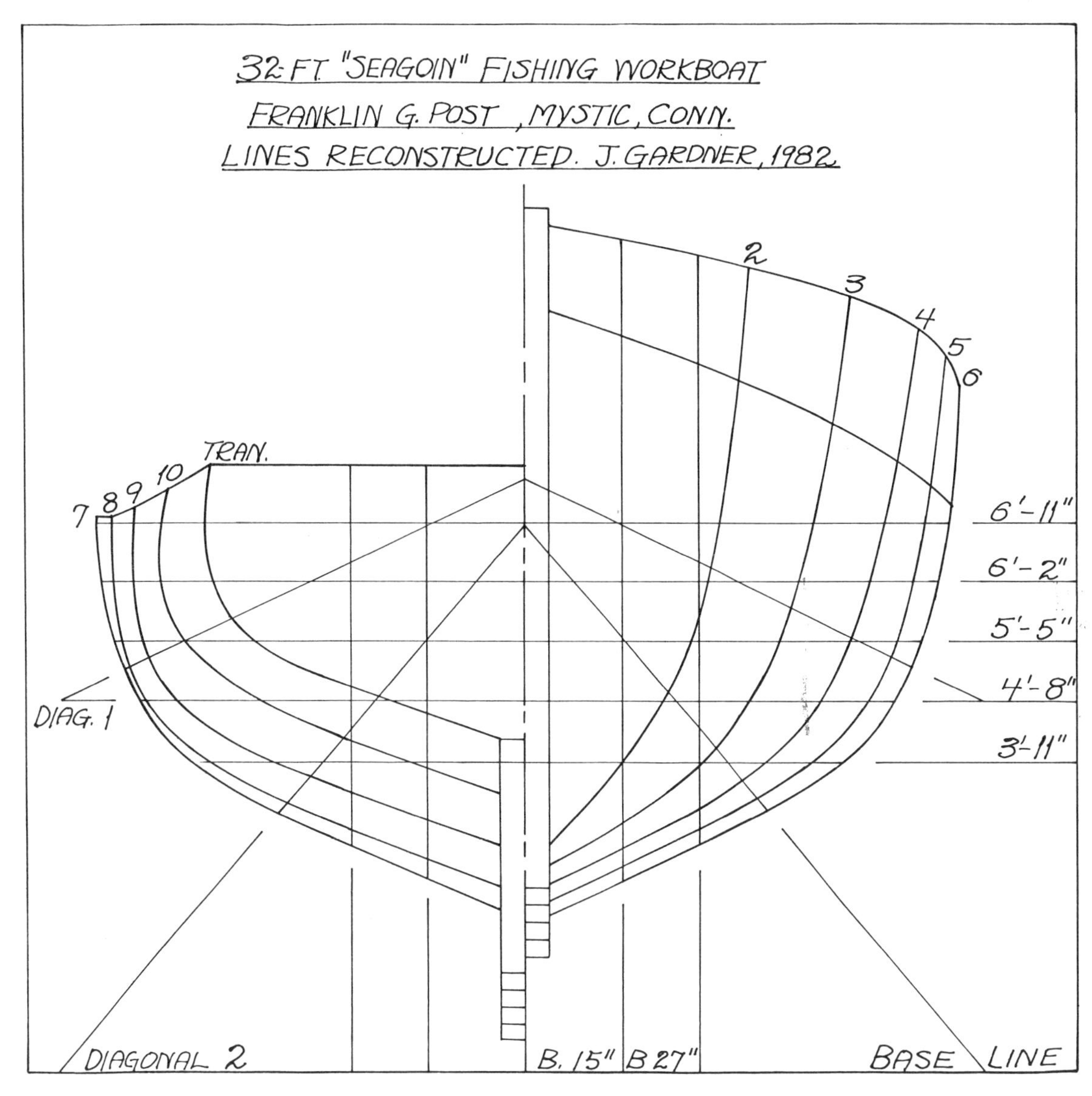

to be a 4-cylinder, 40 h.p., medium-duty motor unless otherwise specified.

The generous size of these scantlings as well as the lines of the boat indicate a heavily constructed displacement hull not suitable, or intended, for high speeds. At the same time, it appears that within its speed range this hull, with its smooth lines, would drive easily and efficiently. And these same smooth, easy hull lines would plank easily, too.

After nearly 50 years, a limited revival of these sturdy, seakindly, low-speed, easily driven hulls would seem to be in the cards. They have much to commend them in some segments of the modern fishing industry where speed is not important but stamina and seakeeping ability are.

In Alaskan waters, for instance, the Seagoin-32 or similar craft would seem well suited to current fishing needs, if reports I am getting are accurate. With only minor changes in scantling sizes, this boat could be built entirely of native lumber, and as already mentioned, construction is straightforward. The heavy scantlings are less critical to fit and fasten.

A boat builder with ordinary skill should be able to put together such a boat more quickly and easily than is possible with much of the glued construction so popular today, which is, of course, quite unsuitable for sturdy, low-speed, displacement craft.

What we see when we examine the construction plan is a sturdily built hull with heavy scantlings. The raised deck forward provides sufficient room below to sleep four with upper and lower berths and clothes locker. There is also a Shipmate

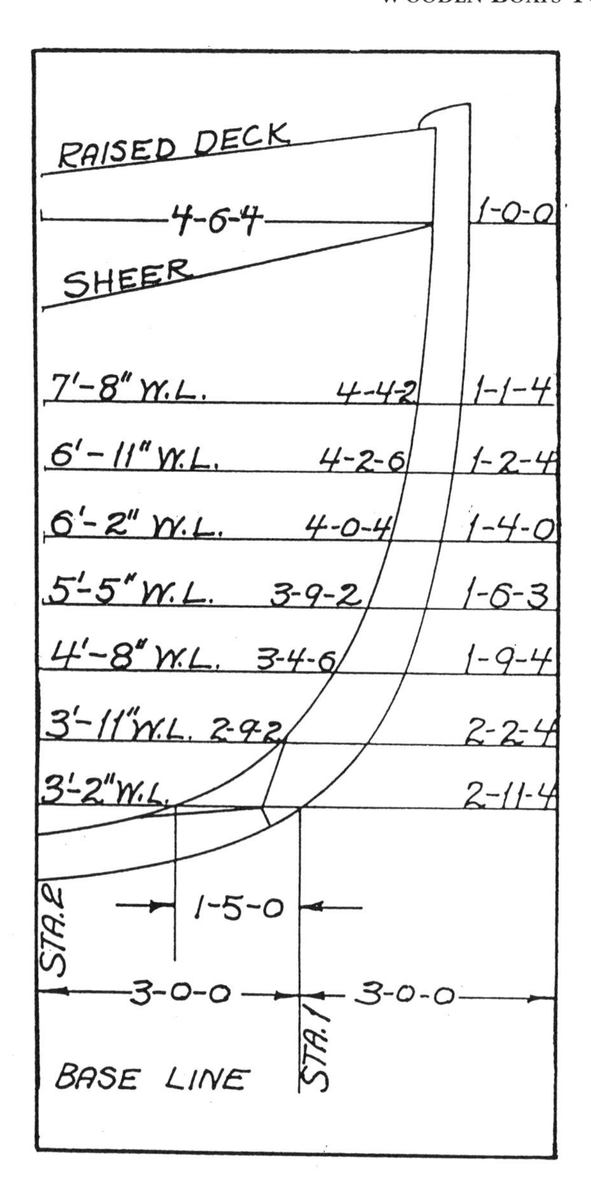

stove. The engine is placed far enough forward to allow for the location of a wet well below the cockpit floor.

Hull construction is straightforward and relatively simple. The lines are such that it would not be a difficult hull to frame and plank. Stem, keel, deadwood, shaft log, stern post, horn timber, transom and outboard rudder are all white oak, which is abundant in southern New England and of good quality. The steam bent frames, floors and deck beams are also of oak. The clamp may well have been of oak, too, but in a slightly larger Post fishing boat the clamp is yellow pine.

While native white cedar is specified for the bottom planking, the topsides call for yellow pine. I assume the reason for this is that yellow, or southern pine is harder than cedar and better able to withstand pounding from traps and other fishing gear, as well as the general hard use that the topsides of a commercial fishing boat must endure. For planking for this boat, my personal preference would be hard pine throughout, but native cedar is just as rot resistant and was probably cheaper locally when this boat was built. However, if the boat were to run aground, the soft cedar bottom would not fare so well.

For decks and cockpit floor, fir was specified. Douglas fir makes excellent decking, provided it is rift-sawn heartwood from old-growth timber, not the sappy, slash-sawn stuff that is widely sold today as construction lumber for floor joists and 2" x 4"s. There is no better decking than northern white pine, and a 35-foot Post commercial fishing boat of the same type and about the same period calls for white pine decking. That was years ago. Native white pine of decking quality is so expensive today that its use on a fishing boat is hardly to be considered.

All fastenings are hot-galvanized iron, but other than that no particulars are given about the fastenings. However, specification sheets for the contemporary 35-foot Post fishing boat, already mentioned, consider fastenings in more detail, and we may assume that fastenings for the 32-foot boat would be much the same.

In that case, keel, stem, deadwood, floors, horn timber, transom and transom knee would have been fastened with 1/2" and 5/8" galvanized rod. Presumably, the 5/8" rod would have been threaded to bolt-up the knee assembly, joining the stem to the keel. Floors would probably have been drifted to the keel with two 1/2" drift bolts, with a 3/4" limber between, as specified.

With the keel sided at 5", there is room for 3" spacing, at least, between these drifts. Half-inch drifts would seemingly have been large enough for deadwood, shaft log, stern post and horn timber, although the 5/8" rod might have been used here, as there was plenty of wood for the larger rod. In fact, with the 5" siding of the shaft log, there is plenty of room for drifts on either side of the shaft hole, which is not often the case. Note that the shaft log is one solid piece, not a split log.

This 26-foot teak-trimmed sport cruiser was the smallest in the Post line of recreational fishing boats, which ranged up to 35 feet. These boats did not deviate greatly from the original commercial hull types (Franklin G. Post & Son Boatyard Collection, M.S.M.).

Steam bent frames in the 35-footer are 2$^{1}/_{4}$" wide and 1$^{1}/_{2}$" thick, spaced 10" on centers, compared with a 2" width and 1$^{1}/_{2}$" thickness spaced on 12" centers in the 32' boat. Floor timbers for the 35-footer are sided 2" and molded 5", and they are located at every frame.

Dimensions for floor timbers are not listed for the 32-footer, but it may be assumed that the two boats do not differ greatly in this respect, although I found when scaling the arrangements plan that their molded depth appears to be greater by a couple of inches at least. This may be because the 35' boat, with an extreme draft of 3', is shallower and flatter in the bottom than the 32' boat, which has steeper deadrise and a draft of 3'6".

Both the deck beams and the cockpit floor beams for the two boats have the same dimensions, with sided thicknesses of 2" and molded depths of 2$^{1}/_{2}$". Other scantling dimensions for the two boats differ only slightly. Decking for the 35-footer is native white pine, 1$^{3}/_{8}$" thick, with the nails bunged. Two bulkheads of 2" fir extend from the keel to the cockpit floor. Fender guards are native oak. Planking and the bilge fender are western oak, finished 1$^{1}/_{4}$" thick and caulked with cotton. Seams are filled with putty and white lead.

Early in the century, so-called western oak, which actually came from east of the Mississippi, was brought in by rail to East Coast yards for planking. The wood was clear, seasoned and reasonably priced. Lengths of up to 60' and 70' were shipped on flatcars to Essex, Massachusetts, where the wood was used to plank Gloucester fishing schooners. Western oak is probably specified here because it was available and attractively priced at the time. Good quality southern hard pine would have served just as well for this, and select rift-sawn Douglas fir almost equally well.

Where timber suitable for steam bent frames is not available locally, heavily built 35-foot and greater fishing boats of this type can be built just as well with sawn frames. When sawn frames are used, there is no need for building molds and forms, which saves considerable time and labor. If care is taken in lofting, planking can be applied directly to the frames themselves as soon as they are set up, simplifying and speeding up the building process appreciably.

It may be said that heavily built, low-speed boats of this type are obsolete today, that they cannot compete costwise with modern types built of plastic or metal, that they are not suited to modern fishing methods and that there are no longer skilled

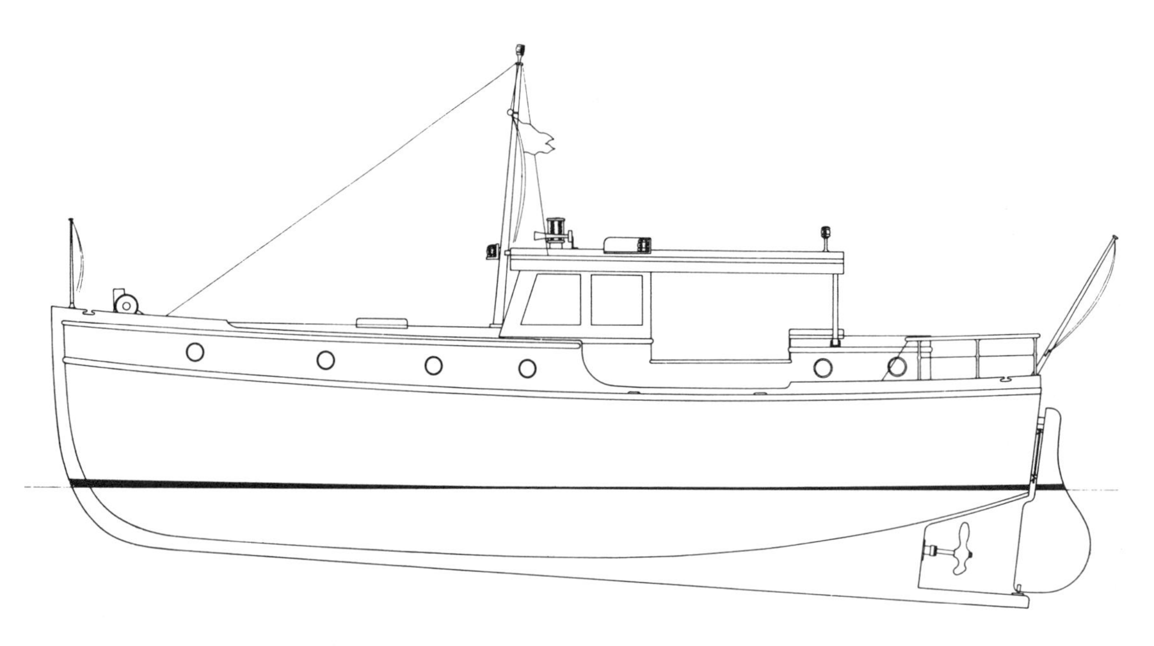

The Post yard turned out a series of Seagoin cruisers built on the same lines as the Block Island commercial fisherman hulls. These ranged from a 26-foot sportfisherman to this 35-foot "off-shore fisherman" type, which features double cabins and a semi-enclosed bridge (Franklin G. Post & Son Boatyard Collection, M.S.M.).

boat builders who can build them. I am not sure that any of these contentions can stand up to examination.

There are definitely places, some far and some near, with fishing conditions and potentials where heavily built seagoing boats like these still have something positive to offer and deserve serious consideration. In regions where sufficient supplies of suitable native lumber are available, these boats would have a definite cost advantage over boats of plastic or metal.

For instance, in parts of New England, the South, the Great Lakes region, the Northwest and, presumably, Alaska, it is possible to buy good-quality boat lumber directly from small local sawmills at very favorable prices. Pine and oak in southern New England, for example, can be had for a reasonable price per board foot, although the cost per foot for seasoned stock from retail lumber dealers may be excessive.

As for lack of boat building skill, this is not quite the barrier that is often supposed. I know a number of self-taught amateurs who are fully qualified to undertake this type of construction. As previously mentioned, the construction of heavily built working craft of this sort is quite simple, straightforward and fast, once the knack has been acquired and the builder is set up for it.

And finally, while it is true that with modern fishing methods there is, for the most part, no use for low-speed, heavily built boats such as these, this would not rule them out completely if fishing methods were to change. There are indications that such changes are already underway.

With energy costs rising and bound to keep on rising, energy-intensive, high-speed fishing methods are less profitable than before. When these costs rise above the break-even point, fishing methods based on energy-efficient, low-speed craft could well come back, at least to a limited extent.

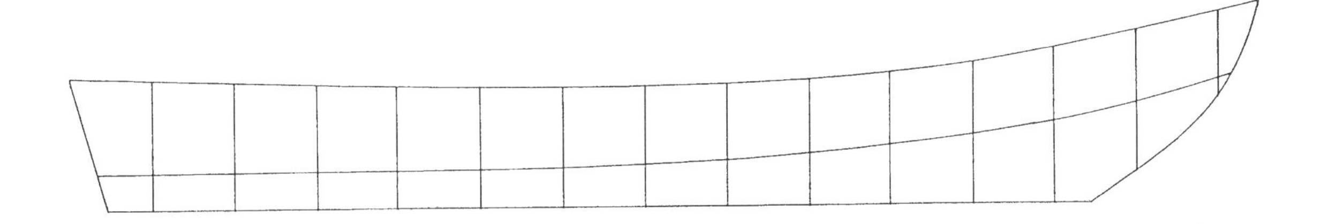

CHAPTER FIFTEEN

22-FOOT OUTBOARD WORK SKIFF

This handsome big work skiff is capable of attaining planing speeds when driven by the powerful outboards that are now available and becoming increasingly popular.

With smaller motors, this boat should do well at the more moderate speeds that are better suited to many fishing operations. How it will perform in the open sea under extreme weather conditions remains to be determined by actual tests, but this handy 22-footer retains enough original dory characteristics to support expectations of a satisfactory weatherly performance. Besides, most fishing operations based on skiffs and other boats in this general size range are not carried on in truly bad weather.

During the 30 or 40 years since they first began to be powered with outboard motors, fishing dories have undergone an extensive and continuing evolution that has finally produced boats such as the 22-foot work skiff under consideration here. Perhaps no innovation has contributed more to this process than the motor well, which brought the outboard motor inboard, where it is more secure and more easily operated.

One of the first of the larger outboard dories was the 19-foot semi-dory I designed for *Outdoor Maine* in 1960 and whose lines were later reprinted in *The Dory Book*. This boat has since been widely built and modified by other builders. The latest of these to come to my attention is from Dick Dunn, Tutka Bay Boats, Red Mountain, Homer, Alaska.

An account of his boats and the modifications he made in his enlarged version of the original 19-footer appeared in the September 1986 issue of the *National Fisherman*. Dunn had high praise for the performance of these dory skiffs in the rough waters of Lower Cook Inlet, Kachemak Bay and the Gulf of Alaska.

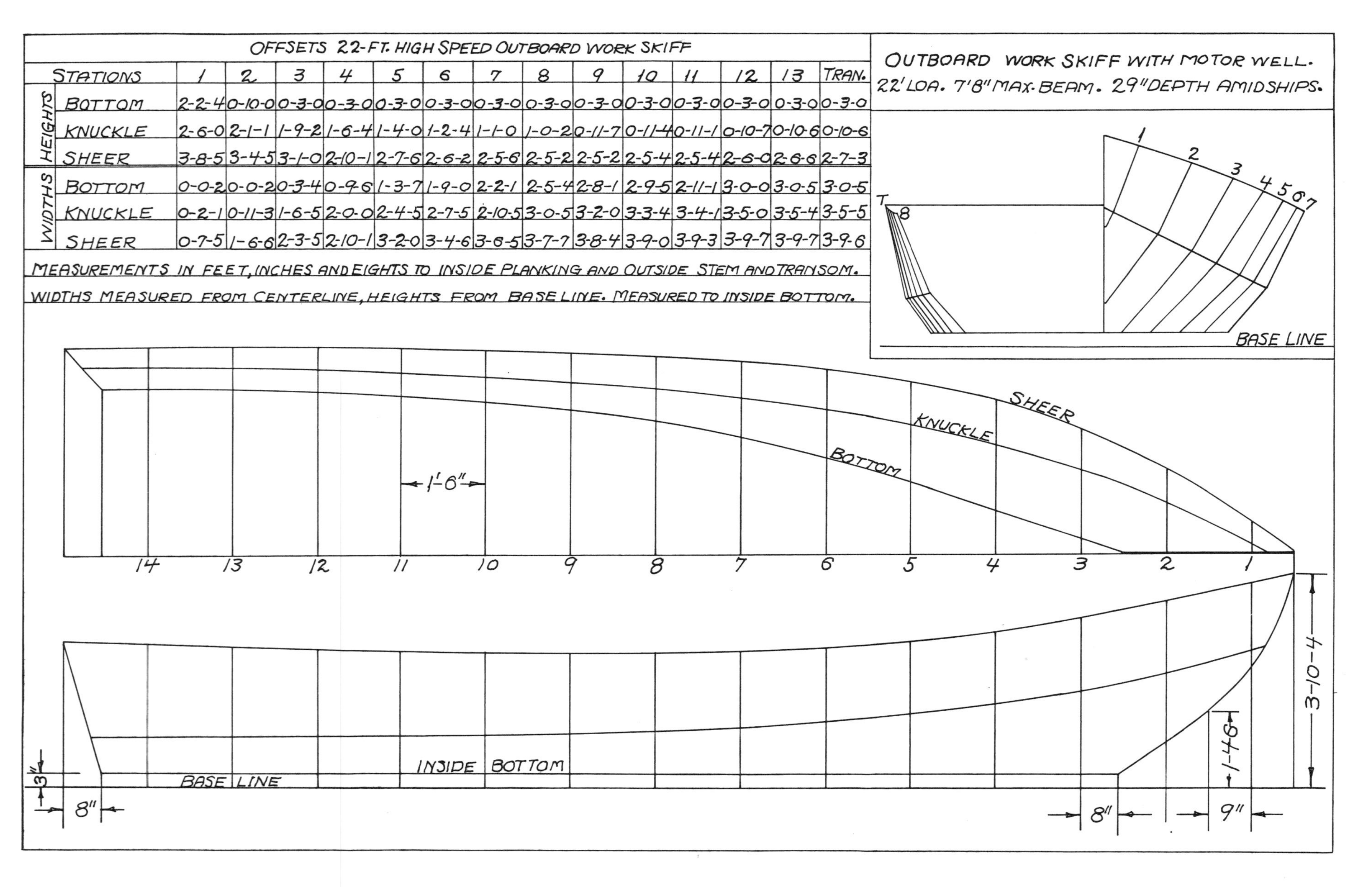

OFFSETS 22-FT. HIGH SPEED OUTBOARD WORK SKIFF

	STATIONS	1	2	3	4	5	6	7	8	9	10	11	12	13	TRAN.
HEIGHTS	BOTTOM	2-2-4	0-10-0	0-3-0	0-3-0	0-3-0	0-3-0	0-3-0	0-3-0	0-3-0	0-3-0	0-3-0	0-3-0	0-3-0	0-3-0
	KNUCKLE	2-6-0	2-1-1	1-9-2	1-6-4	1-4-0	1-2-4	1-1-0	1-0-2	0-11-7	0-11-4	0-11-1	0-10-7	0-10-6	0-10-6
	SHEER	3-8-5	3-4-5	3-1-0	2-10-1	2-7-6	2-6-2	2-5-6	2-5-2	2-5-2	2-5-4	2-5-4	2-6-0	2-6-6	2-7-3
WIDTHS	BOTTOM	0-0-2	0-0-2	0-3-4	0-9-6	1-3-7	1-9-0	2-2-1	2-5-4	2-8-1	2-9-5	2-11-1	3-0-0	3-0-5	3-0-5
	KNUCKLE	0-2-1	0-11-3	1-6-5	2-0-0	2-4-5	2-7-5	2-10-5	3-0-5	3-2-0	3-3-4	3-4-1	3-5-0	3-5-4	3-5-5
	SHEER	0-7-5	1-6-6	2-3-5	2-10-1	3-2-0	3-4-6	3-6-5	3-7-7	3-8-4	3-9-0	3-9-3	3-9-7	3-9-7	3-9-6

MEASUREMENTS IN FEET, INCHES AND EIGHTS TO INSIDE PLANKING AND OUTSIDE STEM AND TRANSOM.

WIDTHS MEASURED FROM CENTERLINE, HEIGHTS FROM BASE LINE. MEASURED TO INSIDE BOTTOM.

OUTBOARD WORK SKIFF WITH MOTOR WELL.
22' LOA. 7'8" MAX. BEAM. 29" DEPTH AMIDSHIPS.

Both the Carolina Dory and the Pacific City Dory, to name two well-known examples of the type, are large outboard skiffs in the 20-foot-plus range whose superior qualities for both pleasure boating and commercial fishing have been demonstrated beyond any question. For years, fishermen and others, from Alaska and Newfoundland to the South Pacific, have been building these large outboard dory skiffs successfully from plans supplied by such outfits as Texas Dory Boat Plans.

But it is probably Tracy O'Brien, Headwater Boats, Chehalis, Washington, who now stands in the forefront of dory skiff evolution. His Headwater V-20, described in the January 1987 issue of *National Fisherman*, still retains aspects of the traditional dory shape, but O'Brien has eliminated almost every element of traditional dory construction with his combination of glued, sewn-seam plywood and glued wood strips.

When I looked over O'Brien's plans for the V-20, I was surprised to find that its hull form was not unlike the shape of a skiff for which I had already made a scale half-model. Moreover, I found that the principal dimensions of the two skiffs were very similar.

Detailed here, that same big outboard skiff is designed to be simply built, inexpensive (as prices go today) and relatively fast. No special boat building knowledge or skills are required. Ordinary woodworking tools will suffice. The necessary materials are readily available almost everywhere and are not high priced in comparison with boat building materials in general.

If the builder has some woodworking experience, and can handle ordinary carpentry, he should have no trouble. This is not to say, however, that the boat can be thrown together; it definitely cannot. Simple as the construction of this skiff may be, building it would be a waste of time and materials without careful and accurate workmanship. In short, a haphazard approach would make the project a headache from start to finish.

The principal material is 3/8" marine plywood, put together with epoxy glue and nails. If some attention is paid to the planning, there need be little or no waste of plywood as the smaller scraps and cutoffs can all be worked into one place or another. There is no need to bevel the frames since these are set normal or "square" with the curve of the bottom and sides, as shown in the accompanying diagram.

It is not necessary to make a preliminary full-size laydown of the lines, as the shape of the bottom and frames can be laid out directly from the table of offsets provided. The perfectly flat bottom is easily assembled and glued. No scarphs have to be cut in piecing together the plywood bottom or in joining the sections of side planking, thus avoiding an operation that is exacting and time-consuming.

No special oversize plywood panels are required; standard 4' x 8' sheets—with a few 8' x 10's for the side planking—will do the job. There are no difficult bends or twists in the side planking. The top edge of the bottom plank is nearly straight except for a slight curve at the forward end. The bottom edge of the top plank is dead straight for its full length.

Access to a small power band saw is highly desirable. If one is unavailable, a Skilsaw and a good saber saw will suffice. One or more panel (hand) saws, 8/10 points, are a must. A power plane would facilitate beveling the bottom but otherwise would not be of any use. We'll talk more about tools, materials, nails, glue and so forth later on, as building operations are explained.

Construction starts with the assembly of the bottom, to which the frames, a stem and a transom are attached. Next, the sides are planked, the side decks framed and laid and the outwales fastened in place. This completes the basic hull, at which point construction moves inboard. Different builders will have their own ideas about interior arrangements that best suit the way they intend to use the boat. The option presented here provides for an open working cockpit that is clear of all structures and obstructions. It measures 8'4" long in the central portion of the boat and is closed off by watertight bulkheads at either end. The motor well is situated in the stern, with good-sized, decked-over storage compartments on either side.

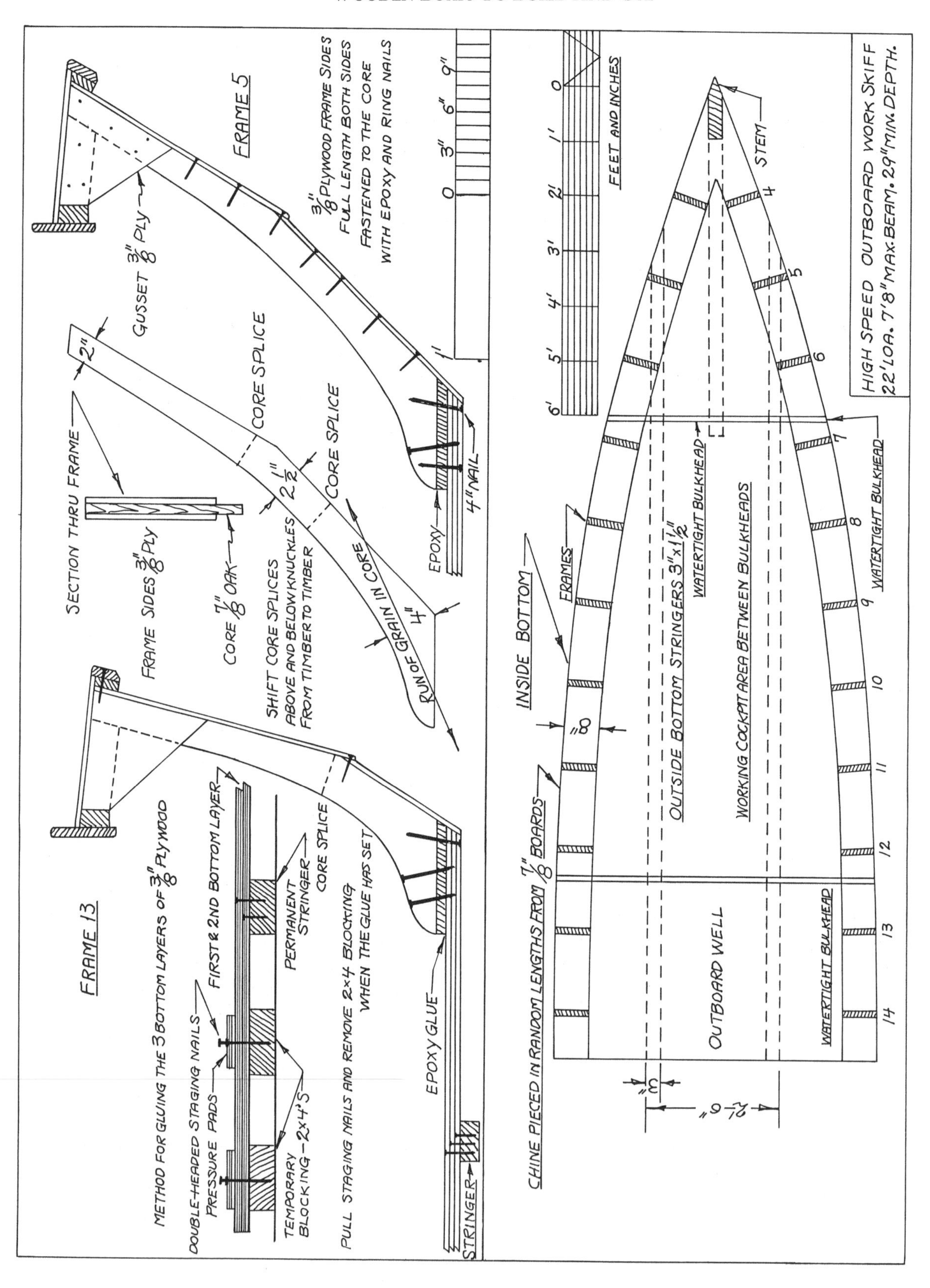
FRAME 5
3/8" PLYWOOD FRAME SIDES FULL LENGTH BOTH SIDES FASTENED TO THE CORE WITH EPOXY AND RING NAILS
GUSSET 3/8" PLY
2"
CORE SPLICE
CORE SPLICE
1 1/2"
SECTION THRU FRAME
FRAME SIDES 3/8" PLY
CORE 7/8" OAK
SHIFT CORE SPLICES ABOVE AND BELOW KNUCKLES FROM TIMBER TO TIMBER
RUN OF GRAIN IN CORE
4"
EPOXY
4" NAIL
FRAME 13
METHOD FOR GLUING THE 3 BOTTOM LAYERS OF 3/8" PLYWOOD
DOUBLE-HEADED STAGING NAILS
PRESSURE PADS
FIRST & 2ND BOTTOM LAYER
PERMANENT STRINGER
CORE SPLICE
TEMPORARY BLOCKING—2×4'S
PULL STAGING NAILS AND REMOVE 2×4 BLOCKING WHEN THE GLUE HAS SET
EPOXY GLUE
STRINGER
0 3" 6" 9" 1'
0 1' 2' 3' 4' 5' 6'
FEET AND INCHES
HIGH SPEED OUTBOARD WORK SKIFF
22' LOA. 7'8" MAX. BEAM. 29" MIN. DEPTH.
STEM
4 5 6 7 8 9 10 11 12 13 14
INSIDE BOTTOM
FRAMES
OUTSIDE BOTTOM STRINGERS 3"×1 1/2"
WATERTIGHT BULKHEAD
WATERTIGHT BULKHEAD
WORKING COCKPIT AREA BETWEEN BULKHEADS
OUTBOARD WELL
WATERTIGHT BULKHEAD
8"
CHINE PIECED IN RANDOM LENGTHS FROM 7/8" BOARDS
3"
2'-6"

Forward of the bulkhead at the bow end is a cuddy with sitting headroom, a 7' berth and a place for a bucket to serve as a basic head. The wheel and the controls are mounted to port against the after side of the cuddy.

The two watertight bulkheads that close off the central working cockpit are important structural members, serving to strengthen and brace the sides of the hull. For that reason, they must be solidly constructed and securely attached both to the bottom of the boat and to the sides. They also support the side decks, which in turn brace the sides and take strain off the frames by acting as trusses.

The builder might well start by making up the frames. These are laid out full size from the offsets on a sheet of plywood serving as a scrive, or lofting, board. Frames are made up with a core of 7/8" oak board to which full-length sides of 3/8" plywood are glued with epoxy. This lamination is further secured by a liberal scattering of 1" ring nails, enough to bring the plywood into firm contact with the core until the glue hardens. In getting out the core pieces, care should be taken to avoid cross grain as much as possible—especially at the foot of the frame, as is shown in the diagram.

To obtain a correct run of grain without too much wastage of lumber, the core may be got out in two pieces, but the location of the core joints should vary so they don't line up from timber to timber. The combination of the nails and the epoxy glue makes a very strong joint, provided the gluing is properly done. The recommended procedure for epoxy gluing will be given special attention later.

In getting out the frames, a sharp power band saw can be a great saver of time and effort. Because they do not require beveling, frames can be sawn out so precisely they hardly need to be touched by any other tool. If oak for the frame cores is unobtainable, Douglas fir, southern hard pine, mahogany or other durable, hard wood would be acceptable substitutes.

Dory construction starts, as a rule, with the bottom. It is one of the most important parts of the dory structure, and among other things, its shape largely determines the form of the hull. Here, the bottom is perfectly flat, with no camber from end to end, which greatly facilitates its assembly. No camber is needed: first, because the bow will lift to some extent when power is applied to bring the boat into planing position and, second, because the forward section is an open V in shape.

This is a work skiff and will be driven hard at high speed, so a solid bottom is an absolute requirement—all the more so because the central portion of the boat's width is unsupported by cross floors. In my judgment, the three layers of 3/8" marine plywood will be adequately strong. The bottom panels are glued and nailed together, stiffened by the outside fore-and-aft stringers, and reinforced by the 8" wide flat chine pieces, which are themselves glued and nailed in place. Besides, the bottom has additional support from the two watertight bulkheads and from cross floors in the motor well and cuddy areas.

In constructing the bottom, the fore-and-aft stringers are first laid down on a flat, level surface or floor. Then the first layer of random-width pieces of 3/8" plywood, cut to slightly more than the required length to allow for later trimming, is glued and nailed to the stringers. The meeting edges of the assembled panels or cross pieces should be coated with glue and brought together with a snug, touching fit.

The second layer of plywood can then be applied. The joints between the pieces in the second layer should overlap those in the first layer as much as possible, which also goes for the joints in the third layer. Indeed, the mated edges in the third layer should not line up with those in the first layer either.

To hold the ends of the plywood cross pieces in close contact until the glue between the layers sets, pressure is supplied by double-headed staging nails that are driven through plywood pressure pads laid on top of the layers being glued. These nails run down underneath the lamination into 2" x 4"s, which are the same thickness as the permanent fore-and-aft bottom stringers.

After the glue between the first and second layers has set, the staging nails are pulled and the third and final layer is put on, with its ends, too, being brought down

Bret Luick of Corvallis, Oregon, built this 22-foot outboard work skiff from plans published in the* National Fisherman *in 1987 (Bret Luick photograph).

tight with the staging nails. In addition to the pressure supplied by temporary nailing, further gluing pressure may be applied with cement blocks or other heavy objects.

Once the bottom is glued, its shape is lined out with a batten set to what the offset table identifies as the bottom widths at the various frame stations. The bottom is then trimmed to the line either with an electric saw or an ordinary handsaw. If an electric saw is used, it is set to the least of the frame bevels taken from the frame layouts—and if the saw is run close to the line, a good part of the work beveling the bottom will be done quickly and easily.

To finish the job, the bottom is turned over and placed on horses or other supports at a convenient height for planing. The bevel angles at the various stations are taken with a carpenter's bevel from the full-size layout of the frames. Finally, the bottom is beveled with a sharp plane to correspond.

When the beveling is completed, the bottom assembly is turned right side up, and the flat chine is put on. This piece is 8" wide and its edge must be made to match the bottom bevels. It can be got out of the same 7/8" boards as the frame coring. Random lengths are all right, and even short pieces will do—though longer ones are better. Joints should be well fitted, and should not fall over those in the bottom. Joining ends are glued when the strips themselves are glued and nailed to the bottom.

The stem, or to be precise, the inner stem, is laminated like the frames, except that the core is made up of three layers (rather than one) of 7/8" oak stock. These, plus an outside facing of 3/8" ply on either side, gives an aggregate thickness of 3 3/8". The stem is beveled for the planking to a width of 1/2" at the outer edge. The bevel angles can be found by running a batten around the frames and clamping it temporarily against the stem, which—after being marked—can be taken to the bench for the beveling operation.

In explaining the construction procedure for this large outboard work skiff, I have now reached the timbering and planking stage. The bottom has been glued together and cut out to shape, and the flat chine has been glued in place and well-nailed. The chine has also been beveled along its outer edges to correspond to the slope of the side frames, as shown in the body plan.

The stem, beveled to receive the planking, and the transom are attached at the two ends of the bottom. This assembly

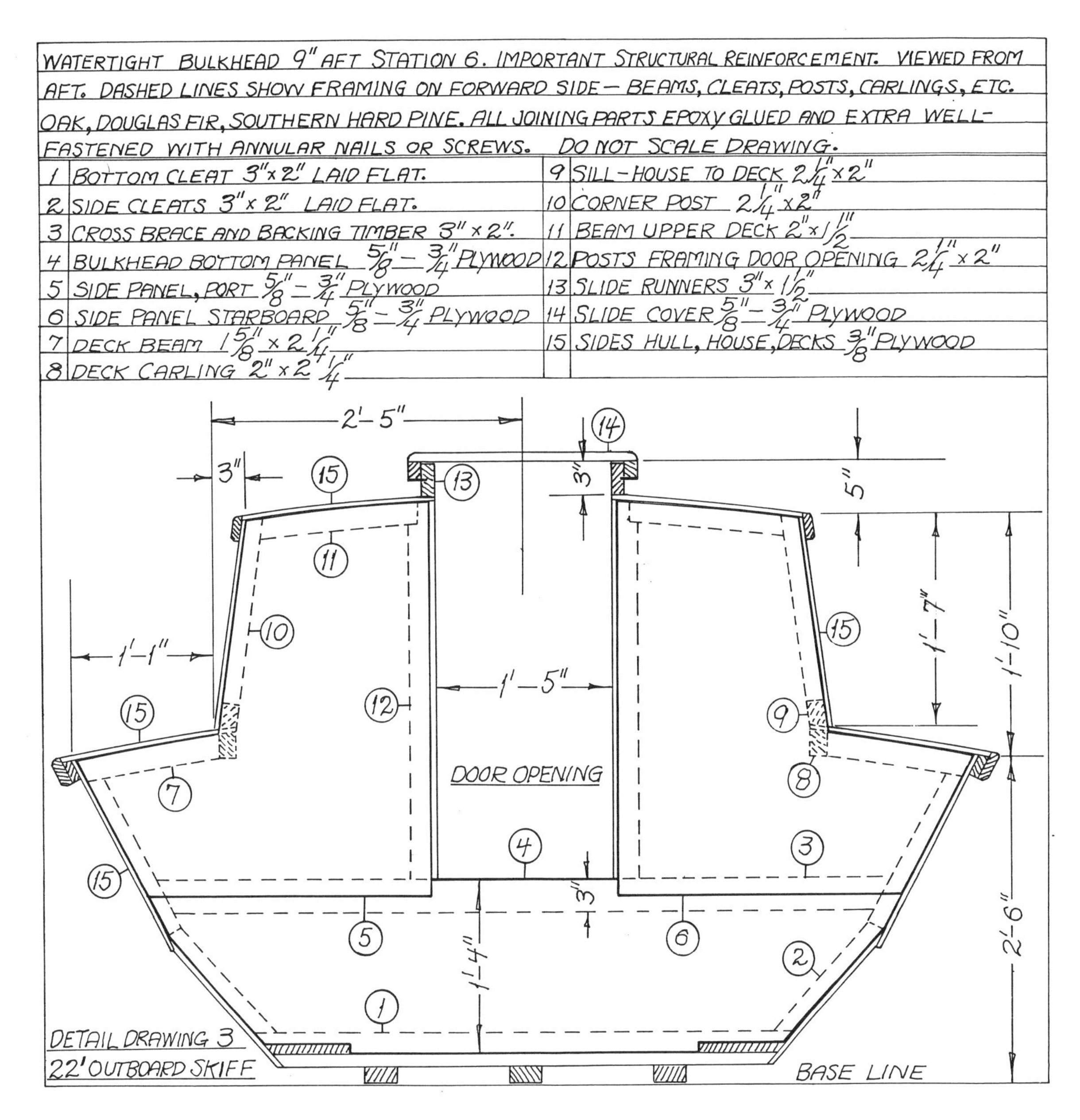

has been raised 18"-24" above the floor and set on horses or blocking to bring it to a comfortable height for working.

Experienced builders, especially if they were building a number of boats in production, might be able to dispense with molds entirely, replacing them with pre-cut frames and applying the planking directly to these. But for a one-off or first-time boat, and particularly in the case of an inexperienced builder, it would be advisable to set up a number of removable molds laid out from measurements given in the table of offsets. Five of these molds should be enough, erected at the even-numbered stations, beginning with No. 2 at the bow.

Made from rough 1" pine or spruce boards, the molds do not need to be beveled. They are set up on their respective stations, braced and temporarily stayed in place. Two ribbands, extending from the stem to the transom, are then bent around the molds and temporarily but securely nailed in place. (Screws with washers under the heads may be needed for fastening the ribbands to the stem because of the bend.)

One ribband should be placed about 2" above the plank knuckle on the side; the other should be set about 2" above the sheer line, the sides of the molds having been made long enough to allow this. The reason for placing the ribbands this high is to provide room underneath for clamps. Made of the same 1" stock as the molds, these ribbands are 2"-3" wide and are free of knots, cross grain or other defects that would weaken them or interfere with a uniform bend. To ensure a fair shape, they should be quite stiff, and they are easily spliced with a butt joint attached to a short piece of backing anywhere aft of Station 6, or where the bend eases up.

When the ribbands are in place, checked for fairness and adjusted as necessary, the permanent side frames are fitted and fastened at the odd-numbered stations. Their lower ends are glued to the flat chine and spiked up through from the bottom. Toward the bow, the frames may require a slight amount of beveling with a sharp block plane before they will lie flat against the inner surface of the ribbands. Temporary staging nails driven through the ribbands into the frames are needed until the side planking is on.

The lower, or garboard, plank is fitted and fastened first. Its upper edge appears to curve in a pronounced upward sweep toward the bow when it is in place on the boat, but the plank is nearly straight over its whole length when it is laid out flat. If this plank, or strake, is made up from pieces cut from standard 4' x 8' plywood panels, two or possibly three splices will be required. If properly made, a strake so spliced is entirely adequate. It is just as good and much less expensive than one cut from a specially made, full-length panel.

Only toward the bow, where the sides show the greatest amount of bend, do planking splices require particular care and attention. Here, tapered splices–accurately scarphed and glued–are a necessity. In the after sections of the boat, where the sides straighten out, the plank sections can simply be butted square and glued to backing pieces of the same 3/8" plywood fitted between adjoining frames. The location of the splices in the two side planks should be staggered, that is, separated by at least two frame spaces.

If the joining ends are accurately scarphed and properly glued, spliced sections of planking will bend in fair curves like uncut sections of plywood. They are theoretically just as strong. Yet for complete assurance that the splices will never let go, they should land on a frame (as shown in Detail Drawing **5**) and be nailed across the splice with a row of small ring nails spaced about 5/8" apart. Where plank splices are square-butted and glued to backing strips, reinforcement with mechanical fastenings such as clinch nails or rivets is not a bad idea.

Three-eighths-inch marine plywood might seem a little on the thin side for planking the sides of this boat, although considerable reinforcement is gained from almost 2½" of gluing surface at the bottom and chine. The assembly gets additional strength from the 1⅛" lap at the side knuckle and from the 2" wide sheer line outwale, to which the sheer plank is glued and nailed from the inside.

Dick Dunn of Tutka Bay Boats in Homer, Alaska, for one, has found 3/8" plywood side planking quite adequate for outboard skiffs in the 20-foot range. One-half-inch thick plywood side planking would be difficult to bend and hold in place on the forward sections of a boat with as much shape at the bow as this model has. The thicker plywood would add appreciably to both the weight of the boat and its cost, although the latter should not be a deciding consideration if the integrity of the structure were at stake.

Besides, in the working cockpit area between the two watertight bulkheads there will be sheathing on the inside of the frames to protect the inner face of the planking. This will also stiffen the sides of the hull appreciably.

The 1⅛" plank lap at the side knuckle is glued and may be either clinch nailed or riveted between the frames. The laps are solidly nailed to the frames with 2" ring nails of no smaller than 12-gauge wire. At the sheer, the inner strip of the laminated, two-part outwale is glued on flush with the edge of the plank. It is nailed into the timber heads and, between the frames, is well fastened from the inside of the planking with 1" ring nails. Nailing from this vantage point, before the glue has set, ensures good contact between the planking and the outwale for a strong glue joint.

Later, when the plywood decking is put on, it is brought out flush with the outside of this strip, which affords good nailing for the outer edge of the deck panel.

And when the outside strip of the outwale is put in place, it is brought up over the outside edge of the decking to cover and protect it. This laminated two-part outwale is simple and easy to put on. Stiff and strong when glued and nailed in place, it completes an assembly that is entirely tight and quite weatherproof.

After the sides are planked, the molds are removed one at a time and the frames, planed as necessary for a snug fit against the inside of the planking, are glued and nailed in place. When this operation has been completed, the upper ribband is removed and the frame extensions are sawn off flush with the sheer line. Little work remains to be done on the outside of the hull, and construction now moves to the interior.

The two watertight bulkheads are a good place to begin. These are important structural members that add greatly to the strength and stiffness of the hull. The bulkhead located between Stations 6 and 7 is especially significant in that it also seals off the forward portion of the hull, confining rain and shipped water to the central cockpit area.

The framing on the forward side of this bulkhead is laid out to brace the sides, and it also allows the plywood facing of the bulkhead to be installed in three sections for easy fitting and economy of materials. Well-glued and nailed, the result, for all practical requirements, is the equivalent of one continuous plywood panel.

Forward of the bulkhead is a cuddy of good size, considering the dimensions of this skiff. Access is through a door near the center of the boat. The steering wheel and controls are mounted on the outer face of the bulkhead on the port side. No provision has been made to seat the helmsman, although a folding seat hinged from a point under the side deck could easily be installed so as to swing back out of the way when not in use. Nor is there any arrangement for seating in the cockpit area, although such could easily be added if the boat were to be used for taking out fishing parties or the like.

Inside the cuddy there is standing headroom when the sliding hatch is open; sitting headroom when it is closed. The platform allows for a 7' berth of ample width, and under its front there is room for a cedar bucket to serve as an improvised head, if there were a need for such. With the cuddy door closed, complete privacy would be assured.

The cross planking of the platform is tightly fitted against the frames port and starboard, and rests on a riser nailed to the frames. One or more lengths of sheathing bear on it from above, and these, too, are nailed to the frames. As a result of this construction system, the platform itself serves to brace the sides of the hull and augments the bracing support of the bulkhead. Also providing structural strength are the heavy, flat bottom; the platform, which sits 13" above the bottom; and the well-fastened, single-piece plywood deck panel at the sheer. Because of the combined contribution of these three elements, the forward end of the skiff will easily withstand the shock of any seas it may be driven into at full speed.

The construction of the side decks, the cuddy, the sliding hatch cover and so forth is more or less routine, offering no special problems. In fact, the greater part of the procedure has already been covered in the accompanying drawings and hardly needs much additional explanation. These drawings and diagrams are intended for close and searching study, and they offer more than can be had from hasty and casual scanning.

Most of what remains to be done is the layout of the outboard well and the watertight storage compartments that lie on either side of it, abaft the watertight stern bulkhead and forward of the transom. Also warranting consideration is the diagram of a so-called dry well for those who might want to mount the motor on the transom. For some, the consequent enlargement of the working cockpit might be a compelling consideration.

Detail Drawing 4 shows a number of referenced structural members. Of these, Numbers **(1)** to **(15)** are identified in the key for Detail Drawing 3. The remaining components are as follows:

(16) Inner layer of the laminated, two-part outwale; 2 5/8" x 7/8"; spruce, southern hard pine, Douglas fir or red oak; special preparation for gluing red oak required;

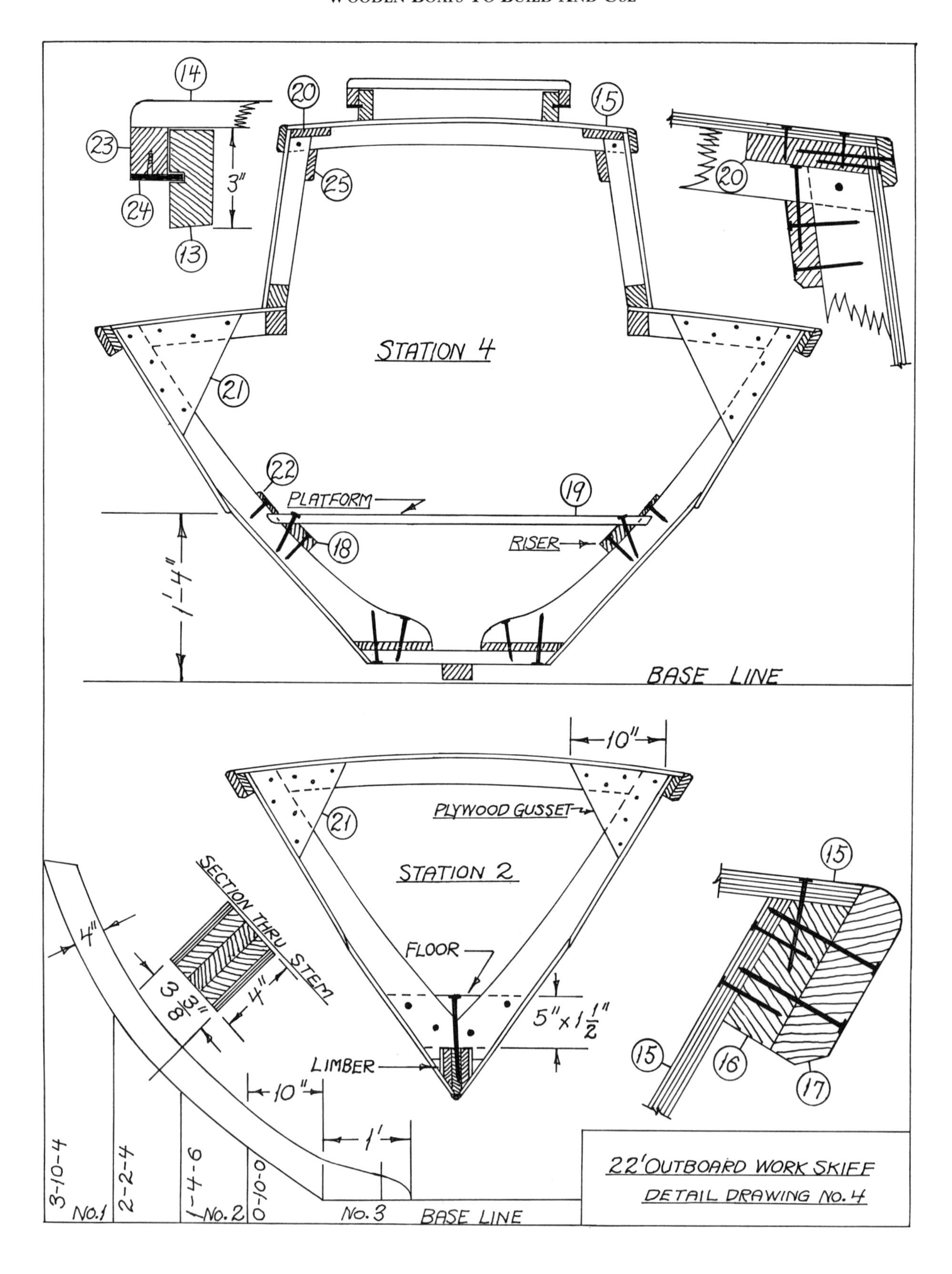

14
20
15
23
3"
25
24
13
20
STATION 4
21
22
PLATFORM
19
18
RISER
1'-4"
BASE LINE
10"
PLYWOOD GUSSET
21
STATION 2
SECTION THRU STEM
15
4"
3 3/8"
4"
FLOOR
5" x 1 1/2"
LIMBER
15
16
17
10"
1'
3-10-4
2-2-4
1-4-6
0-10-0
No.1
No.2
No.3
BASE LINE
22' OUTBOARD WORK SKIFF
DETAIL DRAWING No. 4

white pine too soft; white oak does not glue well.

(17) Outer layer of the laminated outwale; same material as **(16)**; red oak favored because of its wearing quality; nailed as shown with ring nails.

(18) Platform riser; 4" x 7/8"; pine, fir or spruce; glued as well as nailed to the frames.

(19) Platform cross planking; pine, spruce, or Douglas fir; 7/8" thick or thicker; random-width boards, the wider, the stiffer; glued and well-nailed to risers and frames.

(20) Shelf running length of cuddy to provide support and nailing for outside edge of deck and sides of house; Douglas fir, spruce, hard pine or red oak; 4" x 7/8".

(21) Gusset supports for deck beams; 3/8" plywood; approximately 10" on a side at Station 2, 14" at Station 4; glued and nailed to deck beams and side frames.

(22) Sheathing strips; run length of platform; pine, spruce or Douglas fir; 1/2" x 3"; fitted snug to platform; glued and nailed; additional sheathing strips optional.

(23) Slide rails; 3/4" and 7/8"; glued and nailed to slide cover.

(24) Flat brass strip 1" x 3/16"; runs length of slide rails.

(25) Deck beam riser; 3/4" x 3"; glued and nailed to side uprights; supports deck beams, which are nailed to it.

We will now consider the boat's interior, which divides into three distinct and separate areas, or sections.

The arrangement and layout of one of these, the enclosed cuddy forward, was described in some detail earlier. This compartment is large enough to provide an ample 7' berth with sitting headroom. When desired, the cuddy can be closed off for complete privacy and use of an improvised head. This could be convenient for an extended excursion of one sort or another with a mixed party aboard.

When the skiff is put to work as a fishing boat, having a compartment like this forward of a watertight bulkhead provides dry and convenient storage for a considerable amount of assorted gear. This equipment can then be locked up for as much security from pilferers as is possible on a boat of this kind.

The central portion of the skiff offers nearly 10 feet of unobstructed working space. In order to provide as much elbow room as possible, there is no wheelhouse or enclosed structure for the controls. Instead, the wheel is mounted at a convenient standing height on the port side of the forward bulkhead, which does, in fact, give some protection to the helmsman.

A hinged seat that would fold back out of the way under the side deck when not in use could easily be added, and if the interior arrangement shown in Drawing No. 6 as Option 2 were adopted, there would be seating for the helmsman on the port side bench that runs to the forward bulkhead.

It is assumed that the helmsman will be protected in bad weather by his foul weather gear, carried conveniently at hand in the forward compartment. Here, it was a question of which was to receive first consideration: working space or operating comfort. I gave precedence to the former, not being much impressed with some of the smaller working craft we see now with more wheelhouse than anything else. However, if a wheelhouse were desired, one could easily be added.

Drawing No. 6 shows two possible options for the arrangement of the working cockpit in the central portion of the boat. With Option 1, the cockpit is shown clear of all obstructions, except for a temporary cross thwart that can be put up should seating be desired. Ordinarily, this thwart is taken down and lashed out of the way under the port side decking. When it is in place for use as a thwart, it also provides some bracing support for the sides of the boat, although this is hardly needed.

Both the interior sheathing **(26)** and the 12" wide side decks greatly reinforce and stiffen the sides of the skiff. The side decks, in turn, gain rigidity from the plywood gussets **(21)** that support them, as well as from the full-length carlins **(8)** that run from the forward bulkhead back to the stern transom.

For its part, the interior sheathing is made of full-length 3" wide strips of fir, spruce or hard pine. This stock measures 3/4" thick and is fastened to the frames with $2^1/_2$" No. 10 or No. 11 ring nails. Several removable gratings of convenient width for handling (see Drawing No. 6) cover and protect the bottom.

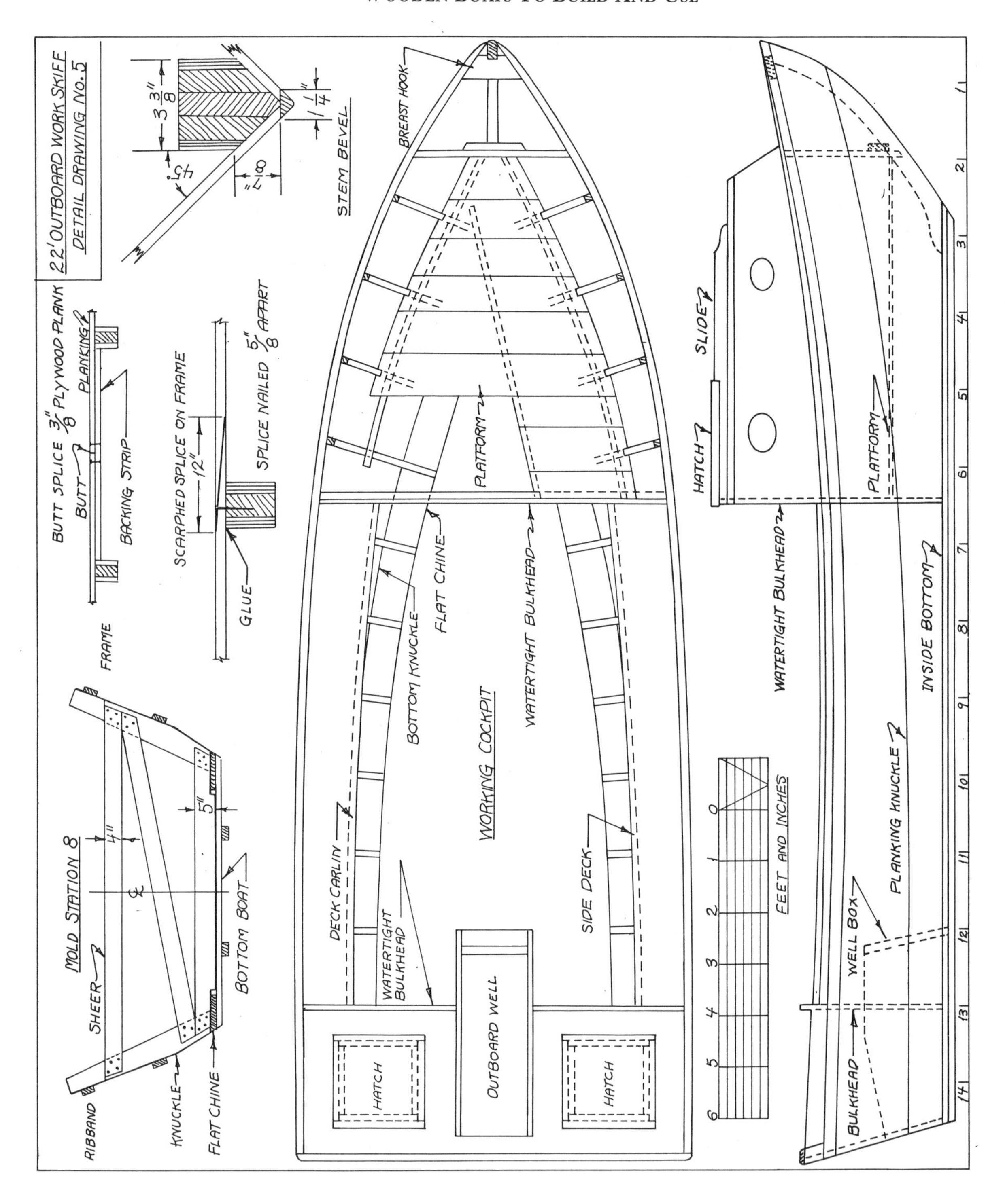

They are made, as shown, from 7/8" fir, spruce or pine boards that run athwartships and are notched to fit the ends of the side frames, which hold them in place. These gratings are easily removed when the bottom needs to be washed out. Although it is not shown, a simple pump of some sort is necessary to clear water from the open cockpit. An old-fashioned, manually operated pump will do nicely.

If the boat is to be used principally for transporting passengers and as a party boat, Option 2 for the interior arrangement is worth consideration. Full-length seating is permanently built-in around the sides, and the side decks have been cut back to a width of 6" to give adequate seating room. Even with these side benches in place, there is still considerable unobstructed space in the central portion of the cockpit. And, though

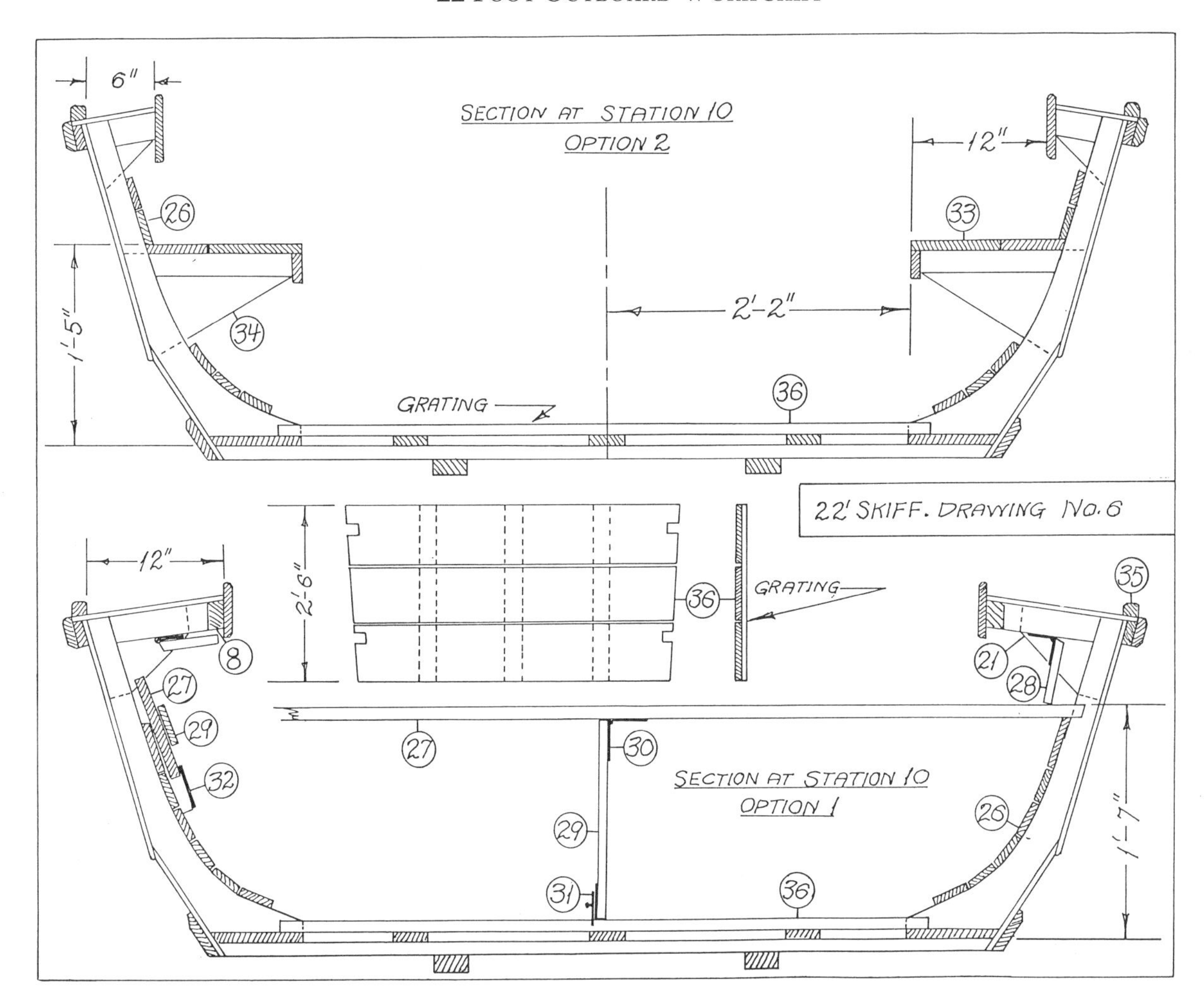

the additional strength isn't really necessary, the benches also greatly reinforce and stiffen the sides of the boat.

The motor well, located in the stern, is flanked by two watertight storage compartments fronted by tight-fitting plywood panels 3/4" thick. These extend from the sides of the boat to the edges of the motor well, combining with the well enclosure to form what amounts to a continuous watertight bulkhead. Running across the boat at frame station No. 13, it measures 20" high (18$^{7}/_{8}$" above the inside of the bottom). Tight decks of 3/4"-1" plywood sit level with the top of the motor well and cap the storage compartment.

These covers are solidly supported by, and fastened to, the sides of the boat, the stern transom and the top of the well. The effect of this boxed-in enclosure at the stern is to provide immense strength and rigid reinforcement for the motor well, as is required by the powerful motors for which this boat is intended. There is also the secondary bonus of ample, easily accessible, watertight storage for tools, line, fuel or whatever may be desired.

Provision has been made to ensure that the motor well is big enough and rugged enough for large outboard motors of 100 h.p. or more. Measuring 20" deep inside, its opening through the bottom and transom is 16" wide, while the transom aperture is 14" high. The sides of the motor well are cut from plywood, either 1" or 3/4" thick. They are fastened (epoxy glue and 3" ring nails) to 2" x 3" sills laid flat on the bottom in glue and through-bolted with 5/16" carriage bolts.

The after end of the well serves as the motor mount and is inclined at an angle of 76 degrees. It is made up by gluing together two thicknesses of 1" plywood and is glued and bolted to a 2" x 3" cross sill **(40)**. This piece, in turn, is glued and bolted to the bottom and set against the mount's after side. Upright 2" x 3" end posts **(39)** on either side provide additional bracing for the motor mount.

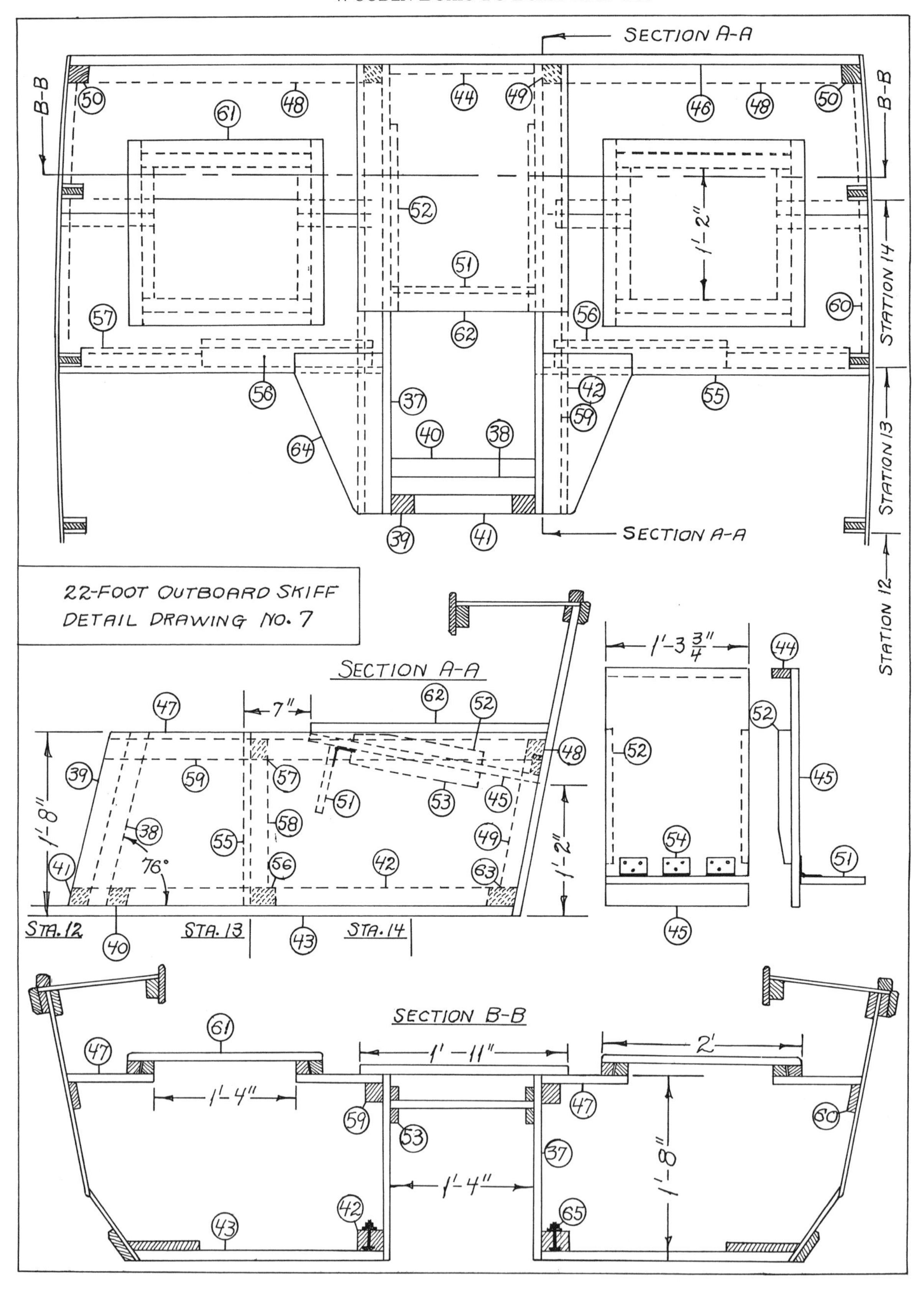
SECTION A-A
B-B
B-B
STATION 14
STATION 13
STATION 12
1'-2"
22-FOOT OUTBOARD SKIFF
DETAIL DRAWING NO. 7
SECTION A-A
7"
1'-8"
76°
1'-2"
STA. 12
STA. 13
STA. 14
1'-3 3/4"
SECTION B-B
1'-4"
1'-11"
2'
1'-4"
1'-8"

Bret Luick added a roomy cabin and 4 inches to the topsides to better handle heavy seas while trolling and leaning outboard to land large salmon over the side (Bret Luick photograph).

Hinged to the underside of the removable internal well cover is a baffle offering protection against water that could otherwise surge through the stern aperture and spill over into the boat. In addition, there is an external well cover, also easily removed, whose principal function is to reinforce the well structure by tying together the decks of the storage compartments on either side of the well.

This outside cover is fastened down with screws running through the edges of the side decks and into the 2" x 2" full-length, fore-and-aft stringers **(59)** that support the decks and attach them to the sides of the motor well. The outside cover can thus be easily removed to get at the inside cover, which slides into place and is supported by cleats **(53)** permanently fastened on the inside of the motor well.

When positioned, the inside cover **(45)** is held in place by a couple of removable screws driven into the well sides through cleats **(52)** attached to the top of the inside cover. An end cleat **(44)** permanently fastened to the inside cover fits snugly against the inside of the transom and lies flush with the upper edge of the well aperture through the transom. It is secured in place by easily removable screws running into the transom.

The inside well cover must be precisely fitted to the well aperture. It is critical that there be no lip or other projection below the opening of the well aperture to impede the smooth flow of water. Nor can there be any exposed transom surface above the aperture to catch and hold water. This might produce a build-up that could overflow the opening at the forward end of the well and spill over into the boat.

One reason for making the well covers removable is to permit fitting and an adjustment of the baffle to suit the outboard motor used. Different motors may require baffles of different lengths, placed in slightly different locations. Baffles have to be hinged to hold back surges into the motor well, and yet they must also move freely in the other direction, corresponding to the flow of the slipstream.

For ultimate strength, the motor well and the storage compartment on either side might be made of 1" marine plywood. Yet, aside from the stern transom and the double-layer motor mount, for which 1" thick plywood is none too heavy, 3/4" marine plywood would no doubt serve quite adequately, with a savings both in weight and in the cost of the material

Timber for framing—sills, posts,

stringers and the rest—should be thoroughly seasoned wood of a species that is suitable for gluing and is hard enough to hold nails and screws well. White pine and cedar are too soft. Good quality Douglas fir, if it can be obtained, would be excellent. (This last recommendation, however, definitely does not apply to the inferior, sappy fir now widely sold for house construction.) Good quality southern hard pine—longleaf yellow pine—or red oak would also be quite satisfactory.

Seasoned red oak will last adequately well for this use, and it is easier to work and fasten than the harder white oak. The latter is prone to twisting and warping, it is more difficult to obtain and it is more expensive. The notion has got abroad recently that white oak is the only kind to consider for boat work. Although it is by far the best for bent timbers, deck beams, the construction of large vessels and various special uses, red oak, from my experience, is the wood of choice here.

Plywood, of course, should be nothing less than a marine grade. It need not, however, be one of the more expensive, imported varieties. First-quality native marine fir would be quite adequate.

Glue should be used, if possible, wherever wood surfaces are joined, even if metal fastenings, nails, screws and/or bolts are also employed. The combined strength of metal fastenings and glue is much greater than the individual strengths of either when used alone. Ultimate durability and integrity in boat work are achieved by a judicious combination of metal fastenings with a good formulation of epoxy glue.

There are many varieties of epoxy glue to choose from. Two that I have used with excellent results many times are Chem-Tech T-88 and System Three. So far as I know, neither of these formulations offers any hazard to the health, although some other epoxy formulations do.

I prefer an epoxy that does not set up too fast, is not of thin, watery consistency, is easy to proportion and mix, hardens at relatively low temperatures and does not require precisely fitted joints or excessive clamping pressure. Both Chem-Tech and System Three meet these requirements. For builders inexperienced in the use of epoxy additives, Chem-Tech has prepared a concise manual that covers the basics, and it is well worth perusal.

Most of the metal fastenings for this boat can be nails, with annular, or "ring," nails preferred. As already specified, a few bolts will be needed for the sills at the bottom of the motor well sides. Only a small number of screws is required for the removable covers of the motor well, for hinges and possibly for a few other places.

Galvanized steel could be used, but in view of the slight extra cost for bronze (slight, that is, in comparison with the total cost of the other materials used for this boat), I would definitely go for the bronze. Perhaps some would want to consider stainless. In my estimation, this is more chancy, and the cost differential is relatively insignificant.

The parts numbered on the drawings are as follows:

(26) Sheathing, inside cockpit; Douglas fir, spruce or pine; 3" wide, 3/4" thick; nailed without glue.

(27) Removable thwart; Douglas fir, spruce or pine; 9" x 1"; could be 7/8".

(28) Hinged shores to hold down thwart; work well when wedged in place.

(29) Hinged leg to support thwart; 4" x 7/8" (or 3/4").

(30) Hinge.

(31) Sliding bolt to secure supporting leg in position.

(32) Support to hold thwart out of the way under port side deck when not in use.

(33) Side benches; 3/4" or 7/8" pine boards, random widths.

(34) Plywood gussets supporting side benches; glued and nailed to frames.

(35) Toe rails; 1 1/2" x 1 1/4" red oak; can be bedded and put down with screws for easy removal.

(36) Gratings; Douglas fir or pine; 3/4" or 7/8"; nailed to cleats the same thickness as flat chines on which ends of gratings rest; notched around ends of side frames; made up in removable sections.

(37) Sides of motor well; 3/4" or 1" marine plyood.

(38) Forward end of motor well or motor mount; two thicknesses of 1" plywood glued together; inclined 76 degrees.

(39) End posts bracing motor mount;

Douglas fir, hard pine or red oak; glued and nailed to (37) and (38).

(40) Horizontal cross brace and sill for (38); located aft of inside well; glued and bolted with 5/16" bolts to bottom; (38) spiked to it.

(41) Horizontal bottom cross brace; located forward of (38) and between end posts (39) that extend to bottom of boat; glued and bolted to bottom; spiled into (38).

(42) Bottom side sills for motor well; extend full length of well; 2" x 3" Douglas fir, hard pine or red oak; put down flat with glue; fastened through bottom with 5/16" bolts; sills serve to anchor nails through well sides.

(43) Bottom of skiff; 1 1/8" thick; three layers of 3/8" marine plyood, glued together.

(44) Strip attaching internal well cover to transom.

(45) Internal well cover; removable; 3/4" plywood.

(46) Transom; 1" marine plywood.

(47) Tight end decks either side of motor well; made up in two pieces for economical use of plywood; 3/4" plywood heavy enough.

(48) Cross cleats; glued and nailed to transom either side of motor well to support decks; 2" x 2".

(49) End posts; glued and nailed to the inside of the transom at either side of the well opening, to receive the ends of the well sides which are glued and nailed to them; 2" x 2".

(50) End posts; glued and nailed to inside of transom to reinforce attachment of side planking; 2" x 2"; beveled as required.

(51) Baffle; hinged to internal well cover; downward reach must be adjusted to suit motor.

(52) Strips for attaching internal well cover to sides of well with screws for easy removal; these strips permanently attach well cover.

(53) Cleats supporting internal well cover; permanently fastened on either side of motor well.

(54) Hinges attaching baffle to internal well cover; must be extra strong.

(55) Watertight bulkheads; extend across the boat on either side of motor well and form front sides of storage compartments; 3/4" marine plywood; could be 1".

(56) Sill backing bottom watertight bulkheads; 2" x 3"; put down on the flat with glue and nailed; does not require bolts.

(57) Cross pieces to back watertight bulkheads at top and support storage compartment decking; 2" x 2"; glued and nailed.

(58) Posts; 2" x 2"; reinforcement at juncture of bulkheads and sides of well box; glued and nailed.

(59) Fore-and-aft stringers supporting joint between compartment decks and sides of motor well; 2" x 2"; glued and nailed.

(60) Filler pieces between frames; serve to link compartment decks and sides of boat; glued and nailed; watertight.

(61) Hatches; dimensioned as shown; covers, 3/4" or 1" marine plywood; frames, 1 1/2" x 2"; Douglas fir or red oak, laid on the flat; glued and nailed.

(62) Removable outer well cover; 1" marine plywood; put down with 3" long No. 14 wood screws.

(63) Cross sills; support for seam between transom and bottom on either side of motor well; 2" x 3"; put down flat with glue and 5/16" bolts; transom spiked to sills.

(64) Carriage bolts; 5/16"; used to secure all sills to bottom of boat.

Bret Luick of Corvallis, Oregon, built the 22-foot outboard skiff shown on page 195 in five weeks in June, 1988, for less than $2,000. This was Luick's first boat building project. He modified the plans I published in 1987 in the February, March and May issues of the *National Fisherman*, adding a large, comfortable cabin, and 4" to the topsides. Luick used AC grade plywood throughout and his principal power tools were a Skilsaw, a 1/2" variable speed drill and a disk grinder. Finished in time for the salmon fishing season, Luick used his new boat for commercial hand trolling in southeast Alaska. Running a 4-cycle trolling motor alongside a 45 h.p. outboard, the *Zephyr* cruised at 14 knots, and fuel economy was 5 nautical miles per gallon. Luick found her very steady in heavy seas. "The broad bottom and stern seems to carry it like a duck over the water," he wrote, adding that *Zephyr* "was uniformly admired on the trolling grounds and a well-received addition to the dory trolling fleet."

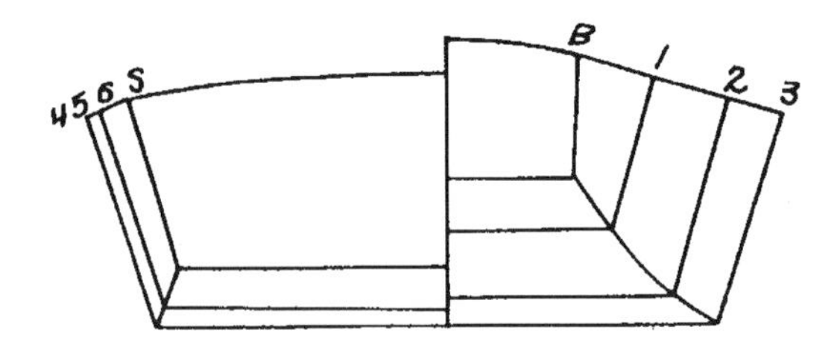

CHAPTER SIXTEEN

18-FOOT 10-INCH GARVEY

The garvey is one of the easiest boats to build. In its simplest form, it is hardly more than a rough, open box crudely proportioned and nailed together.

But for some uses, such boats are all that is needed, and they give satisfactory and long-lasting service. Great carrying capacity, shallow draft (even when loaded), and exceptional initial stability combine to make the garvey an able, roomy transport, as well as a very steady work platform in sheltered and semiprotected waters.

In their traditional form, garveys are not suited for the open ocean. With their wide, flat bottoms, they are more readily capsized by powerful waves than are round-hulled craft that roll with the movement of the sea. Nor is the basic garvey hull well adapted for high speeds, since the wide, flat bottom pounds badly when pushed into a chop. (Nonetheless, the garvey's classic lines can be modified, as has been attempted here, so that its performance in this respect should be appreciably improved.) When operated at moderate working speeds, horsepower requirements are modest and fuel consumption relatively low.

In the design that is offered, I believe I have retained the desirable characteristics of the type while improving seakeeping ability and performance in a chop, as compared with the more boxy working garvies. Pounding will not be eliminated entirely, but it should be reduced. The narrowed bow is raised well above the water at the termination of an easy, gradual incline—an adaption from the Norwegian pram, a proven deep-water type.

In modeling the hull, special consideration was given to a requirement for efficient and economical powering with what today is generally regarded as a rather small outboard motor—something in the range of 10-15 h.p. No doubt this hull would accept larger motors, with corresponding increases in speed and fuel consumption, but that is not being suggested here.

Special provision has been made for another feature of this design, one that should commend it to those who might be interested in a smaller boat. The shape of this 19-footer is such that it can be readily reduced in size simply by removing the desired amount amidships and without altering the ends whatsoever. At the same time, the beam can be reduced merely by moving the sides toward the centerline and revising the offsets accordingly. However, no attempt should be made to go below 16' in length overall without completely reworking the design.

Materials required are relatively inexpensive and obtainable virtually everywhere. In fact, the boat can be framed almost entirely with ordinary 1" softwood boards (net thickness 3/4", planed both sides) as are stocked by most retail lumber yards. This is ample thickness except for cross-planking the bottom and for the fore-and-aft bottom stringers, where 7/8" or even 15/16" stock would be much better.

Northern white pine, southern pine or Douglas fir can be used, but avoid Ponderosa pine, which does not last when exposed to the weather. Plywood 3/8" thick is heavy enough for the outside planking. Three 8' sheets are sufficient for the sides and ends, with some left over. Inside, the motor well and frame gussets will require one sheet of 1/2" plywood.

Of course, marine-grade plywood is far superior to ordinary exterior grade, but it is much more expensive. If a passable grade of exterior can be found, one without large interior voids or thin and patched surface veneers, it might be made to do. But I doubt such stock can be located and one should think twice before using it.

No steam bending is called for. The very little beveling required is indicated on the drawings. Also there are very few curved shapes. If need be, everything can be done with ordinary carpenter's hand tools. It goes without saying they should be sharp. A power table saw for ripping the 1" boards would, of course, expedite the job immensely. A hand-held circular saw would be a less desirable substitute here.

Epoxy glue is indispensable for use throughout the assembly. This boat is a laminated structure that depends upon a powerful adhesive to hold it together. I recommend Chem-Tech's T-88, which I have used for many years with excellent results. Although care should be taken to keep it off the skin, this brand is among the epoxies that don't cause irritation or other problems for most people.

T-88 makes a comparatively thick mix that I like to stiffen still more by adding a thickening agent widely sold under the name of "Cab-o-Sil," which will not weaken the glue like microballoons or sawdust. The result is a loose paste that is easy to apply and stays put. Thin, watery epoxy mixes tend to leak out of the joints, often resulting in a weak, starved bond. In addition, runny glue is messy and gets over everything.

One attractive feature of the construction procedure suggested here is its simple, inexpensive use of laminated softwood lumber. Uniform strips of ordinary stock are glued and nailed together for the internal framing, including the double-thickness chines and the bow and stern transom structures. In this way, much of the construction is reduced to cross-cutting straight strips of uniform width to the requisite lengths, and then gluing and nailing them in place.

As implied, nails can be used for fastenings throughout; there is no need whatsoever for wood screws. Nailing is easier, faster, and less expensive than using screws—and every bit as good for this particular job. Hot-dipped galvanized nails (not electroplated nails, which are worthless) are good for about ten years. Bronze annular (ring) nails will last longer and are somewhat more expensive, but not significantly so. I would go with the bronze ring nails. Nails perform two important functions. They reinforce the glued laminations, and they hold the layers of wood in firm contact while the glue is hardening.

The angle of the sides with the bottom is 107 degrees throughout, as measured on the scale half-model. Because this angle remains the same over the entire length of the boat, neither the plywood side panels nor the chine stringers have to be twisted to

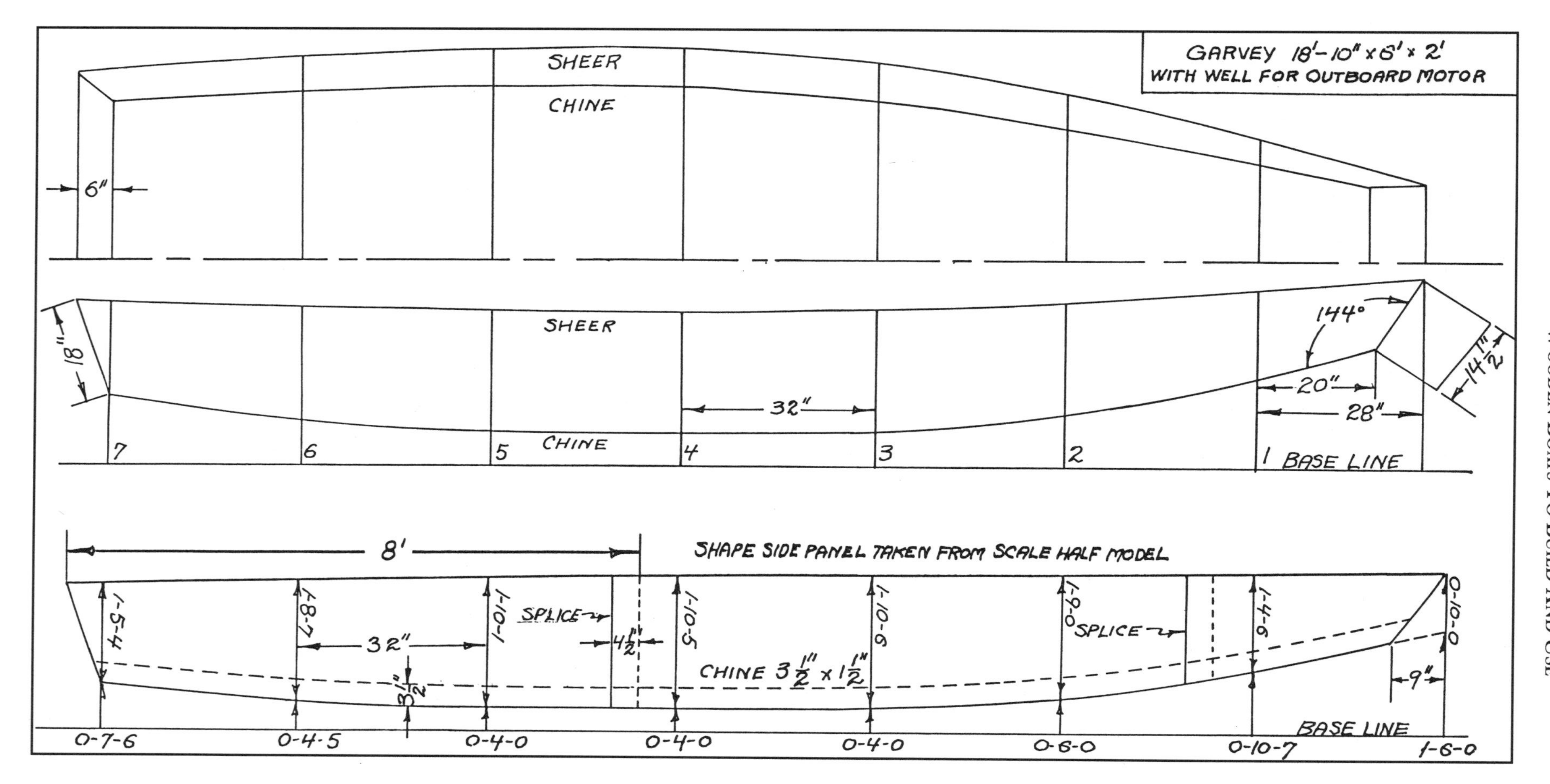
GARVEY 18'-10" x 6' x 2'
WITH WELL FOR OUTBOARD MOTOR
SHEER
CHINE
6"
SHEER
18"
144°
20"
28"
14 1/2"
32"
CHINE
7
6
5
4
3
2
1 BASE LINE
8'
SHAPE SIDE PANEL TAKEN FROM SCALE HALF MODEL
1-5-4
1-8-7
32"
3 1/2"
1-10-1
SPLICE
4 1/2"
1-10-5
1-10-6
CHINE 3 1/2" x 1 1/2"
1-9-0
SPLICE
1-4-6
0-10-0
9"
BASE LINE
0-7-6
0-4-5
0-4-0
0-4-0
0-4-0
0-6-0
0-10-7
1-6-0

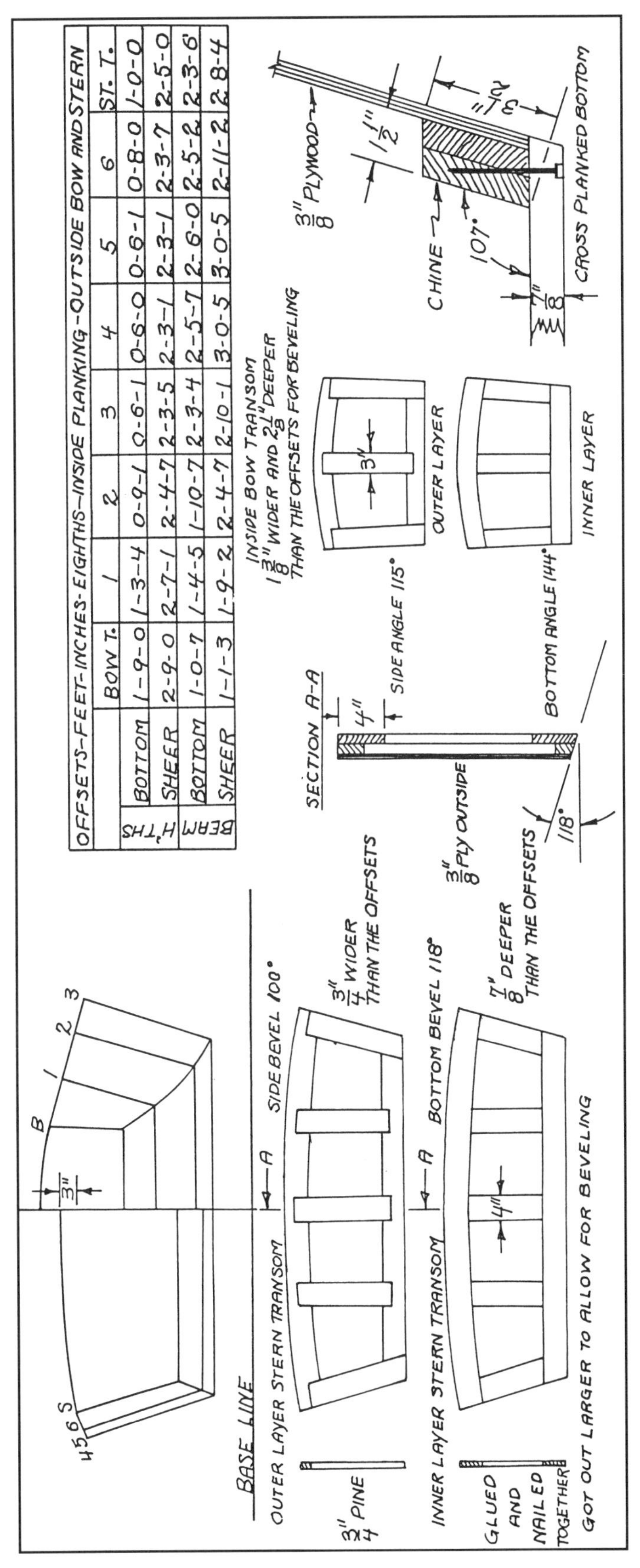

OFFSETS–FEET-INCHES-EIGHTHS–INSIDE PLANKING–OUTSIDE BOW AND STERN		BOW T.	1	2	3	4	5	6	ST. T.
H'THS	BOTTOM	1-9-0	1-3-4	0-9-1	0-6-1	0-6-0	0-6-1	0-8-0	1-0-0
	SHEER	2-9-0	2-7-1	2-4-7	2-3-5	2-3-1	2-3-1	2-3-7	2-5-0
BEAM	BOTTOM	1-0-7	1-4-5	1-10-7	2-3-4	2-5-7	2-6-0	2-5-2	2-3-6
	SHEER	1-1-3	1-9-2	2-4-7	2-10-1	3-0-5	3-0-5	2-11-2	2-8-4

the slightest extent as they are fastened in place. And since the curve of the sides is relatively slight, both the side panels and the chine stringers will go on very easily.

There is no need for steaming the chine stringers, as might have been the case had there been a lot of bend and twist. As mentioned, steaming is not required anywhere in the construction of this boat, which greatly simplifies the building procedure and makes it go much faster and more easily.

The order of construction should be apparent from the drawings. The boat is started upside down on a temporary building form made up of molds for Stations 2, 4 and 6, with the transom frames in place at either end. The molds are supported at the right height by posts; these are set up on level cross-planks, to which they are nailed.

Construction of the boat proceeds on this temporary form until the plywood sides and ends are on; the cross-planked bottom is fitted, nailed, and caulked; and the two outside bottom stringers are nailed in place over the inside fore-and-aft bottom stringers. (At this point, the boat is turned over, the temporary form is removed and the construction is completed right side up.)

Structurally, a great deal depends on having chine stringers solid enough to ensure an adequate nailing surface for the cross-planked bottom. Specifically, they must be sufficiently large to take a continuous row of closely spaced 3" nails without splitting.

The chine stringers which are recommended for this job are made up from two layers of softwood glued together with epoxy and nailed with 1 1/4" nails. The 3/4" thick boards normally carried by most retail lumber yards give a total chine thickness of 1 1/2", which is quite ample for secure nailing of the bottom. Each of the two halves of the chine stringer is got out in

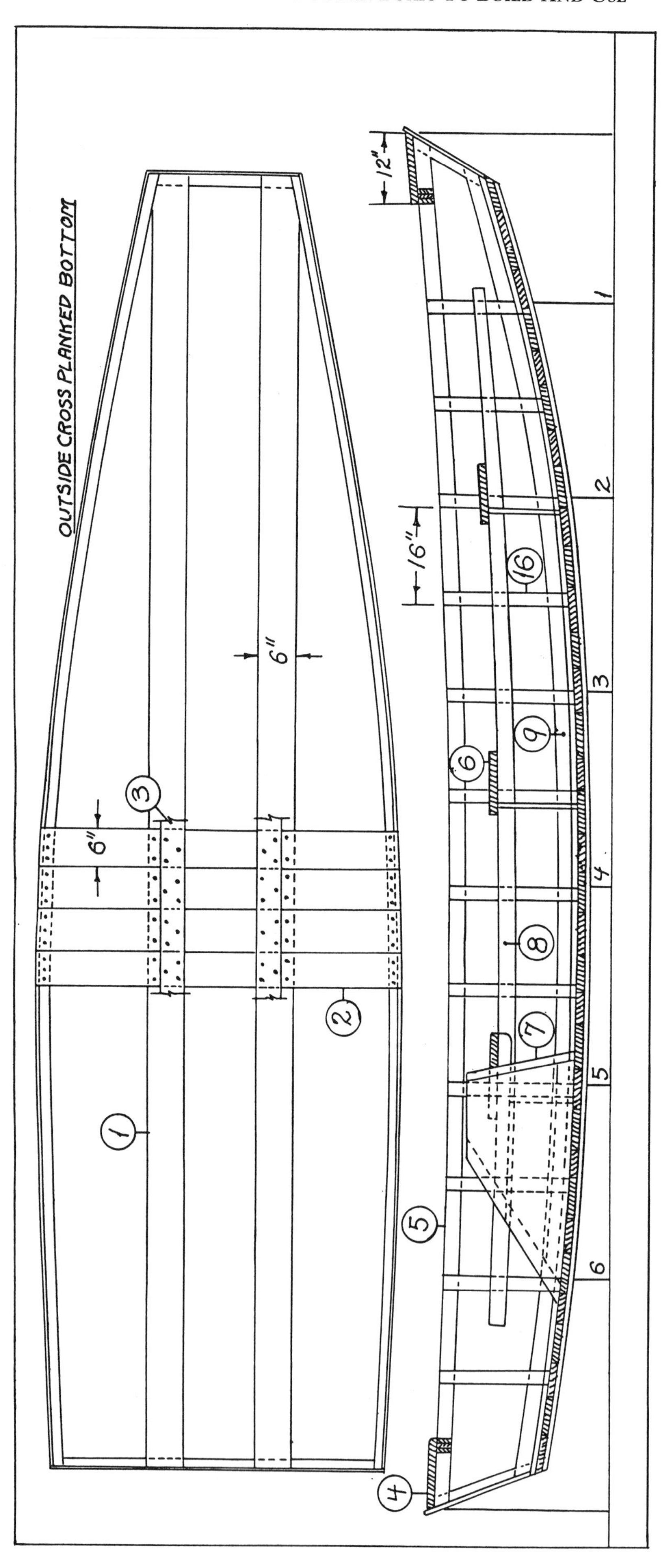
OUTSIDE CROSS PLANKED BOTTOM
6"
6"
12"
16"

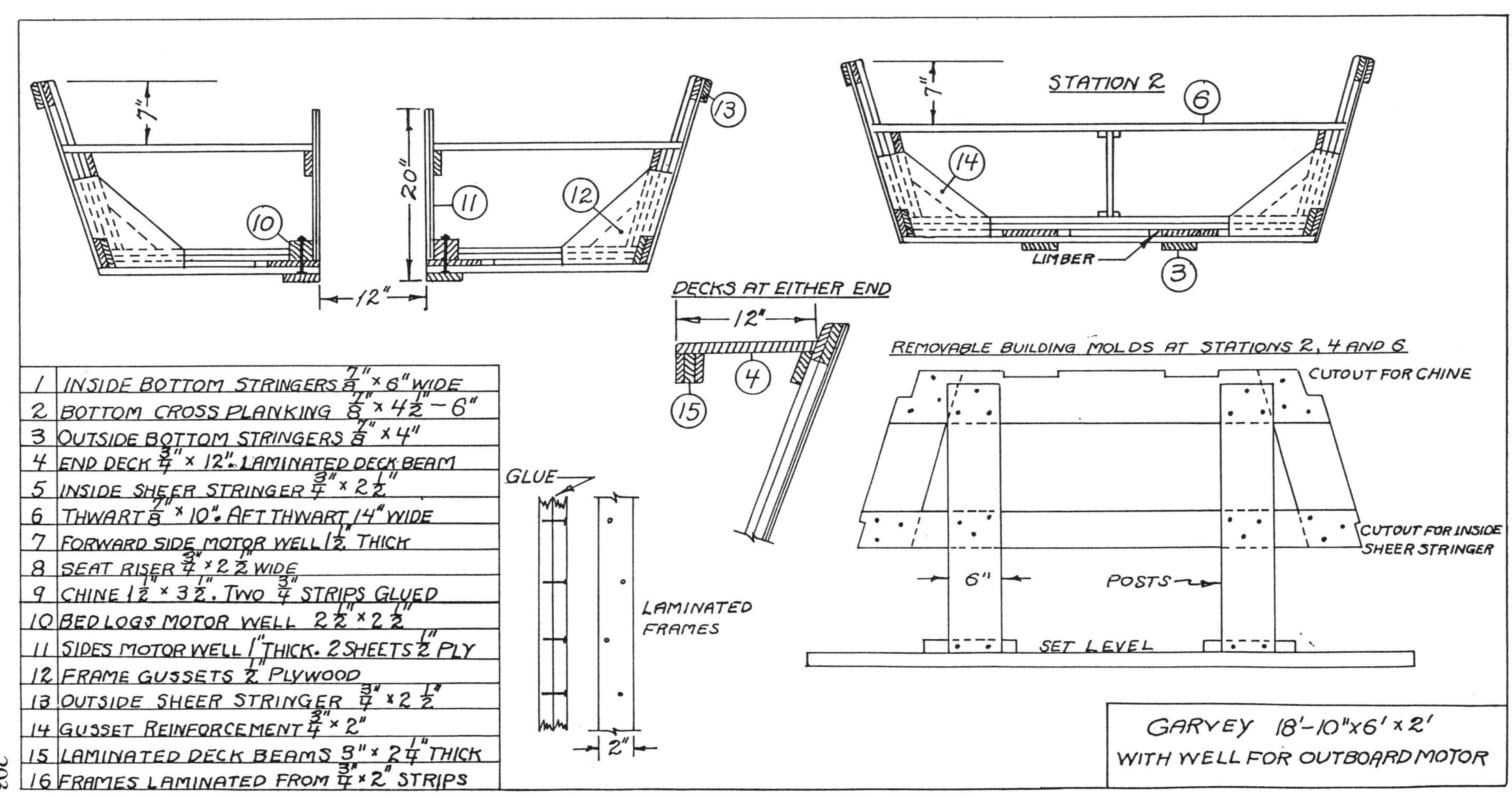
7"
13
20"
11
12
10
12"
STATION 2
6
7"
14
LIMBER
3
DECKS AT EITHER END
12"
4
15
REMOVABLE BUILDING MOLDS AT STATIONS 2, 4 AND 6
CUTOUT FOR CHINE
CUTOUT FOR INSIDE SHEER STRINGER
6"
POSTS
SET LEVEL
1 INSIDE BOTTOM STRINGERS 7/8" x 6" WIDE
2 BOTTOM CROSS PLANKING 7/8" x 4 1/2" - 6"
3 OUTSIDE BOTTOM STRINGERS 7/8" x 4"
4 END DECK 3/4" x 12". LAMINATED DECK BEAM
5 INSIDE SHEER STRINGER 3/4" x 2 1/2"
6 THWART 7/8" x 10". AFT THWART 14" WIDE
7 FORWARD SIDE MOTOR WELL 1 1/2" THICK
8 SEAT RISER 3/4" x 2 1/2" WIDE
9 CHINE 1 1/2" x 3 1/2". TWO 3/4" STRIPS GLUED
10 BED LOGS MOTOR WELL 2 1/2" x 2 1/2"
11 SIDES MOTOR WELL 1" THICK. 2 SHEETS 1/2" PLY
12 FRAME GUSSETS 1/2" PLYWOOD
13 OUTSIDE SHEER STRINGER 3/4" x 2 1/2"
14 GUSSET REINFORCEMENT 3/4" x 2"
15 LAMINATED DECK BEAMS 3" x 2 1/4" THICK
16 FRAMES LAMINATED FROM 3/4" x 2" STRIPS
GLUE
LAMINATED FRAMES
2"
GARVEY 18'-10"x6'x2'
WITH WELL FOR OUTBOARD MOTOR

Steve Roso of Portland, Oregon, built this 18' 10" garvey from plans I published in the 1987 October and November issues of the* National Fisherman. *He used 3/8" marine plywood for the sides and 3/4" for the bottom, Chem-Tech T-88 glue and bronze or stainless steel nails throughout (Steve Roso photograph).

three pieces of varying length, so the butts in the inner and outer layers can be well-staggered when the full-length lamination is made up.

The chine stringers should be slightly longer than the boat itself, so that when they are first put in place their ends will project slightly past the outer surface of the transom framing. After fastening them in the notches cut for that purpose in the transom frames, the ends of the stringers are cut off flush with the outside surface. This, in turn, is covered with an outer layer of plywood, 3/8" or 1/2" thick.

This completes the transoms, and the side panels are not installed until this is done. It should be mentioned that the inside sheer stringers, or inwales, which are also put in place before the side panels go on, are treated in the same way as the chine stringers. It is appropriate to add here that the dimensions for laying out the curve of the chine stringers were taken from the scale half-model and are to be found in the drawings.

The upper edges of these stringers are beveled flat before they are sprung into place. The lower edges can just as well be beveled afterwards, on the building form, when the bottom is faired prior to nailing on the cross-planking.

At the same time the chines and sheer stringers are installed, the 6" wide, fore-and-after center strips are set into the notches cut for them in the building molds. And similarly, the ends of the center strips are nailed to the bow and stern transom frames, and the excess is trimmed flush before the plywood covering the outside of the transoms is put in place.

When a straightedge is placed across the bottom, the center stringers should line up perfectly with the chines. If not, they can be pulled up or forced down with C-clamps until the cross-planking is nailed on. Before this operation is started, the bottom edges of the plywood side planking are beveled slightly to provide a narrow (1/16") caulking seam for driving in a thread of caulking cotton after the bottom is planked.

The 3/8" plywood side panels go on before the bottom is planked. If they are made up from standard sheets, two 8' lengths and one 4' length will be required, each side panel being joined by two splices scarphed $4^{1}/_{2}$" (12 times the thickness of the wood). Because the sides of this garvey are slightly less than 2' wide, three 4' x 8' sheets are enough to cover both sides, with part of a sheet left over.

The three lengths that make up each of the two side panels are put on starting from the stern, so that the completed splices will lead aft. The 4 1/2" scarphs are planed almost to a feather edge but should remain short of sharp. As the lengths of side planking are put in place, they are glued and nailed to the chine and sheer stringers.

To clamp the scarph joints together after glue has been applied, a thoroughly waxed strip of wood slightly wider than the scarph is placed against the joint from the inside. A similar waxed strip of 3/4" thick board covers the splice from the outside.

Several good-sized wood screws are put through this outer piece from the outside and turned up tight to put pressure on the joint and pull it tight. After the glue sets, the temporary clamping strips are removed, and the screw holes through the side are plugged. Neat, tight scarph joints will result if this procedure is followed.

Cross-planking is specified for the bottom, both because it is traditional construction for this type of garvey and because it adds great strength, taking the place of transverse frames. (Plywood, however, could be used, and this alternative will be discussed.)

If possible, bottom planking should be at least 7/8" thick, and slightly thicker stock wouldn't hurt. Widths should range from 4 1/2" to 6", but planks should be no wider than 6" to prevent "cupping." Wide bottom planks invariably curl and cup. In addition, the seams act as expansion joints: the fewer the seams in a cross-planked bottom, the more they will open up when the boat dries out.

The bottom planks should be fitted tight on the inside, but there should be a good 1/8" on the outside for easy caulking with cotton, to be set down tight with a mallet. The operation proceeds with the nailing on of five or six planks, which are set up tight together with either wedges or door clamps. Then, after a space the width of one plank is left open, several more planks are put down in another narrow section. The process is repeated over the length of the bottom.

The empty spaces between the several planked sections are given a taper so as to take wedge-shaped filler planks that are driven in with a heavy hammer or maul. This serves to tighten up the whole bottom. The taper should not be excessive. A 1/2" difference in widths from one end of the filler plank to the other is plenty.

After the seams have been caulked with cotton, they should be thoroughly soaked with an oil-base primer (red lead is best) and puttied over with a non-hardening seam compound. The seam between the bottom and the lower edge of the plywood side panels is caulked and puttied in the same manner as the bottom seams. If the procedure outlined here is followed, a tight, trouble-free bottom should result. But a poorly constructed cross-planked bottom is certain to be a source of unending grief.

This more or less completes the outside of the boat except for painting, but before the hull is turned over to finish the inside, it would be advisable to give the exterior a coat of primer, at least. Pete Culler used to say that expensive marine paints are a waste of money except for luxury yachts, and his recommendation was a good quality oil-based house paint.

Inside, the completion of this boat is more or less ordinary, straightforward carpentry, except for the outboard well, perhaps, which does have one novel feature: its sloping after end is intended to eliminate the drag and scooping action of a vertical wall.

Nothing could be simpler than using glue and nails to laminate the well frames in place from 2" x 3/4" strips. Serving to greatly increase the holding power of the glue, nails should be used generously and their length increased to correspond to the increases in the thickness of the wood.

The frame gussets under the thwarts at Stations 2 and 5, and between 4 and 5, are extremely important bracing members that stiffen the hull. If necessary, the middle thwart may be made removable, but it would be better fastened in permanently. The end decks also stiffen the hull structure. The builder would do well to beef up the sheer stringers under the ends of these decks, adding several more laminations to provide a better nailing and bearing surface.

For those who might want them, gratings could be easily fitted in the bottom. Note the provision for limbers. These should be put in as needed to drain the full length of the bottom.

With the considerable amount of fore-and-aft rocker in the bottom of this 19' garvey, it should move easily through the water at low and moderate speeds.

A draft of 7" amidships with a displacement of about 1,000 lbs. puts the bottom of the stern transom just about level with the surface of the water. At this depth, there will be no stern drag at all. Even with enough of a load in the boat to increase the draft amidships to 10" and the displacement to 1,500 lbs., the amount of stern drag resulting should have no significant braking effect.

In contrast to this garvey, the bottoms of most modern outboard boats which are designed for speed are generally quite straight and flat from amidships aft to the stern. It hardly needs saying that this configuration is required to support the weight of heavy motors. It is also necessary to offset the tendency of the stern to sink deeper in the water as power is applied (thus increasing stern drag) while the bow rises skyward.

It is not unusual for the entire front end of some of these boats to lift clear of the water back almost to amidships. In such cases, there will be practically no depth of displacement amidships when running at top speed; what little displacement there is will be concentrated at the stern. This gives the boat very little grip on the water, making it prone to skid out of control, especially on the turns. The problem will occur in spite of skegs and the like attached to the bottom to correct it.

If a heavy, high-horsepower motor were to be hung on the stern of this garvey, the result would be to stand the boat almost on end when running. To counter any such tendency, the motor is placed in a well located forward of the stern at a distance slightly less than one-third the length of the load waterline when the depth of displacement is 10". This should keep the boat more or less level with its designed load line when running at low and moderate speeds with motors no larger than 15 h.p.

To adapt the boat for larger motors and higher speed, the location of the well could easily be moved forward about 2'. All that would be required is planing a few shavings off the bed logs attached to the sides of the well to make them conform to the slightly changed bottom curvature in the new position.

Should more unobstructed room be desired in the stern section of the boat, the well can be located in this more forward position even with the originally specified smaller motor. This would be my recommendation to the two Virginia Beach readers who wrote to inquire about plans for a skiff with enough working room at the stern for a small gillnetting operation.

The construction of the motor well has been carefully planned to make it easy to build and simple to install. One novel feature that requires explanation is the well's sloping back. Unlike the vertical back of an ordinary box well, it is designed to slide over the water in front of it, offering as little resistance to forward motion as possible.

A well back that stands straight up and down acts, to some extent, like a brake and a scoop when the boat is moving forward at any speed. Water is pushed ahead of it and is forced into the upper portion of the box. To control this flow and hold it back, builders and designers have devised a variety of motor well baffles, but they don't always work as well as they are supposed to. Because of the sloping back in this well, we don't foresee any need for such baffles.

The four principal parts of the motor well are made up in advance. The back is simply a rectangular piece of 1/2" marine plywood that measures 1'7 1/2" long x 1'5" wide and is glued and nailed to the side cleats. The bottom is beveled to fit tight against the filler piece **(8)** between the inside fore-and-aft stringers **(7)**, and it is covered with the cross cleat **(9)**, which is glued and nailed down tight.

The front of the well serves as the motor mount and, like the back, is a rectangle 1'7 1/2" long x 1'5" wide. This piece, however, is made up of three layers of 1/2" plywood glued together to a total thickness of 1 1/2". The ends are beveled 78 degrees to conform to the required angle of rake for

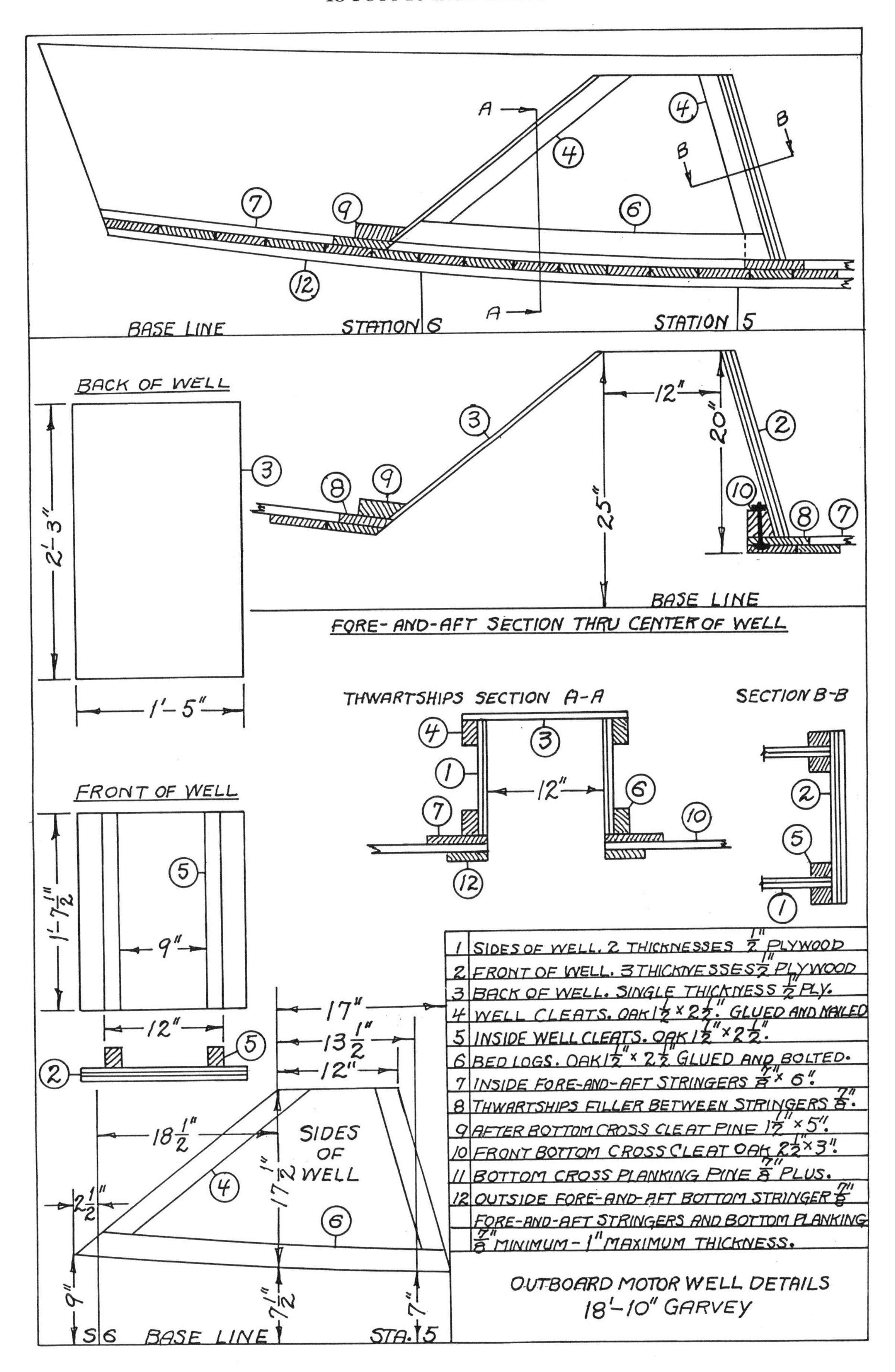
A
B
BASE LINE
STATION 6
STATION 5
BACK OF WELL
2'-3"
1'-5"
12"
20"
25"
BASE LINE
FORE-AND-AFT SECTION THRU CENTER OF WELL
THWARTSHIPS SECTION A-A
SECTION B-B
FRONT OF WELL
1'-7 1/2"
9"
12"
17"
13 1/2"
18 1/2"
SIDES OF WELL
17 1/2"
2 1/2"
9"
7 1/2"
7"
S 6
BASE LINE
STA. 5
1 SIDES OF WELL. 2 THICKNESSES 1/2" PLYWOOD
2 FRONT OF WELL. 3 THICKNESSES 1/2" PLYWOOD
3 BACK OF WELL. SINGLE THICKNESS 1/2" PLY.
4 WELL CLEATS. OAK 1 1/2" x 2 1/2" GLUED AND NAILED
5 INSIDE WELL CLEATS. OAK 1 1/2" x 2 1/2".
6 BED LOGS. OAK 1 1/2" x 2 1/2" GLUED AND BOLTED.
7 INSIDE FORE-AND-AFT STRINGERS 7/8" x 6".
8 THWARTSHIPS FILLER BETWEEN STRINGERS 7/8".
9 AFTER BOTTOM CROSS CLEAT PINE 1 1/2" x 5".
10 FRONT BOTTOM CROSS CLEAT OAK 2 1/2" x 3".
11 BOTTOM CROSS PLANKING PINE 7/8" PLUS.
12 OUTSIDE FORE-AND-AFT BOTTOM STRINGER 7/8"
FORE-AND-AFT STRINGERS AND BOTTOM PLANKING 7/8" MINIMUM - 1" MAXIMUM THICKNESS.
OUTBOARD MOTOR WELL DETAILS
18'-10" GARVEY

the motor mount and to fit tight at the bottom. The inside well cleats **(5)** are spaced exactly 12" from outside to outside.

They are glued in place and fastened through from the outside of the motor mount with 3" nails. Use 10d hot-dipped galvanized wire nails if the boat is galvanized fastened, otherwise choose bronze or stainless "ring" nails of equivalent size. If the cleats are hard, dry oak, it may be necessary to bore for the nails, but the pilot holes should not be large enough to cut down on the holding power of the fasteners.

The sides of the well consist of two layers of 1/2" plywood cut to the shape dimensioned in the drawing, glued together and cleated on the outside as shown, with the bottom cleats serving as bed logs. Cleats are put on with glue and nailed through from the inside.

The aperture for the well is not cut through the bottom cross-planking until the well is ready to be installed. When the time comes, it is sawed out flush with the inside edges of the fore-and-aft stringers. At either end, filler pieces **(8)** are fitted between the stringers. These serve to support the ends of the well and are exactly the same thickness as the stringers.

They must be fitted perfectly tight and be glued and fastened in place to forestall any chance of leakage. This is one area where a few 1½" No. 12 flat-head wood screws could be used to advantage in fastening the ends of the filler pieces to the cross-planked bottom.

Within the well aperture, the ends of the cross-planks will be exposed where they were sawn off flush with the stringers. So will the joints of seams between these planks. Leakage is apt to develop here unless the joints at the plank ends are made perfectly tight. They could be caulked, but another way to handle the problem is to snug them together by driving in thin wedges of the same material as the bottom planking. These can be cut from the pieces of planking removed in sawing out the well aperture. Besides being driven tight, these wedges should also be glued.

In addition to being leakproof, the well box must be solidly built to support the motor. Before it is bedded and bolted in place, it should be precisely fitted all around. If the joints are tight, I would prefer to glue the well in place in addition to bolting and nailing it. But if the fit is sloppy, perhaps it would be safer to use a thick, flexible bedding compound to fill in the voids.

The half thwarts, shown on one of the drawings, extend from either side of the well box to the sides of the boat. They not only brace the well, but when fitted and fastened in place they become important structural members reinforcing the boat as a whole.

Could this garvey be built with a plywood bottom? Yes, of course, but I am not sure enough would be gained thereby to make up for the trouble and expense. Because of the width of the bottom, there might be considerable waste in cutting the required pieces out of standard panels. Moreover, additional labor would be necessary in splicing these pieces together.

If the bottom were fiberglassed, as is generally done with plywood bottoms, that, too, would add extra labor and expense. Costs would begin to rise right from the start, for this job would require marine-quality plywood, which is quite expensive. Ordinary exterior panels are now of such poor quality that they should not be considered for boat construction. Finally, the substitution of plywood for the bottom would call for additional interior bottom framing to provide sufficient support.

In a cross-planked bottom using sawn boards, the planks themselves serve as bottom framing. That is why a thickness of 7/8" is to be considered the minimum for bottom planking in a boat of this size. In my judgment, 15/16" or even a scant 1" of thickness would be better.

There should be no particular difficulty in obtaining suitable lumber that is native to the region where this garvey is to be built. It can frequently be had from small, local sawmills at very reasonable prices. Among the choices: white pine in the Northeast, yellow pine or cypress in the South and Douglas fir, Alaskan yellow cedar or Port Orford cedar in the Northwest.

Western red cedar and northern white cedar last well but are softer and weaker than the first-named species. If the latter

Steve Roso installed a one-cylinder Bukh diesel in this garvey which he completed in July 1989 and used in working an oyster lease on Netarts Bay, Oregon. The boat handles well and is stable enough that two men can work off one side at a time (Steve Roso photograph).

were used, I'd want 1" of thickness. Perhaps it should be added that a cross-planked bottom does not gain from fiberglassing and, in fact, should not be so sheathed.

Some might want to consider powering this garvey with a small inboard engine. Miniature diesels that are not too large for boats like this can now be had. These may have advantages over outboard motors, but whether such benefits are worth the extra cost of purchase and installation is debatable. However, there is another possibility in this connection that should not be overlooked; powering with a small air-cooled engine.

One drawback of the inboard engine for small craft like this garvey should be considered. Ordinarily, boats with flat bottoms are easily landed on the beach, and they conveniently stand upright when they ground out. But, of course, this is virtually impossible if a skeg and propeller are projecting below the bottom.

This problem was solved by the French fishermen on the islands of St. Pierre et Miquelon. Powered with slow turning inboard Acadia make-and-break engines, their big dories had to be pulled up on the beach every night and so were fitted with a hinged shaft that permitted the propeller to be hauled up into a well in the bottom of the boat. I'm not recommending such an arrangement for the garvey, but it is worth mentioning.

When they are hung over the stern, outboard motors are easily swung up out of the way. If they are mounted in a well, however, there must be enough room for the motor to pivot clear of the bottom. The fit will be tight in the well shown here for the garvey, but there will be enough clearance for most motors of the recommended size. If there isn't, narrow twin skegs can be added to the outside fore-and-aft bottom stringers to lift the boat sufficiently off the beach when it is dragged ashore.

Reducing the length of this garvey by up to 3' could not be simpler. No lofting or fairing is required, and no changes in the layout or scantling dimensions are needed. To shorten the boat to 16', let us say, merely eliminate a 2'10" section from the middle of the 18'10" boat (from Station 4 to 2" aft of Station 5) and reunite the bow and stern sections. When the garvey is modified in this manner, the developed shape of the side planking and chine is shortened to correspond.

To narrow the boat, subtract the determined amount from each of the given beam widths for the bottom and the sheer. The garvey can be widened in similar fashion, but if the addition is substantial, it would be well to add another pair of fore-

and-aft stringers spaced midway between the existing stringers and the outside of the boat. Like the first pair, they should be made up of an inside strip and an outside strip nailed through from both sides with 2 1/2" long 10d nails.

If more freeboard is desired, the sides are simply extended by the required amount. Of course the dimensions of the transoms at either end will have to be altered to correspond. There is one thing to bear in mind, however. In order to be able to cut opposite sections of the side planking from a single sheet of plywood, the maximum width of the panels must not exceed 2'.

Excessive freeboard is not advised. In most cases, sides much higher than those shown would likely prove less than advantageous and would certainly not make a safer boat.

Some builders may want gratings in the bottom to stand on. These are easily nailed together from strips of 3/4" white pine or cedar. They should be made in several convenient sections for easy removal when the boat is washed down. Limbers should be large enough to drain readily into the center of the boat for bailing.

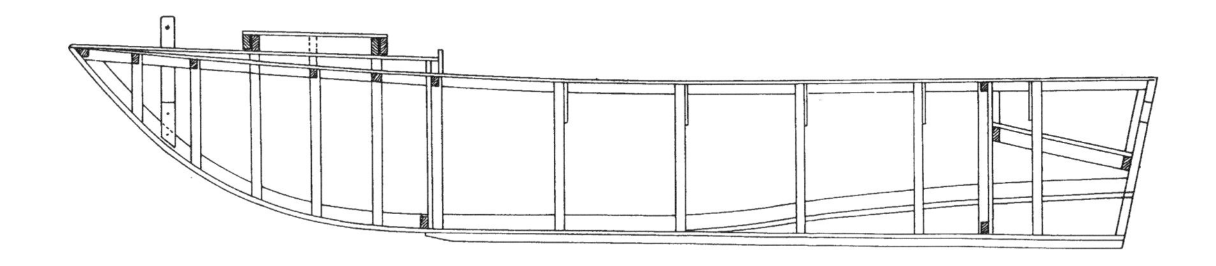

Chapter Seventeen

V-Bottom Garvey Workboat

There are no "instant" small fishing boats. It takes time and effort to put together a hull that will do the job, stand up in service and bring you and your catch home safely at night. Yet, there are simple designs, easy building methods and relatively low-cost materials that make the owner-built boat both very feasible and quite attractive.

Commercially produced craft in either fiberglass or aluminum often tend to be expensive, and they don't always meet the individual fisherman's particular needs or preferences. Frequently, such boats are much better suited for sportfishing and recreational boating. But, while small fishing boats don't build themselves, they can be turned out in good time without special skills or experience, using ordinary materials and simple tools. In short, the job can be done as long as the owner-builder goes about it right.

First comes the selection of a design. For an easily built, relatively inexpensive small fishing boat, standard variations of the time-tested flat-bottomed hull are high on the list to be considered. But in rough water and the open sea, the larger work dories such as the St. Pierre are hard to beat.

For service in more protected areas, garveys have a lot going for them and, in fact, some of these boats will take more punishment in heavy going than they are often given credit for. In actuality, both the well-known Boston Whaler and the old Hickman Sea Sled are variations of the basic garvey hull form.

The garvey considered here is an easy one to build and has great carrying capacity for its 18 feet of length. It should be able to move along at a reasonably good, but not excessive, speed when required—say, for example, when considerable distances have to be covered in the course of the day's work. But, this garvey was definitely not designed as a recreational speedboat for seagoing cowboys who like to drive at breakneck speed, slamming into waves regardless of the weather. Outboard motors larger than 75 h.p. are not recommended, and engines considerably smaller than this would be adequate in many cases.

Boats with wide, flat bottoms spread their displacement over a broad area and thus remain close to the surface even when loaded. Without the grip on the water that greater depth of displacement would provide, they frequently fail to track well and slip out of control in turns. To overcome any such tendency in this garvey hull, a moderate amount of tunnel is built into the stern.

When the boat is moving up to speed, the bow will lift somewhat, and the stern, because of its tunnel formation, will settle a corresponding amount. An additional 1½" of displacement is supplied by the wear strips, or stringers, attached to the outer edges of the bottom. These features should provide a depth of displacement in the after section of the hull sufficient to ensure that the boat tracks well and responds reliably in turns.

Construction is straightforward and fairly simple. Plywood is specified for the sides, bottom, transom, decks and bulkheads, while sawn lumber is used for the internal framing, deck beams, bottom wear strips and laminated inner keel. The boat is fastened together with epoxy glue, nails and a few screws. The entire job can be done with ordinary hand tools, although power saws—a table saw, a band saw and a skill saw, in that order—will greatly facilitate the building process.

This is in no way to imply that accurate and precise workmanship is not necessary. Quite the contrary. Building this boat is no job for a wood butcher.

The first step is to spread out three sheets of plywood and to lay out the shapes of the transom and the sides, spliced as shown on Sheet **2**. The shape of the inner keel is laid out later.

Three layers of 1/2" plywood glued together form the transom. The two inner layers can be made up of two or three smaller pieces butted together, so long as the joints from layer to layer are well staggered. No two should fall together, or even close together. Firm contact between the gluing surfaces is all that epoxy adhesives require in the way of pressure. Such contact can be assured throughout the transom assembly by driving in ordinary nails; these are pulled out and the holes plugged when the glue has set. This method of using removable nails is diagrammed on Sheet **2**.

The procedure shown for butt-splicing the sides is easily adapted for splicing the inner layers of the transom. If C-clamps are used around the edges, care must be taken not to squeeze out all the glue from between the layers, lest weak or "starved" joints result. Prior to gluing together smooth surfaces with epoxy, joining faces had best be roughened by working across the grain with coarse sandpaper. The sides of the transom are left square. The slight bevel required on the bottom is put on when the bottom edges are faired to receive the bottom planking.

After the pieces that make up the sides are cut to the lengths shown on Sheet **2**, they are butt-spliced together (by the method diagrammed) to form two identical sides. They are then temporarily tacked together, outside to outside and the edges planed to make both sides exactly alike.

The next operation is to install the inside chine strips (see Sheet **1**), as well as the vertical frames, or timbers, including those for the transom and the posts for the after bulkhead. These are all put in with glue and nailed through from the outside with 1½" No. 11 annular ("ring") nails spaced 2" to 3" apart.

Finally, the laminated inner keel, or keelson, can be made up. A glance at the table of offsets will show that the bottom of the sides and the inner keel have exactly the same shape from Station 4 forward. This is of critical importance, and to ensure that they are identical, use one of the completed sides as a template for marking, cutting out and checking the shape of the forward part of the inner keel. Aft of Station 4, this timber takes its own shape in forming the tunnel stern.

The six laminations that are glued and nailed together to form the inner keel are cut out of 3/4" thick sawn boards in several shorter pieces that make up into the full-

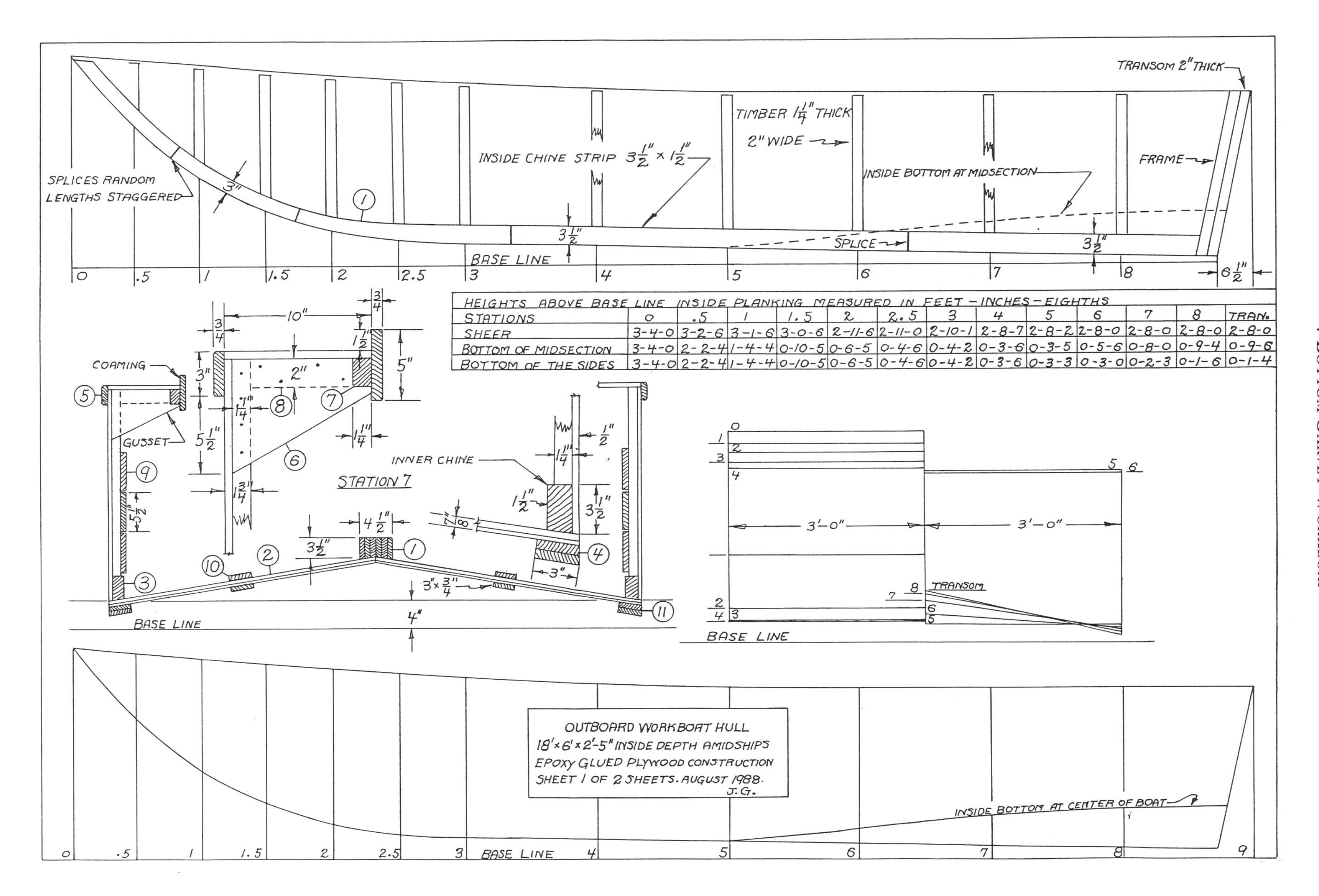

HEIGHTS ABOVE BASE LINE INSIDE PLANKING MEASURED IN FEET-INCHES-EIGHTHS

STATIONS	0	.5	1	1.5	2	2.5	3	4	5	6	7	8	TRAN.
SHEER	3-4-0	3-2-6	3-1-6	3-0-6	2-11-6	2-11-0	2-10-1	2-8-7	2-8-2	2-8-0	2-8-0	2-8-0	2-8-0
BOTTOM OF MIDSECTION	3-4-0	2-2-4	1-4-4	0-10-5	0-6-5	0-4-6	0-4-2	0-3-6	0-3-5	0-5-6	0-8-0	0-9-4	0-9-6
BOTTOM OF THE SIDES	3-4-0	2-2-4	1-4-4	0-10-5	0-6-5	0-4-6	0-4-2	0-3-6	0-3-3	0-3-0	0-2-3	0-1-6	0-1-4

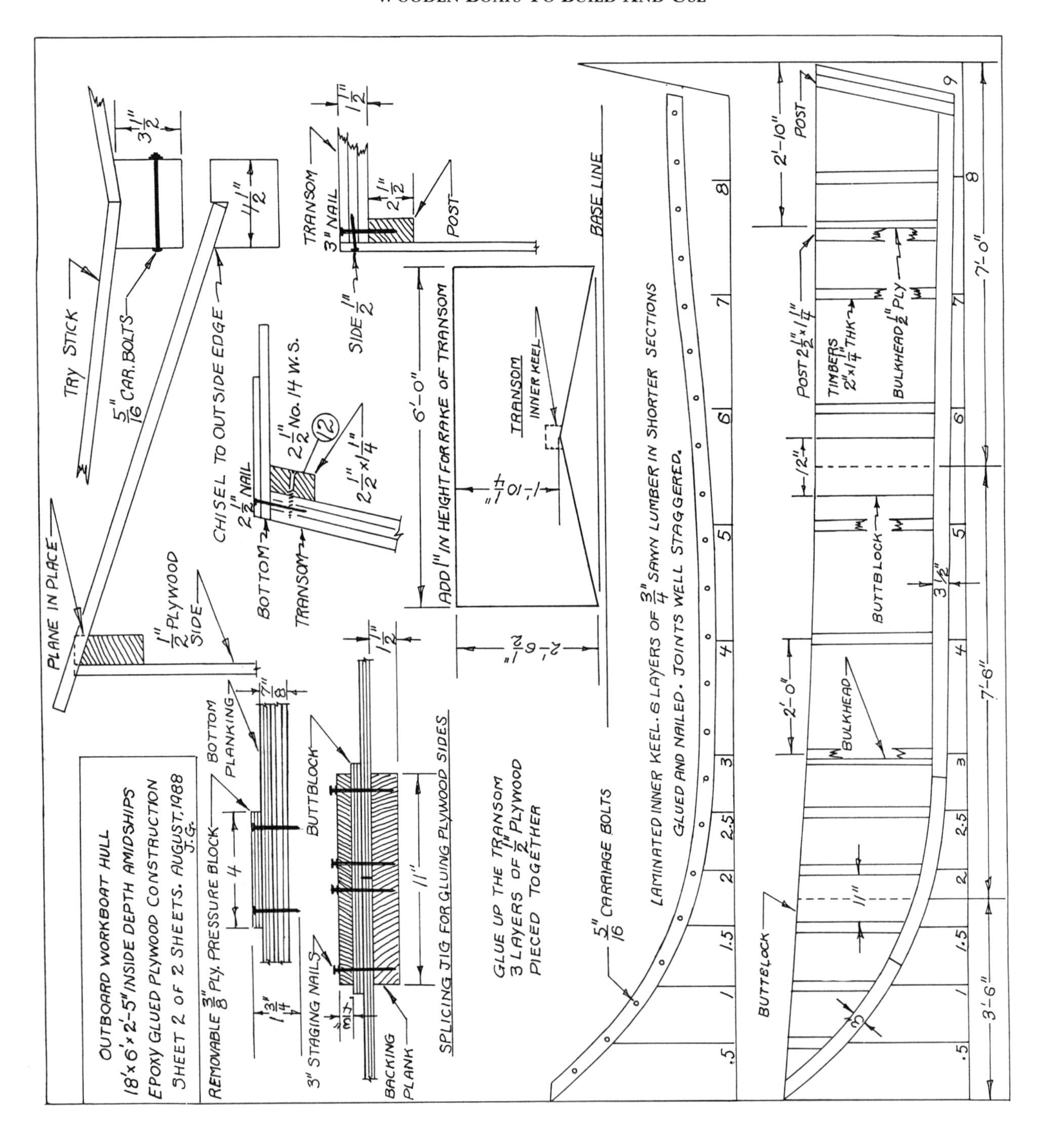

length piece. The length of these sections must vary from layer to layer so that no two joints line up anywhere in the whole assembly; the more widely these joints are distributed, the better.

The only reason for making up these full-length layers, or laminations, from several shorter pieces is to be able to build up the long, curved member from boards of ordinary width and random lengths. Naturally, the longer these sections can be made, the fewer will be needed—and the fewer butt joints there will be to bother with.

The first layer is cut out precisely and checked for exact conformance to the shape of the bottom edge of the sides from Station 4 forward. It can then serve as a pattern for making up the succeeding five layers. These are glued and nailed together with 1½" No. 11 annular nails, one layer at a time. The completed assembly of the inner keel is not through-bolted until after the glue has set. Nonetheless, the location of the bolts must be marked on each layer and lined up before they go together in order that the nails will not be in the way when the holes are bored for the bolts.

It will be prudent to give the after end of the inner keel assembly an extra inch of length to be trimmed later as may be required to ensure a tight fit against the transom when the keelson is put on.

Following the completion of these three sub-assemblies, the boat can be set up for planking the bottom. No form is needed. If a flat building floor is not available, the sides can be set up on two cross skids, leveled from side to side.

The two side panels, set side by side and bottom up, are held apart at the right distance by several cross pieces that are temporarily nailed in place. One of these, placed at Station 4, should be stout enough not to sag when the internal keel is put in place and drawn up snug with a C-clamp. When the keelson has been accurately centered between the sides and clamped to the temporary cross pieces forward of Station 4, the after end is trimmed for a tight fit against the inner surface of the transom when the latter is in place.

Now the transom can be installed. It is glued at the sides and is nailed through (from the outside) into the posts against which it rests on the inside. Nails are also driven through the outsides into the ends (see Sheet 2). Before the transom is nailed into the end of the keelson from the outside, the latter must be positioned just high enough to accommodate the notch that receives the ends of the cross planking (Sheet 2).

By use of the try stick (Sheet 2) at different locations, the amount of changing bevel on the sides and bottom side cleats can be cut away at these selected spots. At the same time, with the try stick as a guide, the V in the bottom of the inner keel is chiseled out to receive the ends of the bottom planking. This V is cut from spot to spot, then made continuous and faired back to the transom.

Likewise, the changing bevel on the bottom edges of the sides and bottom side cleats is planed continuous and fair over its whole length, from the transom to the point where the V dies out completely at Station 4. Before the bottom cross planking is fastened on, the try stick must rest with a touching fit wherever it is placed on the bottom surfaces of the beveled sides, the side cleats, and the rabbeted inner keel.

The final operation before planking is to fasten the two sides of the 2$^{1}/_{2}$" x 1$^{1}/_{4}$" reinforcing cleats that back up the inside bottom of the transom, running from the inner keel to the bottom side cleats. These are glued in, as well as secured in place with screws, and the bottom edges are beveled until the try sticks indicate fairness.

Planking starts at the after end and moves forward. The entire bottom is covered with an inner layer of 1/2" plywood before the outer layer of 3/8" plywood is put on. Forward of Station 4, the planking runs full length across the bottom of the boat. Aft of Station 4, the cross planking goes on in two equal lengths, butting on the centerline of the inner keel.

The planking itself can be of various widths. In general, wider planks are better than narrower ones—so long as they are not too wide to conform to the twist in the after part of the bottom and to the curve of the bottom forward. From Station 3 to Station 5, a single section of cross planking 4" wide would be ideal. At Station 2, 8" of width might be just a bit too much because of the curve at that point.

The edges of the first layer of 1/2" planking should be slightly beveled so that when the planks are pushed together as they are put on, their inner edges will come together tightly, while their outside edges will stand open about 1/16", very much like caulking seams on a conventionally planked boat. When the inner layer of planking is fitted in place—and fastened at the ends and along the bottom of the transom—the seam openings are filled with a stiff epoxy mix that is forced in with a flexible (but not too flexible) putty knife.

After this adhesive hardens, the edges will be solidly glued together, and the bottom planking will be strengthened and stiffened—almost as if it had been put on as one single, continuous sheet of plywood. The builder should use an epoxy resin that is not too thin to begin with, one like Chem-

Tech T-88 or, especially, Fiberglass Evercoat two-part Super Glue. Either should be brought to the consistency of a loose putty by the addition of Cab-o-Sil, or an equivalent thickening agent.

In preparing the outer surface of the inner layer of planking to receive the outside layer of 3/8" plywood, a sharp scraper and several sheets of extra-coarse sandpaper are required. The scraper removes any ridges, bumps, hardened glue and the like. Then the whole surface is sanded fore-and-aft using a block whose length matches that of a sheet of the sandpaper. This will true up the surface as well as roughing it for a good glue joint. Do not sand excessively, however, and avoid cutting in deeply.

The cross planking of the outer layer should lap the under planks about halfway. Epoxy adhesive thick enough to stay put is liberally applied between the planking layers. Fiberglass Evercoat Super Glue is just about the right consistency as it comes from the can. Chem-Tech T-88 or System Three epoxy will require the addition of some Cab-o-Sil. Two-inch No. 11 annular nails spaced 2" apart at the ends and along the bottom of the transom serve to pull the two layers of planking together until the glue has hardened. In addition, nails spaced 3" apart and running fore-and-aft in three rows are driven temporarily through the two thicknesses of planking.

These can be either small, two-headed staging nails or ordinary nails with pressure pads or thin strips of pine under the heads. The nails will be pulled later, after the glue has cured. The holes made by the middle row of nails will be covered by the fore-and-aft bottom stringer, shown as **(10)** on Sheet **1**. The other two rows of holes can be filled with either small wooden plugs or stiff epoxy mix forced in with a putty knife.

At this stage, some might want to fiberglass the bottom. Although I don't think this is necessary, it would add some extra stiffness and strength, and some protection against wear if the boat were to be hauled out on the beach much. The fiberglass would also add weight, the cost for materials and the extra labor of putting it on.

The middle bottom stringer—again, **(10)** on Sheet **1**—measures 3" wide by 3/4" thick and is put on with glue and 1½" No. 12 flat-head wood screws. These pull the stringer down tight on the twist at the stern, one of the few places where screws are called for. The double-thickness stringers, or wear strips, attached to the outer bottom edges of the sides go on in two lifts, each 3" x 3/4". They are glued on and nailed in sequence with 2½" No. 11 annular nails.

These stringers, as well as the single-thickness middle stringer, run from the transom to Station 3 (or slightly farther), where they end in an easy taper. When the stringers are in place, the bottom and outside of the boat are complete except for painting. Now the hull can be turned right side up for work on the interior.

One afterthought: instead of fiberglassing the entire bottom, some might want merely to reinforce the ends of the outer bottom where they meet over the inner keel. This could be neatly done with two layers of fiberglass tape—first a 4" strip, then an outer, 6" strip.

After the hull has been turned over, we can proceed to finish out the interior or, rather, to carry it to the stage where the builder can complete it to suit his own ideas and requirements.

But first, we must consider the materials to be used in the construction of this boat. If a suitable epoxy adhesive is applied in conjunction with metal fastenings, an extremely strong hull structure will result. Thus, whenever two parts are fitted together, glue should be applied to the meeting surfaces. This will require more time and effort than if the pieces were merely nailed or screwed together dry, but the joint will be ever so much stronger with the glue, making the extra pains taken worthwhile.

There are a great many competing formulations of epoxy adhesive now on the market, and these vary markedly in their characteristics. A lot depends on picking one that is well suited to the project in hand. In addition to having superior adhesive qualities and posing relatively few health hazards, the epoxy formulation selected should be easy and foolproof to measure and blend, and the mix should be stiff enough to stay where it is put and not leak out of vertical joints. It should also set slowly enough that the builder need not rush the job for fear the glue will harden before he has completed his assembly.

I previously recommended three widely used brands of epoxy adhesives, all of which meet these requirements. Although I have used all three with excellent results, I have to give a slight edge to Evercoat. The thick, pasty consistency of the 1:1 mix makes application easy. Thus, there is relatively little mess or wasted glue, and few starved joints. As previously mentioned, to attain the same pasty, stay-put consistency with Chem-Tech or System Three, I found it necessary to add a stiffening agent like Cab-o-Sil to the mix.

For those with little or no prior experience with epoxy adhesives, I recommend *The Epoxy Book*, obtainable from System Three, and/or Chem-Tech's *Epoxy Manual*.

If I have paid more attention than seems warranted in considering the epoxy glue to be used in building this garvey, it is because I consider the use of adhesive such an important part of the construction process here. The boat has been designed for nail fastening, which probably wouldn't be strong enough without the glue. Screws are specified only for pulling the outside intermediate bottom stringers up tight to the bottom where there is twist in the inverted V of the after bottom. Even here, 1 1/2" No. 10 ring nails might do the job. The greater part of the nails needed will be ordinary galvanized wire nails in lengths up to 3".

As this is a workboat and not a fine yacht, and because it is intended for hard, rough and more-or-less year-round use, the garvey can hardly be expected to last more than 15 years; and galvanized nails should be good for that length of time. With no inboard engine installation and no battery-powered equipment with through-hull connections, electrolysis will not be a problem. That is true even when galvanized nails are used in combination, but not in contact, with bronze ring nails. The shorter but stouter ring nails are indicated where the combined thickness of the pieces to be fastened together is insufficient to prevent the longer, slimmer galvanized nails from going all the way through. For example, fastening the 1/2" plywood side to the 1 1/2" inside chine calls for a 1 3/4" nail, with the stouter ring nail in this length having appreciably greater holding power than the slimmer galvanized nail of the same length.

The choice of lumber for this boat is somewhat restricted, but should not be too much of a problem. For plywood, I believe we may forgo marine-quality fir (which would be unduly expensive) even though the quality of the ordinary exterior fir now available leaves much to be desired. By placing defects on the inside of glued-up composites like the transom—and by increasing plywood thickness as the builder's judgment dictates—it will be possible to compensate, in large measure, for the less-than-perfect material now available.

As an example here, the bottom might be glued up from two 1/2" thicknesses instead of the combination of 3/8" and 1/2" plywood shown on the drawing. I am not even considering mahogany plywood. A luxury-grade mahogany plywood like Bruynzeel would be far too costly, and some of the cheap grades made with Philippine mahoganies aren't worth considering.

I cannot too strongly advise the prospective builder to give some effort to studying how he will lay out and cut up his plywood panels to best advantage. If he just haphazardly slashes into them, wastage is bound to be high, with a corresponding increase in building cost. In the past, I have gone so far as to make scale drawings of all plywood parts, arranging them to best advantage on representations of 4' x 8' and 4' x 10' plywood panels drawn to the same scale. In this way it was possible to cut the amount of plywood required to a bare minimum and to eliminate waste almost entirely.

When it comes to sawn lumber required for the internal framing, selection is also somewhat restricted. My first choice would be good-quality, air-dried Douglas fir. It glues well and takes and holds nails well. In addition, such fir is stiff and strong, it doesn't weigh too much and it resists rot moderately well. On the West Coast and in

Alaska, Port Orford cedar and Alaskan yellow cedar might qualify as substitutes. Spruce is too susceptible to rot.

In the Northeast, white cedar is too soft for reliable nailing. Southern pine without its sapwood is long-lasting, and it takes and holds nails well. However, it is apt to be quite heavy, though this is perhaps not as objectionable in working boats as it would be in fast, lightly built pleasure craft.

Oak requires consideration. Contrary to what you may read in many of the boat books, I would avoid white oak here. It is unexcelled for steam bending and for heavy timbers in large vessels, where it will outlast most other species. But, white oak—besides being hard to get, hard to work and expensive to buy—tends to twist and move a lot when sawn into strips and boards. This often results in pulled-apart joints, strained fastenings and broken glue lines. As the old boat builders used to say, the wood is "too strong" and doesn't stay put.

These drawbacks are not shared by the red oaks, with the exception of a few like the notorious "piss oak," which rots readily and is easily identifiable by its smell. Indeed, some may prefer red oak to Douglas fir for the framing in this garvey. It is harder and somewhat stronger than fir, but it is also somewhat heavier and is probably better for the deck beams.

The interior construction of this 18-footer is simple and straightforward. After the boat has been turned over and leveled in an upright position, several temporary cross spalls are nailed across the top of the hull. These hold the sides straight and keep them parallel to each other and the right distance apart until the two watertight bulkheads have been fitted and fastened in place. The bulkheads are essential structural members that brace and stiffen the sides of the hull, as well as seal off the ends of the cockpit. The after bulkhead also provides support for the dry well, in which the outboard motor is mounted.

To save materials, these bulkheads can be spliced together from two or more smaller pieces of plywood by the method diagrammed on Sheet 2 for use in the sides. To achieve a tight fit, the bulkheads are scribed to the sides of the boat and to the inside of the bottom. Posts that are glued and nailed to the sides of the boat, and floor sills that run across the bottom and are likewise glued and nailed in place, make up a frame that backs up the bulkheads, which are glued and nailed to it. This makes a watertight closure all around and greatly stiffens and braces the sides of the hull. Such reinforcement is needed because of the long, open work space or cockpit, where there is little cross bracing between the ends.

As soon as the after bulkhead is in place, the dry well for the outboard motor can be built in. First, the horizontal bearing cleat **(28)** supporting the forward end of the dry well is glued and nailed to the after side of the after bulkhead. Then, a matching bearing cleat **(15)** supporting the after end of the well structure is glued and nailed across the inside of the transom.

Between these two bearing cleats, four longitudinal beams **(16)** are installed to support the plywood bottom of the well, which goes in next. With the bottom fastened in place, the framework for the sides of the well is installed. This consists of two posts **(38)** that are attached to the face of the bulkhead at the forward end of the well. Matching posts **(14)** at the well's after end are attached to the transom. Between them there is a longitudinal cleat **(40)** backing the bottom of the side of the well and, at the top, a carlin **(39)** to which the after deck is attached later.

With this framework in place on either side, the well's plywood sides are fitted, glued and nailed to give a watertight cover throughout. The construction of the dry well as diagrammed in the drawing deserves some study. If the step-by-step sequence outlined is followed, the builder will find that it goes into place readily.

Deck framing comes next, starting out with the fitting of blocking around the inside of the sheer. This fills in between the side frames and provides a solid bearing and nailing surface for the outside edges of the plywood decking. The blocks also provide a good bite for the nails used in fastening the sheer trim, or outwale.

Forward of the forward bulkhead, full-length deck beams are crowned 3" in 6' and are spaced 12" apart on centers. Although not shown in the drawing,

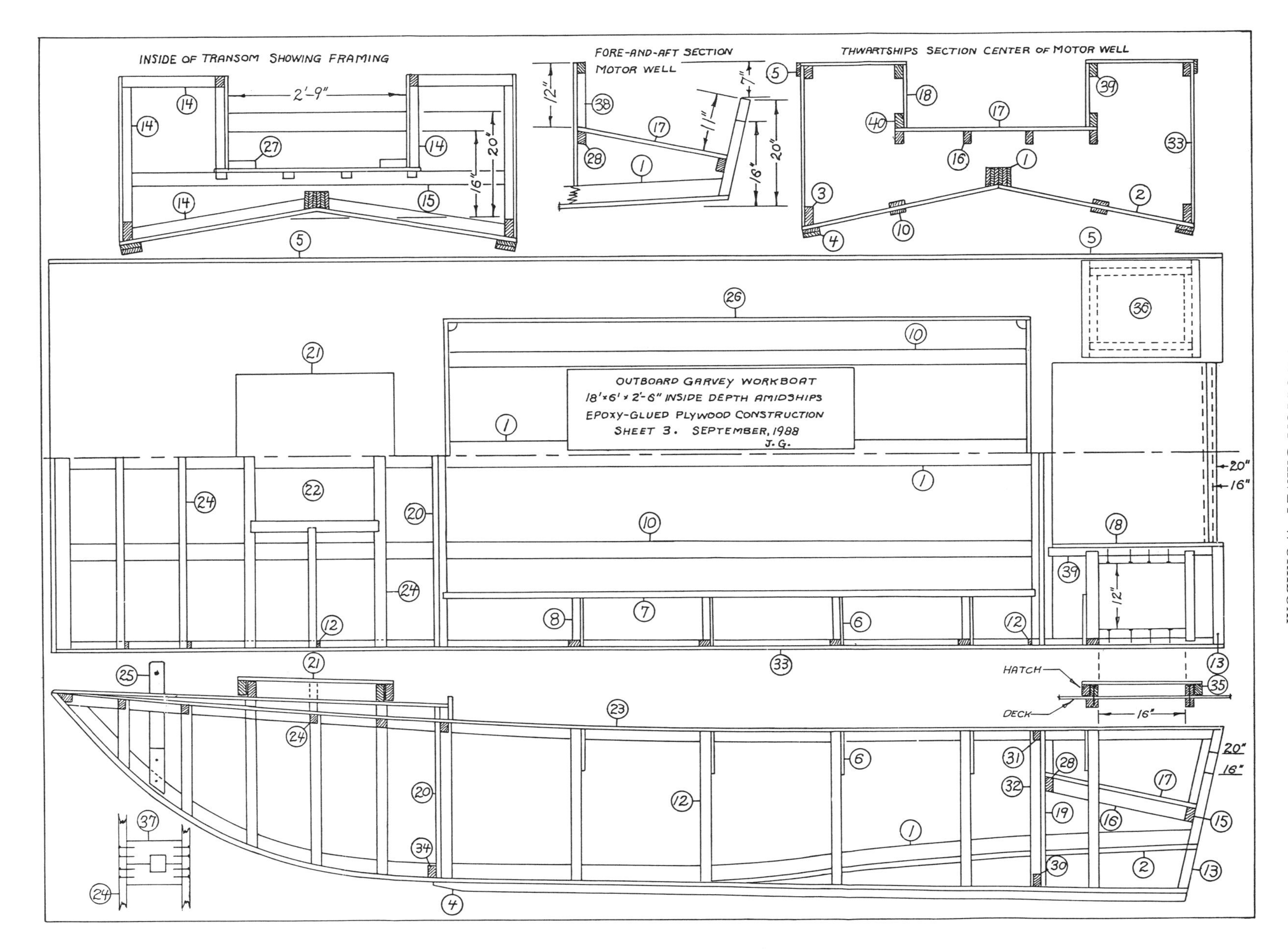
INSIDE OF TRANSOM SHOWING FRAMING
FORE-AND-AFT SECTION
MOTOR WELL
THWARTSHIPS SECTION CENTER OF MOTOR WELL
OUTBOARD GARVEY WORKBOAT
18'×6'×2'-6" INSIDE DEPTH AMIDSHIPS
EPOXY-GLUED PLYWOOD CONSTRUCTION
SHEET 3. SEPTEMBER, 1988
J.G.
HATCH
DECK
2'-9"

longitudinal blocking of 4" or so in width can be fitted between the deck beams along the centerline. Sections of plywood decking too short to extend all the way across the boat can be butted, glued and nailed on these center blocks.

Transverse seams in the plywood decking are laid out to land over the center of the deck beams, with both meeting edges of the seam being glued and nailed to the beams. Note that the aftermost of the full-length deck beams is attached to the after side of the bulkhead, which is trimmed off to correspond to the curve of the beam. The deck will come flush with this deck beam's after face, to which the forward face of the cockpit coaming will be fastened.

As soon as the partners for the Samson post are fastened in, the post itself is installed; the hatch opening is framed; and the plywood deck is laid, glued down and well-nailed throughout. Some will want to cover it with fiberglass. This is not necessary, although it will ensure a tight, smooth deck.

With the deck on, the hatch frame—fastened together with watertight joints at the corners—is put down with glue and 4" galvanized nails, for which lead holes are bored. These must not be so large as to take away from the holding power of the nails. They should be spaced about 4" apart. The hatch cover is made like the hatch frame and is covered with 5/8" or 3/4" plywood. If the builder wants access to the forward compartment without opening the hatch, he can easily make an opening of some sort through the bulkhead, but it should not be so large as to weaken the bulkhead structurally.

The side decks are important for bracing and stiffening the sides of the boat, again, because of its long, open cockpit. Their construction has been fully diagrammed in Sheet 1 of the drawings. The sheathing **(9)** on the sides of the cockpit also stiffens and strengthens the sides of the hull and protects them against the internal wear and tear to be expected in a working boat.

Aft of the cockpit and after bulkhead, the decking runs along both sides of the open dry well, terminating at the transom. The hatches on either side of the well have been made as wide as permitted by the space between the well and the outside of the hull. Hatch frames and covers are made in the same way that is specified for the forward hatch. Here again, the builder may choose to make openings through the bulkhead to get at the interior storage without removing the hatch covers. As was the case in the bow, these openings must not be so large as to weaken the bulkhead to any significant extent.

The following is a list of component parts as numbered on the drawings, with descriptive notes and comments for each. Unless plywood is specified, use sawn lumber.

Parts List

(1) Inner keelson; sides 4-1/2" in its entire length; molded 3-1/2" as far as Station 4, tapering to 3" on upsweep to bow. See directions for lamination in the text.

(2) Plywood bottom; made of two thicknesses; 1/2" and 3/8", glued together as explained in text. Both layers could be 1/2" for a 1" bottom; 7/8" is minimum.

(3) Chines; sided 1-1/2", molded 3-1/2"; glued and nailed to plywood sides. Can be put on in two or three shorter lengths. Principal function is to provide nailing and gluing surface for bottom and for outer chine strips.

(4) Outer chine strips; 3" x 3/4"; glued and nailed in place.

(5) Sheer trim and outwales; 3" x 3/4"; glued and nailed in place.

(6) Gussets supporting side decks; 1/2" plywood; glued and nailed to side frames and deck beams.

(7) Carlins; 2" x 1-1/4"; cut into cross beams fastened to bulkheads at either end of cockpit; fastened into ends of deck beams with 4" galvanized nails.

(8) Side deck beams; 2" x 1-1/2".

(9) Sheathing; 5-1/2" x 3/4"; runs full length of cockpit and nails into side frames with 2" No. 10-gauge bronze ring nails. Supplies protection and needed additional strength and stiffness to sides.

(10) Intermediate stringers; 3" x 3/4"; one lift outside bottom, one lift inside. Put on with nails and glue; use 1-1/2" bronze wood screws to pull outside lift tight to bottom if twist requires it.

(11) Same as **(4)**.

(12) Side frames, or timbers; 2" x 1-1/4";

put in on the flat. Best made of oak, if available.

(13) Transom; three layers of plywood, glued together and nailed. Layers—especially inner two—can be made up from smaller pieces to save stock.

(14) Transom framing; 2" x 1½"; goes around edges on inside of bottom and sides; runs under decking.

(15) Support for after end of dry well; 2½" x 1½". Extends horizontally across inside of transom, where it is glued and nailed.

(16) Four beams; 2½" x 1½". These run fore and aft between **(15)** and **(28)**, and support bottom of dry well.

(17) Bottom of dry well; 1/2" or 5/8" plywood.

(18) Sides of dry well; 1/2" or 5/8" plywood.

(19) After bulkhead; 5/8" plywood.

(20) Forward bulkhead; 5/8"plywood.

(21) Hatch cover; frame is 3" x 1½", topped with 5/8" plywood.

(22) Hatch opening; 22" x 24"; framed 3" x 1½".

(23) Side decking; 1/2" plywood.

(24) Deck beams; side decks, 2" x 1½"; forward deck, 2½" x 1½". Crowned 3" in 6', except for two beams either side of hatch opening; these are sided 2" instead of 1½".

(25) Samson post; rises 8" above deck. Bottom end rests on keelson and is cleated to it on both sides.

(26) Cockpit coaming; 5" x 3/4".

(27) Drain ports for dry well; 5" x 3/4". Can be enlarged, but not to extent of weakening transom for large motors.

(28) Horizontal cleat; 2½" x 1½"; fastened to after bulkhead. Supports bottom of dry well.

(29) Same as **(32)**.

(30) Bottom sill for after bulkhead; 2½" x 1½".

(31) Deck beam; 2" x 1½"; fastened to top of after bulkhead.

(32) Side posts; 2" x 1½"; installed on either side of after bulkhead.

(33) Side planking; 1/2" exterior fir plywood.

(34) Bottom sill; 2½" x 1½"; backs bottom of after bulkhead.

(35) Framing for after hatches; 2" x 1⅜".

(36) After hatch covers; framed 2" x 1⅜"; plywood tops.

(37) Partners for Samson post; each made in two parts; 4" x 1¼"; must be long enough to fit tightly between deck beams.

(38) Posts; 2" x 1½"; fastened to the bulkhead; these provide backing for sides of well at its forward end.

(39) Carlins; 2½" x 1½"; these provide attachment point for dry well and for after deck.

(40) Transverse cleats; 2½" x 1½"; these back bottom of drywell sides.

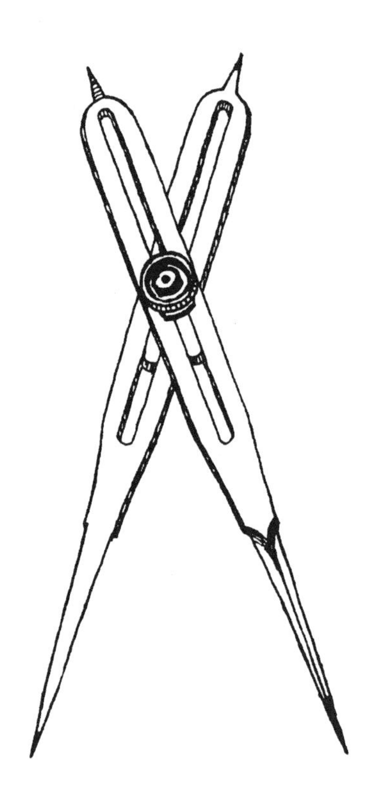

Chapter Eighteen

Scale Half-Models

Anyone planning to build a boat, or have one built, might do well to first make a scale half-model. Since it is difficult to visualize hull shape from either two-dimensional plans or the spoken or written word, the preliminary model would allow him to see the exact shape of the design he is contemplating.

Knowledge of hull form can be gained from personal experience in using a variety of boats and observing them both in and out of the water. In the course of using boats, one learns how different craft perform in all sorts of situations—their strengths and their weaknesses.

The experienced boatman acquires knowledge of the fine points of design. Occasionally that sense is developed so acutely that merely by looking at a boat hauled out of the water, he will have a better idea of how it will perform than the naval architect working at the drawing board and making countless mathematical calculations. This faculty takes experience and close observation, and, of course, all persons working with boats do not develop it to the same degree.

Until early in the present century, most of our traditional boats were built from half-models whittled into shape by their builders. Some notable examples of such boats are Wilbur Morse's Friendship sloops and Will Frost's Jonesport lobsterboats.

Often the experienced boatman can see in his mind's eye what he feels sure would be the ideal boat for him. He has thought about it long and often and has built it piece by piece in the depths of his imagination. Now the question is how to bring it to life, as it were, how to get it out into the real world?

He cannot describe it fully in words, and, not being a practiced marine draftsman, it is unlikely he can draw it either.

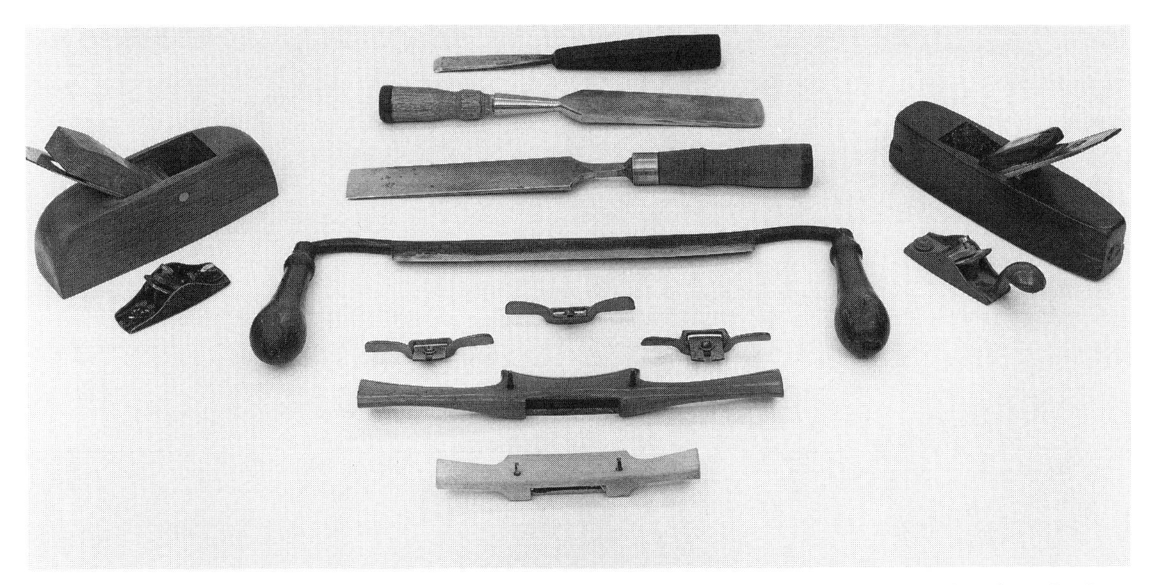

Tools for making half-models, including carver's gouge, paring chisels, drawknives, spokeshaves, thumb planes and smoothing planes (Mary Anne Stets photograph, M.S.M.).

Perhaps the best way, if not the only way, for him, is to carve it out of wood. This will take patience more than special skill, and directions from someone who has done it will help. This is what we are attempting here by outlining procedure and suggesting materials, tools and equipment.

A completed half-model has a wide variety of important uses. The planking lines can be laid out on it in advance with thin strips of pine serving as battens. These should be small enough in section to bend easily around the hull and can be held in place with large dressmaking pins. If the planking lines are marked with a soft pencil, they are easily erased for relocating and re-marking.

The half-model can also be used for spiling the plank shapes in order to determine the widths of the planking stock required. This feature can be very helpful to builders who are not experienced plankers, which is often the case with the novice constructing his own boat. Also, bevels for the frames and molds can be taken off the half-model.

A few words concerning scale should be mentioned. If too large, the model will be massive. If not large enough, the undersized model may be too small for accurate measurements. For a 20-foot boat, a scale of $1^1/_2$"=1' would be about right; for a 40-footer, 3/4"=1' is about as large as can be readily managed.

The wood used for the model must be thoroughly dry and well-seasoned. It should also be fine-grained, soft and easily worked. It is important to select wood that is free of sap, knots, cross grain or other blemishes of any sort. My first choice is northern white pine, without sapwood. Sapwood showing any blue stain should be avoided. Sugar pine would be my second choice.

There are two main types of half-models, lift models and solid models. Lift models are shaped from blocks built up in parallel layers or "lifts," fastened together with removable screws or pegs, but not glued. Generally the lifts correspond to the waterlines, so that when the lifts are taken apart the shape of the waterlines can be traced from them. It is possible to make a model with vertical lifts corresponding to the buttock lines, but this is rarely done.

From the waterline shapes obtained from the lifts, a body plan of sectional shapes is obtained, which, enlarged full size, gives the shapes of the building molds.

Some builders choose to carve their half-models from solid blocks of wood. There are methods for taking off their lines without taking them apart, methods that give more accurate results than can be obtained by tracing the lifts of a lift model.

Sometimes a solid half-model is made from the lines of a boat intended for construction, the better to see what the boat looks like in three dimensions, or perhaps

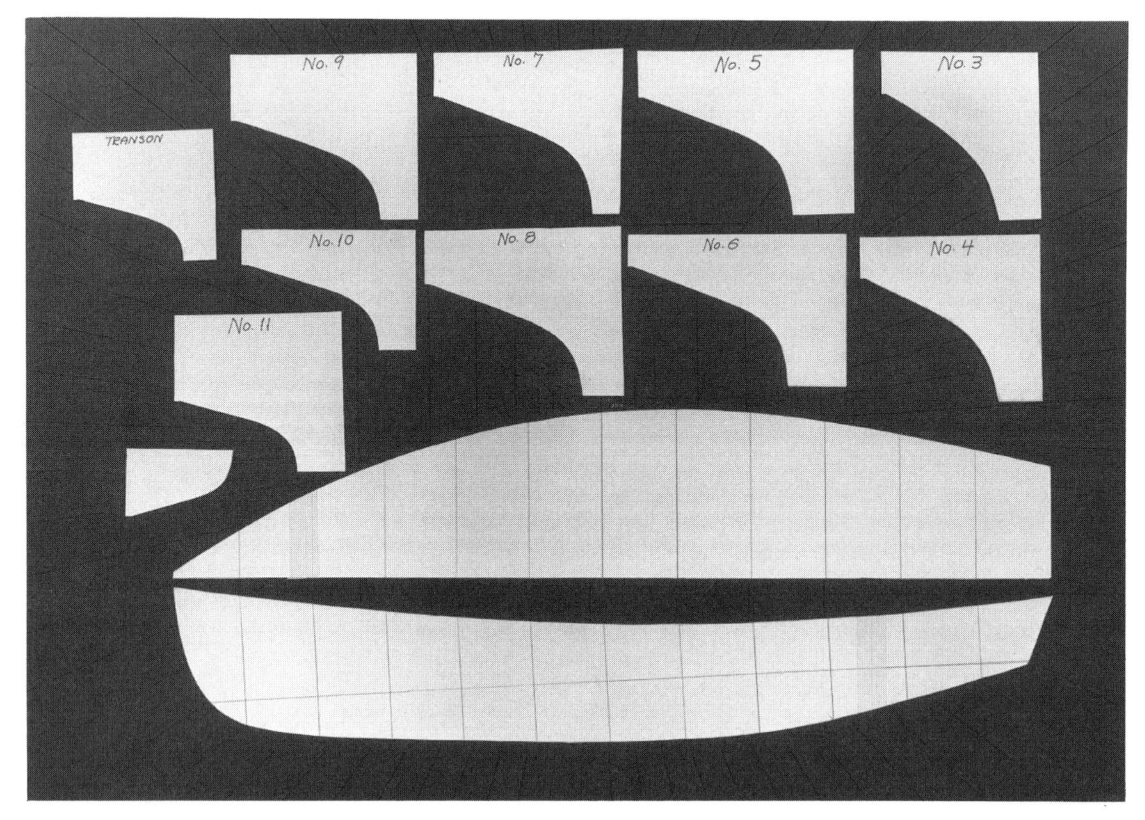

Posterboard templates for shaping a solid half-model. From top to bottom: numbered sectional templates; sheer half-breadth with section lines marked; profile template with section lines and load waterline indicated (Mary Anne Stets photograph, M.S.M.).

some changes in the hull are being considered that would be much easier to visualize from a model than from a two-dimensional lines drawing. Models can be especially helpful when considering minor, and even major, changes in an established design to adapt it for some special use or to meet the requirement of a particular owner. In carving the model it is easily made wider, narrower, fuller, shorter or whatever, by taking off a little more here, or a little less there and it can just as easily be made longer by gluing on additional wood before it is shaped.

To build a solid half-model of a design for which there are lines, the following shapes are required for templates, the profile view projected on the vertical, fore-and-aft center plane, the plan view at the sheer and the sectional curves of the hull surface at each station. These can be traced on stiff cardboard from the lines plans using carbon paper and they are cut out with a sharp pointed knife drawn towards one. Frequently it is desired to build a model from lines published in a magazine or book. Invariably these are too small and must be enlarged to an appropriate scale. For this, proportional dividers are invaluable, the larger the better.

It hardly needs to be said that an indispensable prerequisite for making models from boat lines is at least minimal knowledge of what lines are and how to draw and fair them, and there is no better way of gaining this experience than making enlargements of lines from books and magazines.

In preparing the rectangular block for cutting a model it should be trued by planing to ensure that its opposite surfaces are parallel throughout, and that its sides are square with each other all around. If it is glued up from several pieces a colorless glue should be used and one that will not dull sharp tools when it has hardened. It need not be waterproof.

Begin by laying out the load waterline and the section lines on the back, or inside surface of the block. Section lines are perpendicular to the load waterline. Square the section lines around all four sides of the block, and carry the load waterline through to the front of the block from the back making sure that the horizontal plane generated by the load waterline is perpendicular to the vertical centerline plane represented by the inner face of the model. Apply the profile template to both sides of the block in turn, marking the outline in pencil. Next place the half-breadth plan-view template of the sheer on top of the block, making sure section lines on the template correspond with those on the block and trace around the outside. This line corresponds with the sheer line looking down from above.

The block is now ready to be cut to the lines as marked. The easiest way is with a sharp band saw, provided one is available, otherwise hand tools will do the job.

If a band saw is used, the block is first turned on its side and cut in profile, saving the pieces cut off the bottom and the top of the block to be temporarily tacked back in place. This done, and with the block standing upright, it is cut along the sheer line that was marked in plan view with the half-breadth template. This completes the sawing and the pieces that were temporarily tacked back on are removed and discarded. A word of caution in sawing—do not cut into the lines. Leave enough wood to allow for cleaning up with the plane without getting inside the lines. The saw must be sharp, otherwise it may "run off" and cut under, leaving the block scant on the outer side. This is also likely to happen if the wood is fed into the saw too fast in making a deep cut.

Shaping with hand tools begins by working down the curved surface on the top to conform to the sheer line in profile. This is best done with a plane with a slightly curved or rockered bottom, or with a spoke-shave set fine. As the outside sheer line, as first marked, was on the pieces that were removed after sawing, we are obliged to work from the sheer line marked on the back or inside of the block. As we plane the upper surface we keep squaring across with a try square from the inside or back to make sure we do not get too low. This operation is finished when this upper curved surface is smooth and fair and exactly square—that is, perpendicular to the centerline plane which forms the back of the block. The section lines are squared across it and marked and the plan view template is again applied and the sheer line is remarked, this time permanently.

Half-hull scale models generally do not include the outer portions of the stem, keel and deadwood, which are made and applied separately should the model be mounted on a backing board. Models for larger vessels, made to the outside of the planking, go as far as the rabbet line which

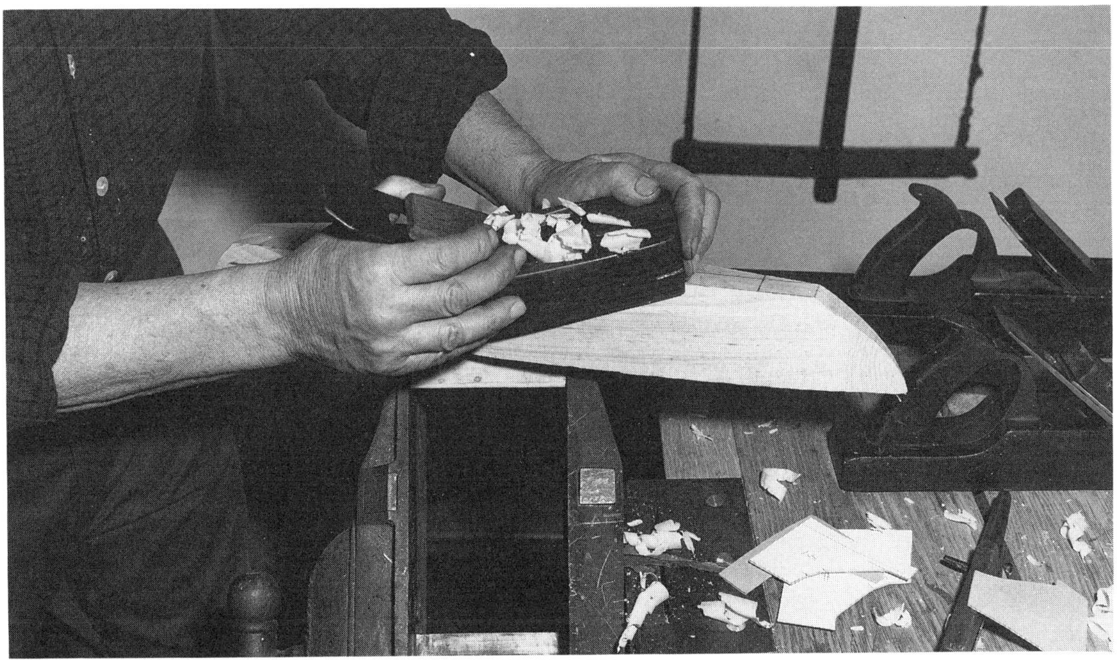

The author at the bench shaping a solid half-model of his first Mystic-built boat, the* Lawton *(Louis S. Martel photograph, M.S.M.).

Solid half-model being shaped on an iron swivel vise. Note the octagonal wooden gripe screwed to the top of the model and the template used to test the shape (Mary Anne Stets photograph, M.S.M.).

squares in to the center plane. Half-models of small craft made to the inside of the planking, end at the bearding line, likewise squared in to the center plane.

What remains to be shaped is the outer curved surface of the hull between the sheer and the stem and keel. Templates made from the lines for each section to fit the finished surface of the hull are required for each of the hull sections.

Working down the hull surface to the shape required by the lines is a slow, fussy, painstaking job calling for close attention and sharp tools. These may include chisels and gouges, the latter of various curvatures, small planes, some with slightly curved soles, spokeshaves, as well as special small model-makers shaves. And all must be kept razor sharp. A leather strap helps for this. Some model-makers depend on fine wood rasps and make extensive use of sandpaper. But I have found that I get better results shaving the wood than by abrading it.

To hold the block solidly in a convenient position for working, a sturdy iron vise that swivels is recommended, similar to the one shown in the photographs. A wooden gripe octagonal in shape is screwed to the block so it can be clamped upright or at a 45 degree angle in the vise. The gripe is attached with two large wood screws either to the top or back of the block, whichever affords the most convenient position for working. It can be changed from one to the other as desired.

At first, considerable excess wood can be trimmed away with a drawknife, spokeshaves and large paring chisels and gouges. Care must be taken not to remove too much, particularly at the turn of the bilge. To guard against this, the templates of sectional shape should be frequently tried at their respective stations.

As the shaping process proceeds, an effective way to control the removal of excess wood is by using a very sharp flat gouge of good size, similar to what carvers call a grounding gouge. This is used across the grain along the line of the section running from the sheer to the keel rabbet. Again, the template for that section is frequently tried until it fits throughout.

The wood remaining between the stations is removed with paring chisels and gouges, spokeshaves (the old wooden ones work best) and small planes set fine. The final shaping is done with even smaller metal planes and shaves. These should be very sharp and set for tissue-thin shavings. The resulting surface will hardly need sandpaper. Some builders find using small flexible battens bent around the model lengthwise, or the way the planks will run, effective for checking fairness in the final stages. Touch will sometimes reveal irregularities that the eye fails to detect.

The procedure just outlined will produce a solid half-model identical in shape to the lines laid down at the beginning. As the shaping process proceeds, the designer/builder is able to make changes as they seem to be required. It is easy to leave wood here or cut away a little more there. He may even decide to glue on pieces to get the shape he wants.

After the builder has finished and has achieved a shape that satisfies him, there

Posterboard inserted in saw kerf to take section shape (Mary Anne Stets photograph, M.S.M.)

But if this is done the model will no longer be intact. The sectional shapes can also be marked on thin pieces of stiff posterboard inserted into narrow saw kerfs that have been cut into the model along the section lines. Enough wood is left uncut at the back to prevent the model from falling apart.

In making these cuts, the saw should be fine enough for posterboard to fit snugly in the saw kerf. The sectional outline is accurately marked on the posterboard with a sharp hard pencil. For most design models up to 30" long, a carpenter's large adjustable miter box will be the best means for making these sectional cuts in the model. For larger models, rigging a special jig for this should not be difficult to improvise.

remains the job of taking off the revised lines, or merely getting sectional shapes for building molds. There are various methods of doing this. The simplest is to cut the model all the way through at the stations and trace the section lines from each piece.

If for any reason sectional cuts cannot be made in the half-model, there is an accurate method for taking off lines which utilizes a device called a bridge. A bridge for this purpose is what amounts to a movable,

Bridge for taking off sectional shapes. The model is resting on its back on its profile view, which has been traced out on the horizontal drawing board. The bridge is set up on one of the extended section lines and the baseline is marked on both the drawing board and the bridge face (Judy Beisler photograph, M.S.M.).

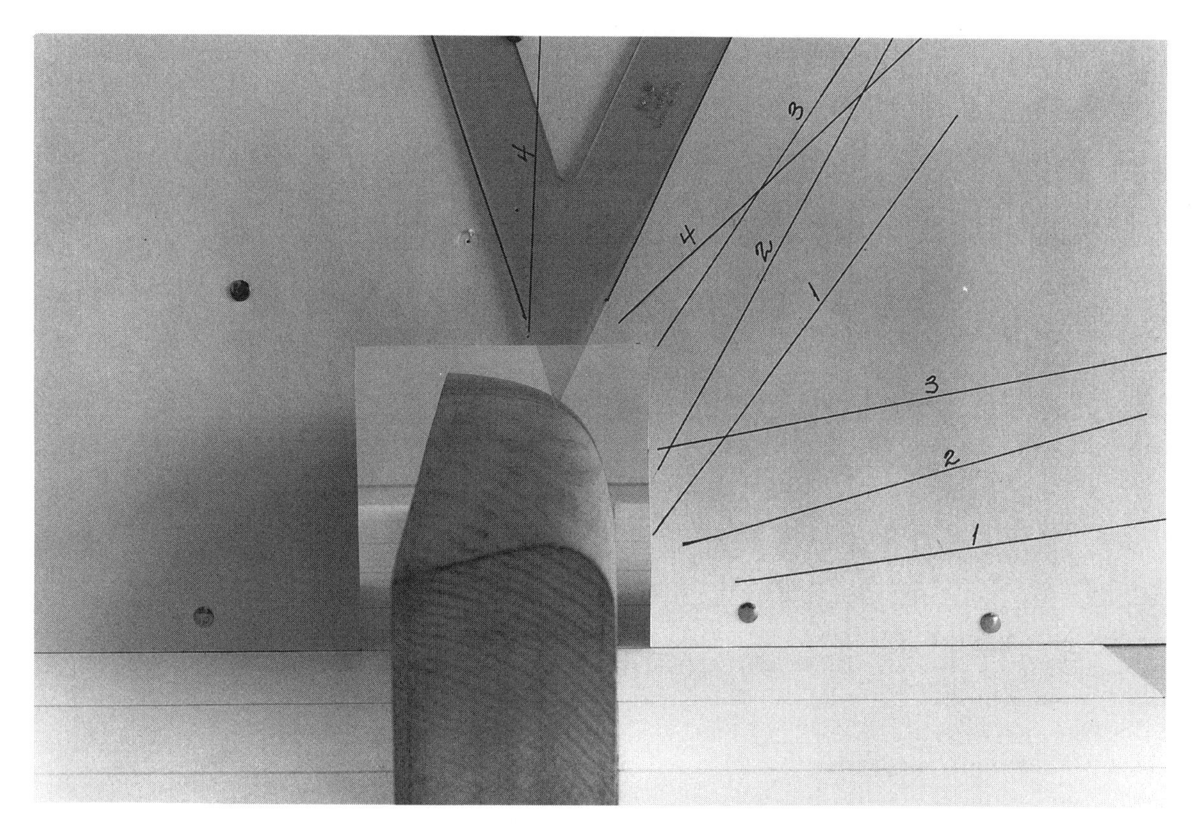

Close-up of bridge used for taking off lines from a solid half-model. When sufficient points are taken at this station and the location of the baseline marked, the posterboard is removed and the points transferred to the body-plan layout (Mary Anne Stets photograph, M.S.M.).

vertical drawing board of sorts with an aperture cut out of the bottom large enough to allow it to bridge over the half-model being measured when it is lying on its back on the horizontal drawing board on which it is standing. The bridge should remain where it is put without moving or slipping out of position, with its working face perpendicular to the horizontal surface on which it is standing. For this reason the bridge should be solid and heavily made, and if it can be clamped in place it should be. Drawing lines on its face, even if they are lightly done, will tend to move it off the mark and out of position.

In taking off lines by this method, enough level drawing surface will be required to lay out the model for measurement with the bridge, and for drawing a body plan of the model from these measurements taken with the bridge. One large board may suffice. If not, two separate, smaller boards will be required.

Preparatory to positioning the model for take-off, the surface of the board is covered with a sheet of detail paper on which the profile outline of the model is drawn with its section lines extending beyond the profile outline five or six inches or more. A baseline parallel to the LWL and some distance below the profile outline is also drawn. The half-model is then placed back down over the profile outline with its section lines matching those on the detail paper exactly. Some provision should be made to prevent the half-model from slipping out of position. A few pieces of masking tape are usually all that is needed.

The working face of the bridge is covered with special drafting media, temporarily pinned in place with thumbtacks. Its surface should be smooth enough to take fine pencil lines readily. It should be stiff enough so as not to wrinkle or buckle and it should be as thin as possible. Sheets of ordinary thin white posterboard meet these requirements. Also the bottom edge must be straight in order to make contact throughout with the horizontal surface on which the model rests. Make the cutout in the bottom to clear the model no larger than necessary.

The curved shape of the hull surface is taken off at each section with the model's bow facing to the right. The sectional shapes of the forward half of the boat are taken off starting at the midsection and proceeding to the right. Then the bridge is reversed and the sectional shapes of the after half are taken off starting amidships and proceeding aft or to the left.

In taking off the shape of a sectional curve, the bridge is positioned on the extensions of the line of that particular section as drawn on the horizontal surface on either side of the model. If an extension of the plane represented by the working face of the bridge were dropped, it would cut through the half-model exactly on the section line.

With the bridge set up in this position, the location of a number of spots on the surface of the half-model where the sectional plane cuts through it, are registered on the sheet of posterboard tacked to the face of the bridge. These spots must be sufficient in number and so positioned, that when they are transferred in making a body plan of the model, a curve drawn through them will reproduce the curve from the model exactly.

If this procedure is carefully followed, very accurate results can be achieved, more accurate than those often obtained by using complicated lines-taking machines with moving parts.

When the bridge is used, all that is needed in taking off the sectional curves and transferring them to a body-plan drawing is an ordinary plastic drafting triangle with a 30-degree angle and a sharp, hard drawing pencil. The 30-degree end of the triangle needs to be perfectly sharp. The triangle is laid flat against the working face of the bridge with its 30-degree end barely touching, yet touching, the surface of the half-model, and light pencil lines are drawn on each side of it on the sheet of posterboard temporarily tacked to the face of the bridge. When the triangle is removed the two lines are numbered to identify them as a pair. This process is repeated at frequent intervals along the entire length of the curve, including its ends. Each time before moving the triangle to a new location the two lines just drawn are numbered to identify them as a pair.

Finally, before removing the tacked posterboard from the working face of the bridge, the location of the baseline is marked on it, squaring up with a large triangle from the horizontal surface on which the model is resting.

On the other drawing surface preparations for laying out the body plan of the model are made. The baseline is drawn and a perpendicular representing the vertical, fore-and-aft center plane of the model is erected. The posterboard sheet marked with the data for the first of the sectional shapes, is removed from the bridge and transfefred to the body-plan layout. It is correctly located by lining up its bottom edge with the vertical centerline for the body plan drawing, and by matching the baseline location marked on the posterboard take-off with the baseline on the body plan drawing.

What remains to be done before the curve of the section can be drawn is to extend the pairs of lines drawn with the 30-degree triangle until they intersect. Their intersections produce the points through which the sectional curves are drawn. With care, a reasonably high degree of accuracy in duplicating the sectional curves of the hull can be achieved.

By this method each of the sectional curves is taken off in turn and transferred to the body plan, and as soon as the body plan is complete building molds can be made.

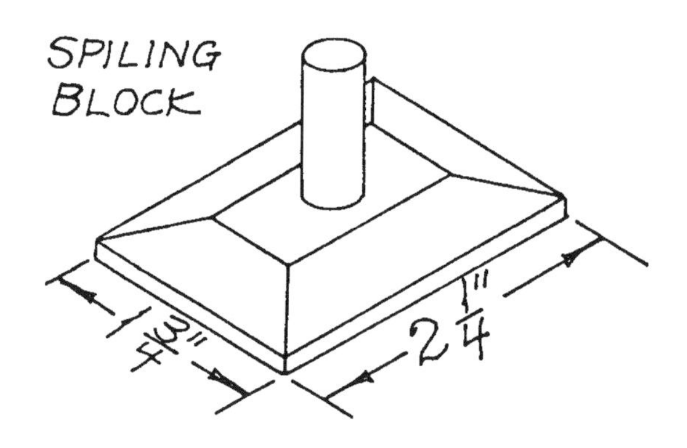

CHAPTER NINETEEN

TAKING OFF BOAT LINES

There is no mystery in taking off boat lines and no special aptitude or skills are required. Almost anyone so motivated can take off an acceptable set of lines, provided he has a basic understanding of what boat lines represent and how they are drawn.

The importance of this proviso is not to be underestimated. Even under the most favorable circumstances, there is much painstaking, not to say tedious, work involved. But more than this, different classes and sizes of boats call for variations in the basic procedure. And, going a step farther, almost every boat presents its own special problems which can occasionally tax the resourcefulness of the measurer.

The complications may include: where the boat is located—in a dark, cluttered shed, on a muddy river bank or in a clean, lighted shop; the boat's condition—badly rotted and falling apart, partly smashed with broken planks or clean and retaining its original shape; the amount of time available for the job—a hurried few hours at the end of a long trip to a distant location, or with the boat close at hand and an unlimited opportunity to check and re-check doubtful or omitted measurements; and, finally, the order of accuracy required or obtainable—rough shapes and dimensions as background for working up a derivative design, or precise measurements for the creation of an accurate replica. In this last regard, absolute accuracy is never possible, even if it were desirable.

These are some of the factors that often complicate the lines-taking process and for which the boat measurer will have to make provision. He will need to be ingenious, even inventive, in adapting his procedure to fit the occasion. And it will not

Small craft staff of Mystic Seaport taking off the lines of a nineteenth-century Sheldon whitehall during field research in Cohasset, Massachusetts. Note the datum or reference line running level fore-and-aft from the top of the stem to the transom. All vertical measurements are taken from this line (Robert A. Pittaway photograph, M.S.M.).

hurt to have some boat building experience, although this is not necessary.

What is absolutely essential, as I have already stated, is a basic knowledge of boat lines and at least minimal facility in working with them. In boat construction, the builder starts with a draft of the lines in order to produce a boat. In taking off lines, the process is reversed—the measurer starts with the boat in order to produce a draft of lines.

The procedure he adopts may vary, but usually it consists mainly of taking the shape of a number of transverse sections at measured intervals along the length of the hull. These sections must be established perpendicular to the load waterline plane and at right angles to the centerline plane of the hull. From the outlines of these sections, a body plan of the hull can be worked up. This, when related to the profile shape of the boat, provides the principal data for completing and fairing a layout of lines for the hull.

The simplest and easiest example of the procedure is taking a number of sectional shapes from the interior of a small, open boat using templates and a spiling block, as is shown in the accompanying diagram. The section depicted is located near the after end of a 19-foot Whitehall.

Now, in the case of a 40-foot decked-over lobsterboat that was partly sheathed inside, with a large, tight, decked-over, self-bailing cockpit, it would be necessary to take the sectional shapes from the outside of the hull. This would require a different procedure and other equipment, as I shall explain later, but the principle and the end result would be the same: a series of sectional shapes through the surface of the hull.

The smaller, open boats are easily blocked up and leveled athwartships and fore-and-aft at a convenient height for measurement. When the boat is so secured, a chalk line or thin wire is tightly stretched between the stem head and the top of the transom to serve as a datum or reference line. It is best if this line runs level fore-and-aft. All vertical measurements are taken from this line.

Moreover, the vertical distance from the datum line to the sheer at each section is marked on the template for that station and used in locating the template at the correct height when the layout of the body plan is made. If possible, the layout should be made full-size on some sort of table or scribe board prior to its reduction to whatever scale is chosen for the draft of lines.

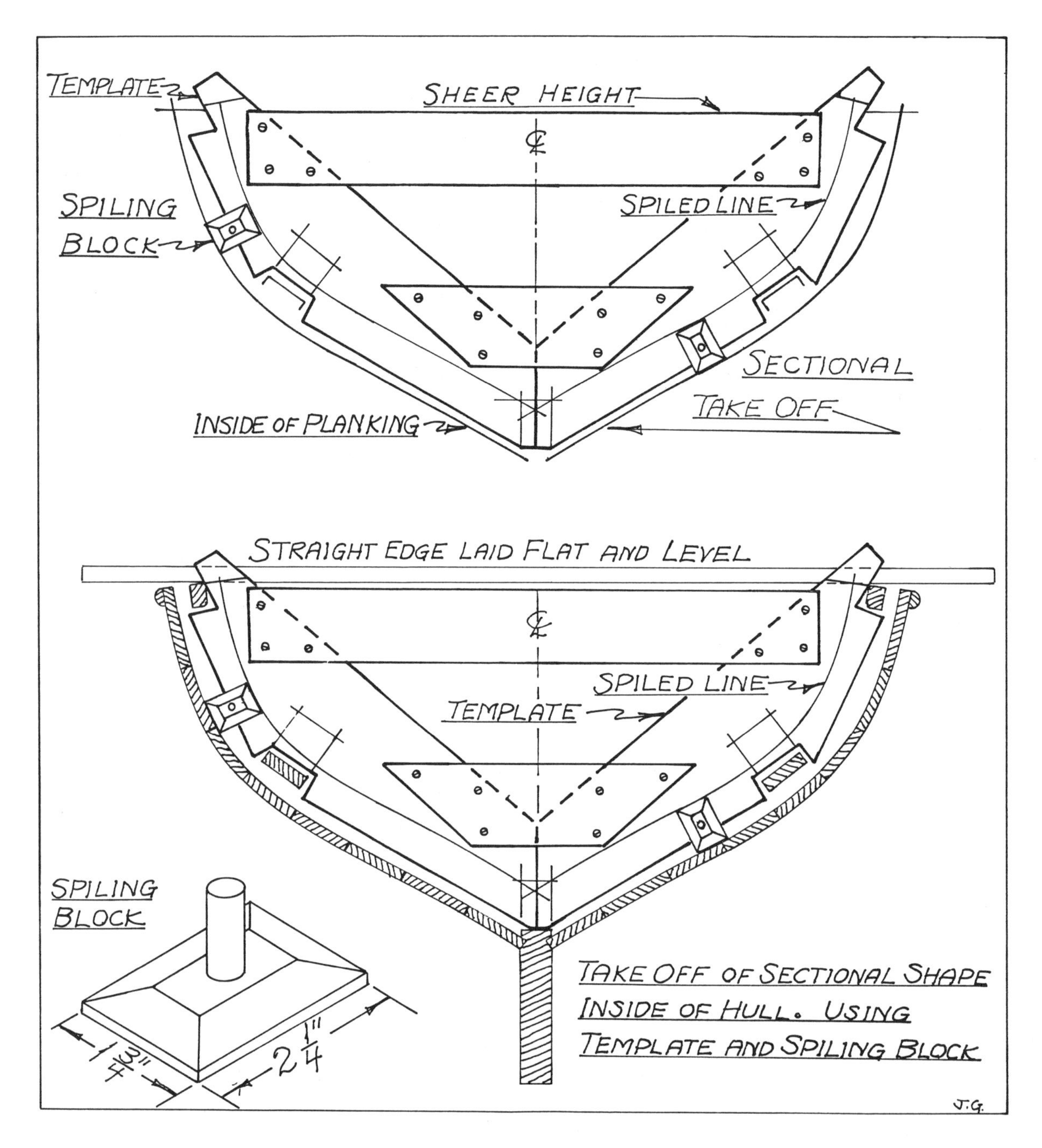

Before the sectional shapes are taken, both their number and location in the boat are decided. This is a matter of judgment based largely on what has worked well for similar boats. Naturally, obstructions in the interior of the hull should be avoided if possible. There is plenty of leeway in choosing these stations, as body sections can be relocated as desired when the final draft of the lines is made. At the ends of some hulls, where the shape of the boat changes rapidly or abruptly, more accurate results will be obtained by locating stations closer together.

In making up templates on which to record the shape of the boat at each of the selected stations, one could use a single piece of thin plywood, scribed and cut roughly to the shape of the hull's interior. Where single lengths of plywood are not available for templates, several shorter pieces of template stock can be fastened together to form a template that will approximate the shape of the side.

Small screws are best for fastening templates together, as they are easily removed, allowing the template material to be used again and again. Where a number of short pieces are joined, several small clamps are useful for holding the components in position until they can be screwed together.

In most cases the shape is taken on one side of the boat only, which is much less work and gives close enough results for most purposes. When a boat is badly out of shape or partly wrecked, it is sometimes necessary to take some of the sectional shapes from one side and some from the other.

The horizontal straightedge shown in the drawing is clamped on to the inwales on either side, and the upper ends of the sectional template are secured to it. Frequently, the bottom of the template is fastened to a small block that has been temporarily tacked to the top of the keel. Of course this is only one of many ways of holding the template in place until the spiling is made.

Note that the upper edge of the cross spall (horizontal brace) at the top of the sectional template in the drawing is put on level athwartships at sheer height. Note, too, that a vertical centerline is marked on the template. This ensures that the sectional shape will be correctly located when it is transferred to the full-size body plan.

The use of the spiling block in taking off lines should be clear from the accompanying drawing. In any case, the procedure is a simple one. Once the measurer has set up and plumbed a given template on the desired section within the boat, he picks up the spiling block with one hand and a good drawing pencil with the other.

With the flat (bottom) surface of the block held flush against the face of the template, the outboard edge of the block is pushed tight against the interior of the hull. Then, with the pencil against the inboard edge of the block and its point against the template, the block is moved along the hull, thus duplicating the section's shape on the template.

One thing deserves mention here. In transferring the scribed line from the template to the full-size body plan, the spiling block will be raised above the drawing surface by the thickness of the template stock. If the pencil is tipped in under the block, the transcribed line will fall short; if the pencil is tipped out, the line will be wide. To ensure accuracy, a small block with a square end may be placed on the layout surface and set against the edge of the spiling block; the pencil line is then marked along the inner, bottom edge of this secondary block.

It is worth noting that the spiling block has other uses in the lines-taking process besides duplicating the sectional shapes of a boat. Using the same procedure explained here, the profile shape of the stem and stem rabbet can be taken off the outside of the hull using the spiling block and a curved template that is set vertical and clamped to a post.

On larger vessels with restricted access to the hull's interior, the measurements required for taking off lines are usually made on the outside. From a series of stations established at regular intervals along the length of the hull, sectional shapes are taken off. This produces a body plan from which a half-breadth plan and a profile plan are developed.

Diagonals are run in, as the fairing proceeds. Of course, in order to be able to do this one must have a basic understanding of how boat lines are drawn and what they represent, as was noted in the opening paragraph. Unfortunately, this procedure is too long and involved to explain here. Adjustments and corrections among the three views are made, and differences are worked over until they are reconciled or eliminated. Eventually, a fair set of lines results. From start to finish, this can be a time-consuming and demanding process, with success hinging on getting an accurate take-off of sectional shapes to begin with.

The basic procedure for making such a take-off is shown in its simplest form in the accompanying drawing. Many variations of this procedure are possible, and modifying it for special circumstances can require much ingenuity. However, if the principles are understood, they can generally be adapted to any situation you're likely to encounter.

The primary measurement base is the load waterline. If a boat were floating on an even trim in smooth water, this imaginary plane would cut through the hull even with the water's surface. All of the horizontal measurements are taken parallel to the load

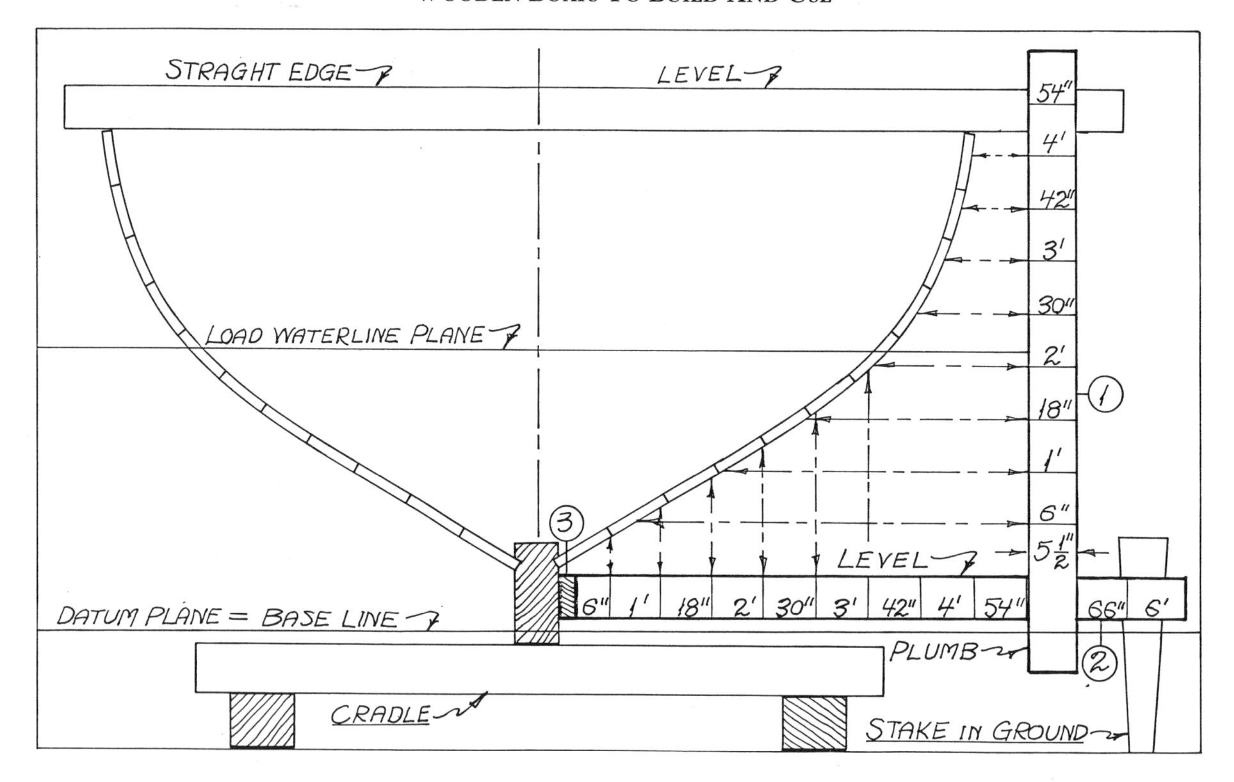

waterline plane. All of the vertical measurements are taken perpendicular to it. For convenience, a datum, or base, plane parallel to the load waterline plane is assumed beneath the boat, and vertical measurements are made from it. The datum plane could just as well be assumed to lie above the boat, and it sometimes is.

Obviously, on two-dimensional, flat drawings, planes must be represented by lines. Thus, the load waterline plane appears in lines drawings as the load waterline, and the datum, or reference, plane is the base line. In addition, the fore-and-aft plane that divides the hull into two identical, longitudinal halves becomes the centerline on the lines plan.

It is highly desirable, although not absolutely necessary, that the boat be leveled athwartships prior to measurement, and usually this is quite easily done. Though it is possible to take lines off a boat that is listing to one side, it is a much more difficult operation, and a special procedure is required. It is not so important for boats to be leveled lengthwise to conform to their load waterline because any compensatory adjustments in the fore-and-aft leveling can be made on the drawing board. Though this means considerable extra work, it's not especially difficult.

The diagram appearing here represents a sectional view of a boat which is leveled athwartships. The extra-long, parallel-sided, wooden straightedge extending across the boat forms a right angle with the centerline plane, and its forward face coincides with an extension of the sectional plane at the station being measured. The straightedge is clamped to the cap rail or bulwarks or is otherwise temporarily secured in position. The vertical measurement arm (1) may be clamped to it, although this vertical arm (which is also a parallel-sided straightedge) is often set up close to the boat and clamped to the toe rail or bulwarks. This vertical arm stands plumb both fore-and-aft and athwartships, and its after face coincides with the sectional plane along which the sectional measurements are made.

A block (3) attached at right angles to the inboard end of a horizontal measurement arm (2) is clamped to the parallel-sided keel, lining up the forward side of the horizontal arm with the sectional plane. If the sides of the keel are not parallel, other provisions must be made for holding the horizontal arm in position. To hold the horizontal arm both level and square with the center plane line, clamp it to a stake, as shown, or to some other secure support.

At regular intervals, measure the distance to the skin of the boat from both the vertical and the horizontal arms. This allows the sectional shape to be reproduced to scale on the drawing board, or at full-size on the lofting floor. The more accurate the measurements, the more reliable is the reproduction. It is absolutely critical that these measurements are made exactly perpendicular both to the leveled horizontal arm and to the plumbed vertical arm. Also, they must lie exactly in the sectional plane, which is at right angles to the center plane of the boat.

Toward the ends of the boat, where the curvature of the sides falls off increasingly between stations, the utmost care must be taken to keep measurements within the sectional plane. This becomes increasingly difficult as the fore-and-aft curvature increases. If measurements are deflected forward, they will be too long and if deflected aft they will be too short. Some means must be found for locating the sectional shape on the hull surface. Sometimes this is done by sighting, using two plumb bobs. Sighting with a transit is ideal, and for accurately measuring large vessels, a transit is indispensable.

Another method especially well adapted for lifting the lines from larger vessels, which is just as accurate and possibly easier and faster than the method just considered, is the triangulation method used by the naval architect, Charles H. Wittholz. He has sent me a full explanation of the triangulation method he has settled on as being best for lifting lines.

"After having performed this exercise with varying results over a period of years," Wittholz states, "I have concluded that a system of triangulation at selected stations is by far the most accurate and convenient method of doing the job." And, he points out, being able to do the job quickly is a definite advantage when working on a tight budget.

Wittholz became acquainted with the triangulation method years ago when he was making calculations for the Coast Guard required for refitting the 91-foot schooner *Western Union* in adapting it for passenger service.

To make the necessary calculations, Wittholz needed a proper lines plan. None existed, so, at the suggestion of the Merchant Marine Inspection Division of the United States Coast Guard in Washington, D.C., he adopted the triangulation method of lifting hull lines. Wittholz found the results were highly satisfactory, as demonstrated in his notes and diagrams.

The steps involved in applying this method are outlined as follows. Naturally, they can be adapted to suit particular circumstances.

First, the vessel is leveled athwartships and also fore-and-aft on its waterline. If this is not possible, the amount of declivity is measured and recorded so that the necessary adjustments can be made later on the drawing board. Then, measurements for laying out the profile view of the vessel are taken off, so that a profile plan can be drawn. Beam measurements are also taken for drawing a plan view of the sheer.

The location of a number of transverse sections at selected stations through the hull are spaced off. These run square with the vertical center plane, represented by the centerline and they are perpendicular to the horizontal datum (or base) plane which is represented by the base line. At each station, the locations of the sheer and the keel rabbet for that section are readily ascertained. They lie at either end of the sectional curve through the surface of the hull.

Completing that curve requires establishing a number of intermediate points spaced at what the measurer judges to be suitable intervals along the curve on the outside of the hull.

The measured locations of these points are taken from the vessel by means of triangulation. The measurements which are thus obtained are then transferred to the drawing board, enabling the draftsman to reproduce the sectional shapes needed for constructing a body plan of the vessel.

How the triangulation method works is illustrated by Diagram **I**. Here, three points—**A**, **B**, **C**—are shown laid off under the hull at measured intervals from the centerline and on a line in the datum plane perpendicular to the centerline plane of the vessel.

Seven points on the surface of the hull are shown, marked **S** (sheer), **2**, **3**, **4**, **5**, **6**, and

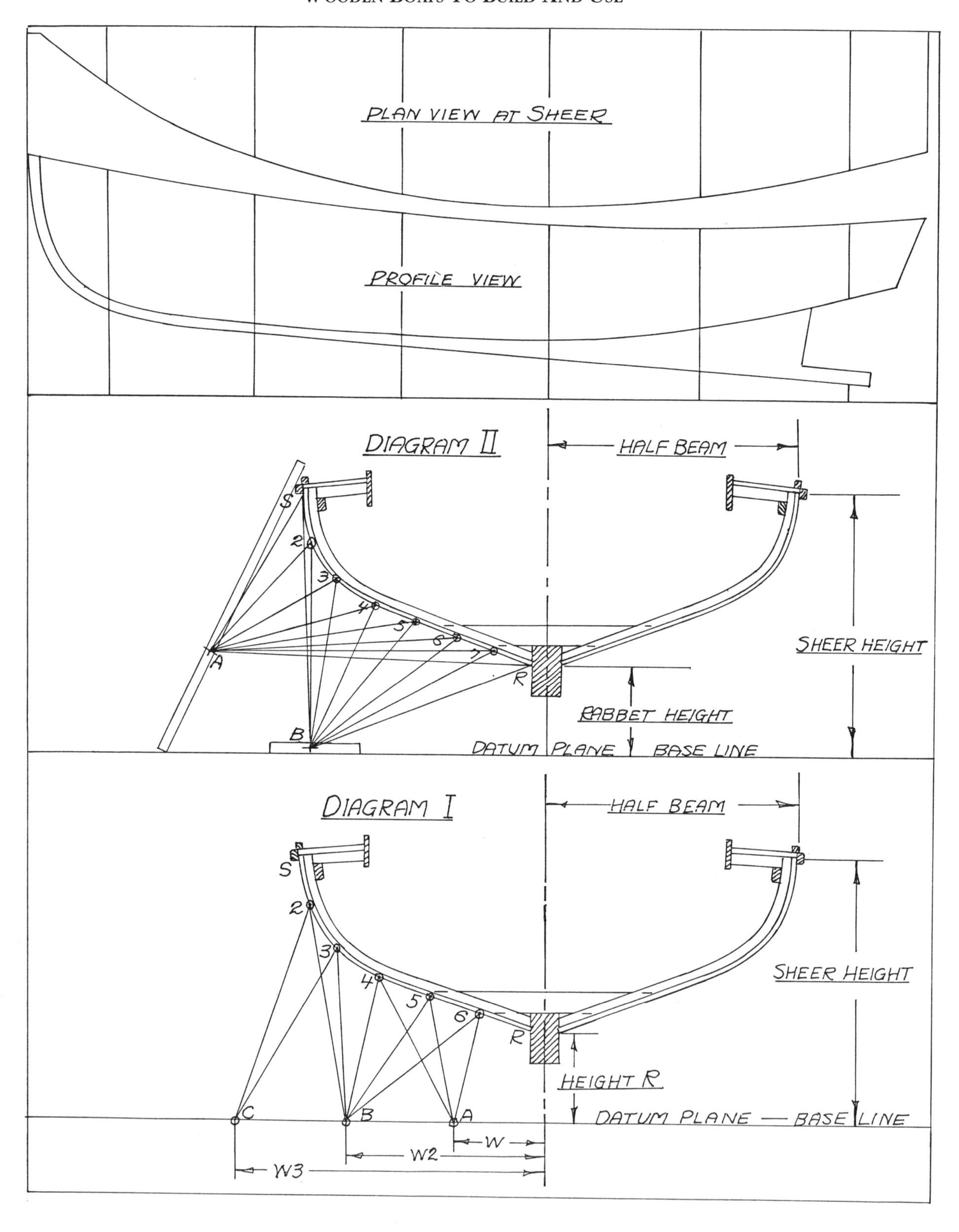

R (rabbet), spaced at intervals along the sectional curve cutting the hull surface. The locations of **S** and **R** are already established by known dimensions of beam, sheer heights and keel siding. What remains to be lifted off are the locations of points **2**, **3**, **4**, **5** and **6**. The measurements for triangulation are taken from the vessel to be transferred to the drawing board using a simple procedure.

The distance from Point **A** to No. **6** is measured on the vessel, and dividers are set to this measurement (with adjustments for the selected scale). At the drawing board, with one of the dividers' legs on Point **A** as the apex, an arc is swung in the likely vicinity of No. **6**. Next, the distance from Point **B** to No. **6** is measured from the vessel, and the dividers are reset accordingly. With one leg on Point **B** as an apex, an arc is swung so as to intersect the previous one.

The intersection of these two arcs completes the triangle and reproduces the location of No. **6** exactly. By using the same triangulation process, the locations of the remaining points are established and then carefully transferred to the drawing board, whereupon the completed sectional curve is drawn through them. Note that Points **B** and **C** are used to locate Nos. **2** and **3**, as Point **C** is in a better position relative to Nos. **2** and **3** than is Point **A**.

It should be mentioned that all measurements of the vessel are usually taken in the field as a single, separate operation, after which they are removed to the drafting room in the form of notes and sketches. Only rarely are the two operations combined and carried on at the same time, as I have shown here for purposes of explanation.

Note that unless the sectional shape is chalked or otherwise temporarily located on the surface of the hull before the measurements are taken, there usually is considerable room for error in measuring either to one side of the section or the other. Toward the ends of the vesel, where the shape changes more rapidly from section to section, even slight deviations of this sort are bound to produce serious mistakes. On large vessels, in particular, errors of this sort can be avoided by using an engineer's transit to line-off sections on the surface of the hull.

By this means, it is easy to locate and mark the sections precisely. On smaller craft, or when a transit is not available, much the same result can be achieved by sighting past the strings of two plumb bobs dropped at some distance apart to a continuation of the section line extending a bit beyond the side of the hull. (A chalk line stretched at right angles to the hull will do for this extension.) I, myself, have used this method with satisfactory results.

Diagram **II** shows a variation of the triangulation method used by Quershi, Magnusson and Jan Olof Traung for taking off sectional curves from West Pakistani fishing craft. They explain the technique in *Fishing Boats of the World, Vol. 1*, p. 45-48. Here, too, a profile view is laid out first, followed by a plan view of the sheer from beam measurements. This permits both the overall height and breadth of each section to be laid out in turn. It also allows the terminations of the sectional curves at each station to be located precisely at the sheer and the keel rabbet.

As shown in the diagram, a series of points, Nos. **2**-**7**, is spaced off on the outside of the hull. Measurements are taken to the two apex points **A** and **B**, conveniently located in the sectional plane outside and below the hull. While these two points must be located in the sectional plane, no vertical or horizontal measurements made in locating them need be taken into consideration. So long as they are in the sectional plane, they can be located where it is most convenient for triangulation measurement.

To locate **B** on the drawing board, set the dividers with a scaled measurement of the distance from **B** to **R** (rabbet). Place one leg of the dividers on point **R**, whose position has been established in the profile view and from the keel siding, and swing an arc in the likely vicinity of point **B**. Next, reset the dividers with the measurement from point **S** (sheer) to point **B**, and with one leg of the dividers on point **S**, swing a second arc intersecting the first. The intersection of the two arcs will complete the triangle locating point **B** on the drawing board. Point **A** is located in the same manner. Once the locations of points **A** and **B** are established, the intermediate points between **S** and **R** (Nos. **2**, **3**,**4**, **5**, **6** and **7**) are readily located by triangulation.

The difference between this variation of the triangulation method and the standard form used by Wittholz is very minor. The greater latitude permitted by Traung's method in locating points **A** and **B** could be a slight advantage when the vessel must be measured in an obstructed situation, as was the case with some of the Pakistani fishing boats, which were hauled out on a river bank.

In reply to comments made by me, Wittholz has since responded in part as follows: "When I took the lines off the schooner *Western Union*, I measured directly from the heavy, horizontal cross timbers on the dry dock under the ship's keel. This was much more convenient than establishing apex points at varying heights, as described in *Fishing Boats of the World.* With my method, I used as many apexes as required, but all at the same height, at the top of the cross timbers on the dry dock.

"You are absolutely right that the fore-and-aft position of the points taken on the hull must be accurate. Rather than going to the trouble of chalking each station line along the hull, I simply raised a plumb bob from the transverse station line up to the hull to insure proper fore-and-aft location of each point."

Boat builder and designer Paul Gartside and assistant Dorothy Jones taking off the lines of the Grand Banks schooner* Robertson II *built in 1940 by W. G. McKay and Sons of Shelburne, Nova Scotia. Using a movable, temporary frame Gartside recorded a series of horizontal and vertical measurements to take the sectional shape at each station. Unique field problems included the building of a drydock deck extension to accommodate the 12' overhang of the vessel and adapting the procedure to account for the list which developed in the drydock trim each day (see also,* The Boatman, *May 1966). An ambitious building project is underway to recreate the* Robertson II *which has been operated as a sail training vessel by the Sail And Life Training Society (SALTS) of Victoria, British Columbia, for the past 20 years (Bill Hayward photograph).

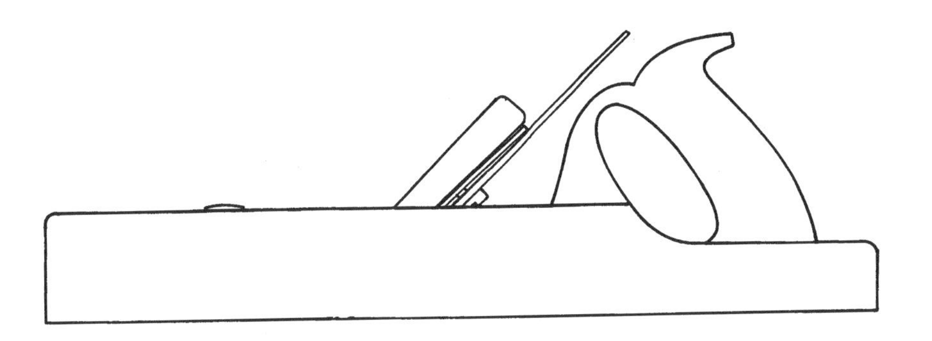

CHAPTER TWENTY

ON BUILDING SMALL OLD-TIME WOODEN BOATS

The observations and directives that follow touch briefly on a few salient aspects of small craft construction about which I have been questioned from time to time in the past by readers of my books and articles. No attempt is made here to present a complete and systematic coverage of the subjects considered. To do so would fill many volumes, yet it is hoped that this information, as limited as it necessarily is, may yet prove useful to the beginner who is considering building a small, old-time wooden boat for the first time for his own use on the water, as well as for the creative satisfaction of putting it together with his own hands.

Building for oneself these old-time wooden boats, almost impossible to obtain now in any other way, is a doubly rewarding undertaking. Over and above the boat itself, the pleasure of accomplishment and the enrichment of the experience that its construction provides are things that cannot be bought with money.

Building and using a small wooden boat helps wonderfully to reestablish and strengthen connections with the natural world that so many of us have lost or are in increasing danger of losing. Wind and water have not changed, and the age-old workings and needs of the human body and psyche remain the same and cry for expression and fulfillment in a cold world of artificial abstractions and flickering images.

Building a small wooden boat for the first time takes the would-be builder into unknown territory. There is so much to figure out on his own. Neighborhood boat shops where he can hang out to see how it is done are all but nonexistent. He is thrown back on books that often do not explain procedures in a way that he can understand. Little things that don't seem worth going into for writers, often prove to be serious roadblocks for those who have never done it or seen it done. What may actually be very simple and easy to do, sometimes brings the novice builder to a standstill, wracking his brain and ingenuity. Reinventing the wheel can be a slow, uncertain and frustrating process, and so unnecessary, except that much that was once common knowledge has been forgotten and all but lost.

What follows are explanations of a number of processes and procedures that beginners may find helpful, things that are frequently taken for granted and passed over or ignored completely in the books and manuals.

Caulking is an all-but-lost boat building skill. The driven caulking keeps out water and also acts to stiffen and solidfy the sides of the hull (John Gardner photograph).

Caulking

Caulking carvel planked boats and vessels with strands of cotton or oakum driven into the seams is becoming a lost art with nothing that adequately replaces it. What is frequently not considered is that driven caulking serves another function besides keeping out water, and a critically important function, especially for larger vessels. Each of the thousands of loops or tucks of caulking, when driven into the seams and set down hard, becomes a wedge imparting pressure that stiffens and solidifies the sides of the hull. A well-caulked vessel has many tons of pressure locked into the hull walls binding them together as if they were fabricated in a single piece. A loosely caulked ship in a violent seaway will soon work herself to pieces, and not a few vessels have been lost in this way.

Modern caulking compounds applied with caulking guns or putty knives, may keep out the water, but do nothing to strengthen and stiffen the hull.

In carvel planked craft except for small tenders and the like, the seams between the planks are V'd, that is they come together tight on the inside, opening on the outside about 1/8" for each inch of plank thickness. Plankers sometimes use a gauge called a "feeler," made from a broken piece of thin hacksaw blade ground off smooth. In "hanging" or putting on the planks, they are forced together with clamps and sets until the "feeler" will not slip through between them. They need not fit tighter than this.

If a plank is incorrectly beveled so that it opens more on the inside than it does on the outside, it will not hold the caulking, and generally must be replaced with one correctly made. Sometimes a short gap of this sort can be covered with a strip of wood on the inside to prevent the caulking from going through. A botch at best.

In caulking a vessel, right-handed caulkers proceed in a counterclockwise direction, filling the seam with short loops of the caulking thread picked up and tucked in with the caulking iron held in the left hand, and tapped in with the mallet grasped by the right hand.

After the seam has been filled with small loops for a length of two or three feet, the caulker stops tucking, leaving a short length of caulking thread "tailing" out, and works back on the section he has tucked, driving the caulking down hard in the seam with hard blows of the mallet, that is, "making it in," as this is called. This compacts the caulking, wedging it firm and solid in the seam and far enough below the outer surface to be covered with putty or something similar. The caulker then moves forward, picking up the "tail" and resuming tucking as before.

In tucking caulking into the seam, the thread or strand of caulking extending across the palm of the left hand is grasped between the tip of the left index finger on the outside and on the inside by the blade of the caulking iron held in place by the pressure of the thumb. The strand of caulking is pinched in this way just far enough below the seam for the bite to be raised in a short loop and tucked into the seam. The pinch on the caulking strand is momentarily slackened, dropped slightly, tightened and another loop is lifted to be tucked into the seam. By varying the size of the loops and the closeness with which they are tucked, the caulker controls the amount of caulking that is put into the seam, less where the seam is tight, more where it opens wider.

This is the basic caulking procedure. As caulking strands are used up, they are replaced by others. Intervals of tucking are made in, and tucking is resumed again. All the while the mallet in the right hand is applied to the end of the caulking iron as needed to tuck in the loops and drive them down hard.

A right-handed caulker holds the iron in his left hand between the thumb and index finger and against the tucked cotton loops in the seam. The caulking, held between the underside of the iron and the index finger, falls between the fingers or lays across the palm (Sharon Brown photograph).

Never paint seams before they are caulked but soak them well with linseed oil paint as soon as the caulking is in. On new work, caulk the tight seams first which will close the wide seams tending to equalize them. Sometimes wide seams will need two strands set down twice. Large vessels with plank three inches or more in thickness may require several strands, say two of cotton followed by three of oakum set down hard with beetle and hawsing iron.

There is much more to the caulking trade than this, but this outlines the basic procedure, and is enough to start with.

Gluing

A strong, reliable, long-lasting, completely waterproof glue was a boon long denied wooden boat builders. The old-timers would have welcomed such a glue with open arms had it come along years ago, and the course taken by traditional boat building surely would have been changed by it. Not until World War II, when war needs turned research to thermoplastics, did plastic glues find their way into the boatyard.

Before that, the best we had was water resistant casein ("Casco") glue which worked fine so long as the glue joint did not become water soaked. This limited its use mostly to hollow spars protected from moisture by varnish.

First of the plastic glues to be taken up by boat builders were the urea resins. One of these, "Weld Wood," is still around today. Like the casein glues, the urea resin glues developed in this country were only water resistant, not waterproof, but they had some advantages over casein glue. They were easy mixing for one, and they continue to be used to a limited extent in boat work today for interior construction where the glue joints will be relatively dry.

A urea resin glue called Aerolite developed in Great Britain for gluing Mosquito Bombers is both waterproof, strong and gap-filling, up to a point. An acid hardener is applied to one surface of the joint and resin to the other. And the two are brought together in firm contact. Strong pressure is not required. Besides being waterproof and strong, Aerolite is easy and relatively clean to apply, colorless and economical, yet it never caught on in this country for some reason, and is not readily obtained here.

Urea resin was followed by resorcinol. Resorcinol glue is both strong and waterproof but exacting in its application. Joints should fit wood to wood, strong clamping pressure is required, as well as higher curing temperatures than are often found in the small boat shop. Besides, resorcinol stains, making it unacceptable where wood is to be finished bright.

Epoxy adhesives became available about the same time as resorcinol. Slow to gain acceptance at first, epoxy's superiority as a boat building glue has since been amply demonstrated. It is strong, durable and gap-filling to a moderate degree. There are epoxy formulations that will cure at temperatures barely above freezing, and others that will cure when submerged under water. By altering the constituents of the mix, setting time can be as short as several minutes, or as long as overnight.

Epoxy has acquired a reputation in some quarters of being hazardous to the health. This is largely undeserved. True, some hardeners and activators used in some of the adhesive formulations are caustic, may be sensitizing to the skin and may even be carcinogenic; but protective precautions can be taken when using them, and, better still, they can be avoided altogether. Formulations that are quite harmless and safe are available that can be used instead.

There are occasional individuals who are allergic to epoxy, just as there are those who are allergic to cat dandruff, bee stings, ragweed or cedar dust. Such persons are advised to avoid epoxy altogether. I may say I have used epoxy intermittently for more than 40 years with no more than minimal precautions and have never experienced the slightest ill effects that I am aware of.

Success in gluing with epoxy requires that a thin film of the glue remain in the joint after it has hardened. "Starved" joints, as they are called, from which all the adhesive has been squeezed out, will not hold. Often this is the result of excessive clamping pressure. Glue will be retained in the joint or scarph if the gluing surfaces are slightly roughened. In ripping up strips on power saws for glued laminations an ordinary splitting blade should be used instead of a planer tooth blade.

Long-length strips such as planking, outwales or guards, when joined with well-made scarphs, and epoxy glued, can be just as strong as uncut lengths of timber. But when these are bent around the hull in curves, the constant tension strain on the outside of the scarph may start the thin end to creep back and separate. This is easily prevented by a row of small rivets or other mechanical fasteners close to the thin end. Epoxy adhesives and mechanical (metallic) fasteners used in combination are stronger than either used separately.

Considering the great variety of components now available for formulating modern epoxy adhesives—resins, copolymers, hardeners, activators and more—there would seem to be no end of possible adhesive combinations.

Large industrial concerns can hire chemists to work out formulations crafted to their needs. The small boat shop and backyard builder have to take what is available on the market.

Epoxy adhesives most suitable for the small shop should have the following characteristics.

(1) Non-sensitizing, non-hazardous, as far as can be determined.

(2) Easy mixing. Equal parts or 2 to 1.

(3) Thick enough to stay put without sagging, dripping, or leaking out of vertical applications.

(4) Slow setting with occasional special exceptions. Most gluing operations take longer than planned. Plenty of working time is eminently desirable.

(5) Low temperature curing, except for special applications.

For some obscure reason the idea of lamination seems to have special appeal for inexperienced boat builders, and so often they have to learn the hard way that lamination is not the breeze it might appear to be. Except when mass-produced under controlled factory conditions, laminated

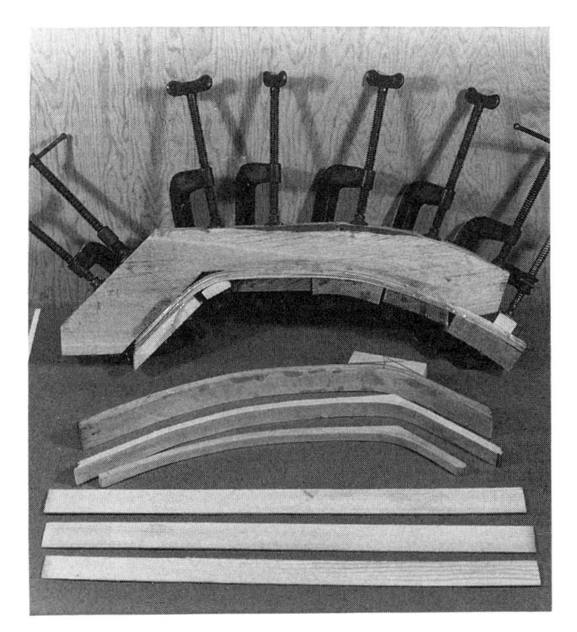

Method used for gluing up laminated frames for the* Green Machine. *Three frames nearing completion are in the middle, the gluing form on top and the Douglas fir laminations below (John Gardner photograph).

boat parts almost always take longer to produce, and cost more, than if they had been made of solid timber, and very often they are not as good. Laminated stems and keels for larger boats generally do not wear or last as well as those made of solid wood.

Yet there is one application in which lamination serves very well indeed, namely for making up curved frames for small, lightly built boats, when suitable bending stock is not available. If properly made, laminated frames can be even stiffer and stronger than steam-bent ones. Other recommended uses are for thwart knees, quarter knees and breasthooks and for light, ultra-strong deck beams for cabin trunks and the like. Generally speaking, and with the exception of bent frames, laminations should only be considered when suitable solid timber is unavailable.

In gluing up laminations, the thinner the strips or laminae used for this, the better. This is because the thinner the strips are, the less they will swell and shrink when the moisture content of the wood changes, and the less tendency there will be to pull apart at the glue line. With strips no thicker than 1/8", any tendency to separate at the glue line may be safely ignored, although for making up frames for small round-sided tenders and the like, I would prefer to have strips even slightly thinner.

Assuming that epoxy glue is used, and epoxy is the only kind to be considered for this, the surfaces of the strips or laminae should be fairly rough to insure the retention of a film of glue, for the reason previously explained. There is danger here of squeezing out the glue completely when a lot of clamps are required to force the strip assembly against a curved mold. With thin strips and plenty of clamps and shaped pressure blocks, any desired shape can be reproduced exactly, as for example, frames for the Adirondack guide-boat, which when properly laminated are stiffer and stronger than solid frames cut from curved spruce stumps. Further confirmation is provided by the laminated frames of the 17-foot modified Herreshoff rowboat featured in Chapter 1 of *More Building Classic Small Craft*, published by International Marine Publishing Company of Camden, Maine. The epoxy-glued laminated frames in the highly successful pilot model of this boat, known as the *Green Machine*, built over 15 years ago and now in the boat livery at Mystic Seaport's Boathouse, are as sound as the day they came off the clamping molds in 1978.

As a precaution against so-called "starved joints" in highly critical gluing operations, it is recommended by some that a preliminary coating of the joint surfaces with a penetrating epoxy be made and allowed to become tacky before a second and final coating of epoxy adhesive is applied and the joint is brought together.

Battens

Common wood battens are essential, in fact, are indispensable boat building tools, yet they are passed over by most boat building manuals, and rarely, if ever, receive the attention they rightly deserve. Battens are something the boat builder must provide for himself. He must make them; they cannot be bought. Without battens boat building as we know it could not be carried on. They are essential for getting fair curves and sweeps when laying down the lines on the mold loft floor, for drawing the curves of the body sections when the molds are made, for lining up the molds when the boat is set up for planking, for lining out the curved shapes of the planking, for laying out the deck beams and much else as well. Obviously without the right battens the boat builder is greatly handicapped, and without battens it is hard to imagine what boats would look like and how they would be built.

The property of wood that enables these long strips of wood of uniform width and thickness to bend uniformly in fair sweeps and curves is the same characteristic that enables the tall trees in the forest to sway back and forth without breaking as they respond to the pressure of the wind.

To be any good, battens require straight-grained lumber that is free of knots, or weakening blemishes of any sort. Years ago clear boards long enough for full-length battens were plentiful and cheap. Not so today. Now long battens usually must be spliced. Fortunately, when this is correctly done, battens that have been spliced bend with the same uniformity as full-length unspliced ones.

First off, a word of warning. Do not attempt to splice together short lengths of battens already cut to size. This is difficult to do without developing kinks or other distortions. It is far easier and better to splice the stock from which the required battens are to be cut, and by doing it this way only one splice has to be made in producing a number of battens, instead of having to splice each one individually.

Let us say that several 16 foot battens are needed, 1¼" to 1¾" wide and 1/4" to 3/8" thick, and all we can secure in the way of a straight grained pine board clear of knots and other defects is a piece 9' long, thus one splice will be required.

Also let us assume that a circular or table saw is available for the job. Nearly everyone today who does any woodworking either owns or has access to a table saw. Of course perfectly acceptable battens could be produced by hand, in the old way with nothing more than a sharp handsaw and a jack plane, but at best this is a slow tedious process, requiring some skill in using the hand tools. With a table saw it is ever so much easier and quicker.

We start by stripping out on the table saw, two lengths from the 9' foot board, 4" or slightly more in width, that is a convenient width for splicing, and wide enough to yield at least four finished battens.

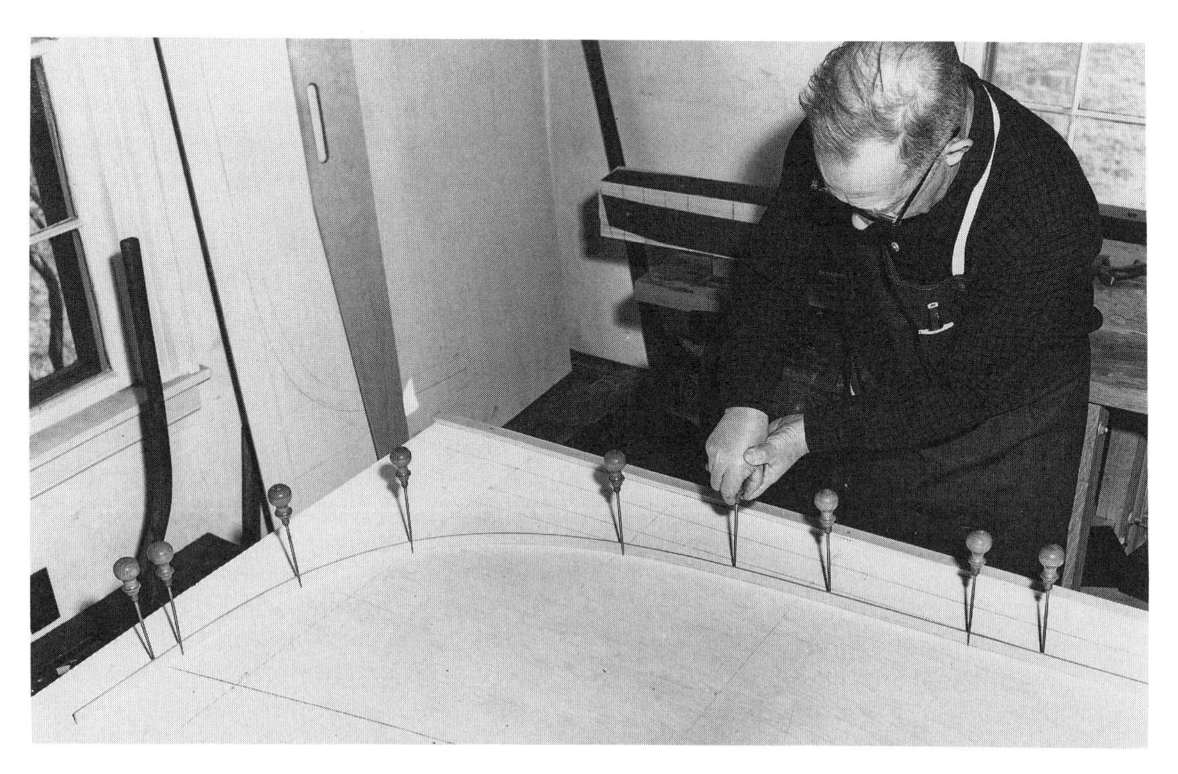

John Gardner using awls and a batten to loft the first boat he built while working at Mystic Seaport, the pulling boat* Lawton *(Louis S. Martel photograph, M.S.M.).

Assuming we are working with lumber 1" thick, the length of the splice should be 12 times the thickness or 12".

In scarphing the ends that will be spliced together, they are finished with a plane set fine to fit together perfectly. Their outer ends, however, are not brought to a feather edge, but are left a scant 1/16" thick to be planed off later after the splice has been glued.

If a power surface planer is available, the glued-up strip can be passed through, first one side and the other, taking off just enough to smooth the two surfaces. Otherwise if there is no planer, a sharp hand plane will do the job just as well. Before stripping out the battens on the table saw, one edge of the strip should be trued with a hand plane. This edge will run against the fence when the first cut is made to true up the other edge, taking off just as little as possible.

In stripping out the battens a planer tooth saw blade is desirable. To hold the work securely against the fence, when the battens are being stripped, pressure is applied by a device called a featherboard. Some modern table saws come so equipped. If not, a featherboard is easily made. A strip of 1" board 3" or 4" wide and long enough to reach across the top of the saw table is slotted at one end that has been slightly rounded with saw cuts 4" or 5" deep and 1/8" or so apart.

The featherboard placed at a slight forward angle, is clamped to the table top, the slotted end, positioned just back of the saw, presses the work against the fence. Its flexible fingers hold the work firmly in place, but allow it to pass by the saw. By changing the position of the fence, and that of the featherboard to correspond, battens of any desired width or thickness are easily stripped from pre-glued lengths of board.

In laying down the lines of large vessels in the mold loft, the battens are held in place by awls stuck in the mold loft floor, not through the batten. In boat work, battens are mostly used flat with small nails driven through them to hold them in place. Slim 1" long common nails are best for tacking down the battens. They are easily drawn with the hammer, and do little appreciable damage to good softwood battens. Battens can be nailed over and over again and without weakening them unduly, often lasting for years of intermittent use before they break or have to be replaced.

Planking

Of all the operations involved in the fabrication of a boat, planking is by far the most difficult and involved, not to say the most puzzling and frustrating oftentimes for the beginner. One reason for this is that the shape planks appear to have when they are fastened in place on the boat is, more often than not, entirely different from their shape when laid out flat.

Volumes could be written on planking, on the many different kinds and methods without exhausting the subject, but the fact is that precious little has so far found its way into print compared with what remains to be recorded. Only quite recently has any systematic attempt been undertaken to explore and record in print the mysteries of planking. One of the first manuals that took up planking was Howard I. Chapelle's *Boat Building, A Complete Handbook of Wooden Boat Construction*, published in 1941 by W. W. Norton and Company, Inc..

In former times boat builders were occupied in building boats, not in writing about them. Many of the top-notch old-timers were practically illiterate anyway, and the idea of divulging their hard-won trade secrets to every nosy Tom, Dick and Harry who came snooping around would have filled them with horror. More than once observers who watched too closely have been accused of trying "to steal the trade," and sent about their business. That was the way it was once. Today there are very few builders to watch, or yards where old-time wooden boats are still being built. So what is the aspiring beginner to do after he has read what little there is to read, and often with scant success in visualizing what he was reading? The answer is simple and obvious, turn to the boats themselves. That is where the action is. Learn from the boats. Observe, photograph, study and measure them whenever the opportunity permits. Occasionally, with luck, the remains of an old boat can be acquired, too far gone to be repaired and restored. By painstakingly taking it apart piece by piece, and strake by strake, much can be discovered about the construction of boats and how they are planked, much not easily learned in any other way. Another prime source of planking information are the small craft collections of large maritime museums when these are arranged and exhibited to facilitate study. Their many examples of historic small craft permit comparison of various styles and methods of planking. Indeed museums that take the preservation of historic small craft seriously will make provision for such study, encouraging it and providing the opportunity and means for it.

There is no more effective way of discovering how plank shapes are proportioned and lined than by laying them out on the surface of a half-hull scale model. The model should be large enough so that the plank shapes can be taken off by spiling, just as if it were a full-size boat.

Let us say the craft to be studied is an 18-foot carvel-planked, round-hulled launch with a slight amount of bottom rocker and a moderately raked transom stern—a displacement hull that might either be power driven or rigged to sail. A half-model scaled 1" equals 1' would be on the small side, but a scale of 1½" equals 1' would be large enough for spiling the shape of the planks.

Working to scale, the shapes of the planks are lined out using a strip of white pine to serve as a batten, 5/16" wide by 1/16" thick would be about right. This strip, or batten, if you will, is held in place by large pins which do not weaken it. I have found that the easiest way to insert and to remove these pins is to grip them with needle-nose pliers.

The outlines of the planks are laid out on the half-model with a soft lead pencil so that they can easily be erased in adjusting the plank shapes. They may need to be changed several times. When they have been adjusted to look right, the shapes can be taken off by spiling and laid out flat, exactly as if they were full-size planks. In this way should any of the planks show excessive curvature, it will be detected, and adjustment can be made to correct it. When everything is satisfactory, the plank widths can be enlarged full-size and lined out with long battens on the frames or molds set up for planking, and marked. If these marks are adhered to as planking proceeds, there is little chance of going wrong.

Experienced plankers familiar with the shape of the hulls they are working on have no need to follow this procedure, relying on their lining batten to give them fair shapes and making minor alterations in plank widths as they go along, if need be, in order to come out right in the end. But this takes experience, and for the inexperienced builder to set about planking a boat hit or miss, and especially a round hull, is an almost certain recipe for failure.

It has been assumed that the builder has a visual image of what his boat should look like when it is correctly planked. Such judgment is acquired by observation and study of representative examples of the boats themselves. The maritime museum's small craft collection can be of great assistance here. The close attention required in the photography of boats in and out of the water will reveal much that otherwise might be missed.

The inexperienced builder can look at a boat without really seeing it. Which is to say that if he does not know what to look for, he can miss a lot of what makes all the difference between a sloppy planking job and one to be proud of. It is not enough that a boat be tight enough to keep out the water, and strong enough to withstand the buffeting of the waves. It must look right. For those who know boats and take boats seriously, looks really matter. For them the classic forms of well-built boats are sources of esthetic pleasure and satisfaction, nor are they backward in expressing their disdain for those inferior craft that fail to measure up to their high standards.

As much as anything, perhaps, regularity in the shape and proportions of the topside strakes are what please and satisfy the discriminating eye. These planks should follow the sweep of the sheer, gradually and gracefully diminishing in width as they approach the stem. Especially to be avoided are planks that appear to grow tired and droop at the ends, as well as those that widen at the ends into what is known in boat building parlance as "fish tails."

Appearance does not have to be taken into account in laying out the plank shapes on the bottom where some of them can be

A student in Mystic Seaport's winter Recreational Boat Building Class demonstrates backing out a carvel plank (photograph courtesy of John Gardner).

For backing out planking a special plane, called a backing out plane, is needed. These cannot be bought in stores. Generally what is used is an old-time wooden smoothing plane that has been converted by rounding the sole and the blade or "iron" just enough to do the job and no more. It is easy to spoil a plane by too much rounding. Working boat builders often have several backing out planes with varying amounts of curvature, one of which will be almost flat. Sources of planes for conversion are flea markets, antique shops, secondhand stores, tool collectors and the like, or the boat builder can make his own. There are several recently published books and articles that explain how, including Chapter 16 in my *More Building Classic Small Craft.*

For carvel planks to fit together properly, that is, tight on the inside and open enough on the outside to admit caulking, their edges must be beveled. When there is any curvature in the hull surface, the greater the amount of curvature, the more bevel is required. To take off the bevels and apply them to the planks when they are being "got out" at the bench, a small bevel gauge or bevel square is handy, which plankers often make for themselves. Ideally, carvel planks should come together tight on the inside with just enough opening on the outside to admit caulking, although in some fancy carvel planked yacht tenders, the planks fit together perfectly tight the whole way, so that caulking is not required.

In most carvel-planked boats, however, the planking rarely comes together perfectly tight on the inside when it is first put on the boat. A slight opening on the inside can be tolerated so long as the seam opening between the planks tapers enough to hold the caulking from going through on the inside. However, if the seam is too open it must be set up tighter, making use of a device called a "set," operated by wedges or

quite wide, hence the term "broad strakes" or "broads" for planks next above the garboard. In some boats the end of the garboard should be kept down to reserve ample width on the stem for the ends of the upper strakes.

In carvel construction the inside surface of the plank must be hollowed or "backed out" to fit the curvature of the hull, thus the wider these planks are the thicker they have to be to start with, and the more they have to be backed out. More than one inexperienced boat builder has neglected to take this into account, ordering all of his planking stock the same thickness. When laying out a boat's planking, planks that come where the hull's greatest curvature is should not be made any wider than is necessary both to save material and work in backing out.

Bob Luedtke of Fairmont, Minnesota, built this 12'6" dory skiff from lines published in The Dory Book. *He used 1/4" mahogany plywood for the planking, 3/8" plywood for the bottom and epoxy-glued the laps. Built for his sister in Connecticut, Bob reported that the boat carried 700 lbs on its maiden voyage and never shipped a drop negotiating wakes on the Connecticut River (Robert W. Luedtke photograph).*

screw clamps. In hurry-up production work, planks are often much more open than this when they are first clamped on the boat, but still are forced into place with "sets." Anyone undertaking a job of carvel planking had best secure a few sets and familiarize himself with their use. He will probably have to improvise something that will serve for this himself. This is but one of numerous small challenges he will face in planking an old-time carvel boat.

In Colonial times small boats were planked clinker or lapstrake for the greater part. Plank-on-frame, carvel-built boats did not begin to take over in this country until well along in the last century when more and more small boats came to be planked smooth like yachts. By the middle of this century small clinker-planked boats had all but disappeared. But since then, with the appearance of superior marine plywood and reliable thermoplastic adhesives, lapstrake planking has taken a new lease on life with the limited few still building wooden boats. Plank-on-frame carvel construction is definitely on the way out, with amateurs at any rate.

Lapstrake planking is simpler, if not easier than carvel planking, with fewer building operations. In one respect, clinker planking may be somewhat more exacting, as plank overlap can vary only slightly, while ill-fitting carvel plank can usually be forced up tight enough with sets to caulk. There are no shutters to fit on clinker boats. Steaming is not required, nor do clinker boats have to be caulked and finished off on the outside by planing and sanding before they are painted. Clinker planks on the other hand can be finish sanded before they leave the bench. Furthermore, if plywood is used for clinker planking small boats, as is frequently the case these days, this material is more readily obtainable and often less expensive than the high quality sawn lumber required for a carvel boat.

On the down side, clinker planking is less versatile than carvel in the number and variety of hull shapes it is suited for, which limits its use. Smooth carvel hulls are quieter in the water, of importance for boats used for hunting and fishing, and carvel planks are easier to replace when damaged.

Well-made, plywood-planked clinker boats with epoxy-glued laps are perfectly tight, and will not dry out and leak when taken out of the water and exposed to the hot summer sun. Glued plywood laps will not split between fastenings, and require fewer lap fastenings. In fact some successful clinker plywood boats have been built without lap fastenings, relying solely on the glue to hold them together. In one instance that came to my attention such a boat was being towed behind a car traveling at a good rate of speed when it slipped off the trailer and bounced several times on the highway. Its glued laps held. The boat was undamaged except for a few scrapes and scratches. The owner thought this was a pretty good test for glued lap construction.

Of necessity lapstrake planking starts with the garboard and proceeds up the side until it reaches the sheer. Unless a precise layout of the planking is worked out in advance and rigidly adhered to, small errors of judgment as to planking widths and shapes can creep in along the way and remain unnoticed until the sheer is reached, where the cumulative effect can spell disaster. An ugly, ill-shaped sheer plank spoils everything, and very rarely can anything be done to correct it.

Of course if the builder is working from plans that show the planking knuckles on the body sections, as in plans for knuckle-sided dories, for instance, he does not have this problem. But he might want to adapt a carvel design for clinker plank, or undertake to design a clinker-planked boat from scratch.

There are several ways of laying out clinker planking in advance. One has already been explained, the scale half-model method. Another is to set up molds for planking as if the intention were to build a carvel boat, and then to proceed with a long lining batten to lay off the lines for clinker plank starting at the sheer and proceeding down the side to the garboard, making such adjustments along the way as judgment indicates, perhaps going back several times to change the markings.

A visitor to Mystic Seaport's Boathouse tries out the Herreshoff/Gardner pulling boat,* Green Machine. *The boat is a favorite of many who admire her simplicity and row for speed (Sharon Brown photograph).

A refinement of this method was devised by the late Pete Culler, an acknowledged master of lapstrake planking. His boats were admired for their graceful and shapely plank lines which stand out prominently on a lapstrake boat. After the molds were set up and made ready for planking he would get out a number of battens the width of his laps, and as long as the boat. These he would tack in place on the molds where he thought the plank laps should be. Then he would stand back and view the result from various angles. Any unfairness he detected, anything that didn't please his eye, he would correct by moving the batten however slightly. This done, very likely he would go off about another job, to come back later to have another look. Finally, when everything was right, he would mark the molds, pick up the marks from each mold on a short piece of thin batten bent around the side of mold and transfer the marks to the other side.

Hull shapes that curve abruptly are not well suited to clinker planking. It is best adapted to craft with slack or moderately curved bilges. On quick turns, laps tend to be too short to fasten securely, especially if planks are wide. Laps that are too short can often be avoided by narrowing the bilge planks when they are laid out. It can be helpful, here, to lay out ahead of time one or more full-size sectional views through the side of the boat with the plank and planking laps drawn in place. This will show how wide, or narrow, the plank should be to lap properly. Also, if the upper portion of a short lap is allowed to project 1/16" or even 1/8" beyond the under part there will be more room for fastening, and this will not detract from the appearance of the boat, in fact will not be noticeable if done carefully.

It is not advisable to glue the laps of clinker boats unless they are planked with plywood. When the laps are glued on boats planked with sawn lumber, splits in the plank are liable to develop from shrinkage when the boat dries out.

Clinker boats can be planked right side up or upside down. In the latter case riveting the lap fastenings must wait until the boat is turned over. Most frequently clinker boats are planked on molds to which they can be temporarily secured by a few leatherheads (slim nails with leather washers under the heads), and have their ribs or frames installed afterwards, but in some cases the frames are set up first. Or the two can be combined, as in the Swampscott dory, planked on frames which serve as molds and remain in the boat permanently with the addition later of a number of small intermediate steam bent ribs that go in after the boat is planked.

After the laps are riveted or glued, steamed ribs can be bent in and fastened through the laps. In the best work, steam-bent ribs are clamped in place and allowed to cool and take shape after which they are notched or jogged around the laps so they fit tight and snug to the inner surface of the plank the whole way before they are fastened in.

The Adirondack guide-boat, a clinker-built boat, although smooth both inside and out, was planked directly on pre-shaped frames instead of molds, cut according to a set of patterns, from natural-crook lumber sawn from large spruce tree stumps. Lacking such stumps in recent times, it was found that frames built up from glued laminations served just as well, if not better.

Although the material is wood, the strip-built boat can hardly qualify as traditional, dating back no earlier than the beginning of this century when this method was devised as a simple, easy, quick, inexpensive way of building lobsterboats and other motorized fishing launches. It was frequently the choice of inexperienced builders who lacked confidence in their ability to plank a boat in the regular way. Actually it is not quite as simple as it might appear to be.

Strip planking works best for hulls of simple rounded shape without reverse curves or abrupt changes in surface contour. It sometimes helps on larger boats to start with an ordinary garboard plank before continuing up the side with strips. The upper edge of the garboard takes the initial strip better than the keel rabbet. Besides, a garboard plank is easier to caulk and holds the caulking better.

Quite recently, now that thermoplastic adhesives are available, various shapely and high performance glued-strip canoes and related craft are being built, but such construction is not within the scope of this chapter.

Another well-known and widely distributed craft not to be overlooked is the time-honored "flatiron" skiff with its flat, cross-planked bottom. The one thing to remember here is not to make the bottom planking too wide. Six inches is plenty, five inches is not too narrow. Wide bottom boards open up too much at the seams when the boat dries out, and they tend to curl no matter how well the ends are nailed.

One final observation concerning planking. Builders have been known to finish planking one side of a launch or similar craft before planking the other side, finding it easier to get at, if, after the bottom was planked, they rolled the boat over on its bilge, finished planking up that side, then rolled it back the other way, and planked up the remaining side. I don't advise doing this. The force required in bending the planking on one side, without a counter force on the other, could very well pull the boat out of shape.

Steaming

Steaming is a process for heating wood to make it pliable, whereby it can be bent in abrupt curves and twists that would otherwise break it. Steaming does this by softening the lignin, a natural plastic which encases the long, flexible wood fibers and locks them into place. When lignin cools and hardens after the wood has been bent, the wood fibers are locked in their new position.

Steaming did not come into use by boat builders in this country until early in the previous century when the supply of natural grown crooked limbs and root knees that up to then had supplied the material for the curved parts of boats became harder to get and more expensive, and later on in the century when the new carvel built boats designed by naval architects required extreme bends and twists in their planking that could only be obtained by steaming.

Inexperienced boat builders are often put off by the prospect of steaming, thinking it is an arcane process requiring equipment difficult to set up and operate. Actually steaming is quite easily done, and with a minimum of special equipment. With a little ingenuity on the part of the builder, any number of simple devices can be improvised to heat wood to the point where it will bend easily.

Boiling wood can be just as effective as steaming it, and in this connection many years ago in Eastport, Maine, I recall watching three men framing a 70-foot vessel with 4" x 4" bent oak timbers consisting of two 2" x 4"s, one placed on top of the other and spiked together. The timbers were bending easily and going in fast. They were not steamed but boiled in a length of secondhand culvert pipe partly filled with sea water and set at an incline on top of a pile of beach stones on the beach. The lower end had been stopped up and buried in the sand. Heat supplied by a fire of driftwood and odds and ends of scrap lumber, kept the water boiling.

In one boat shop that comes to mind the steam box was supplied with steam from a five gallon gasoline can partly filled with water and sitting on top of a low wood stove from which a cover had been removed. A short length of large diameter copper tubing with its lower end capped and its upper end soldered over a hole in the bottom of the can, projected into the fire below. So long as there is water in the can the solder will not melt. Steam passed up into the steam box above through a short length of ordinary garden hose attached to the can's spout. This was an arrangement that worked, and that saw a lot of use over a goodly number of years.

And yet another simple but clever steaming or rather boiling device for bending small timbers comes to mind—a short length of iron sewer pipe standing upright with its lower end plugged, filled with water kept boiling by a propane torch standing beside it.

Boat Building Lumber

Because of the increasing difficulty of obtaining sawn lumber of boat quality, in addition to its mounting cost, more and more would-be boat builders are turning to clinker construction which permits the use of plywood with laminated frames. Nevertheless, there is still some suitable oak, black locust, white cedar and white pine to be had in the northeast, some cedar, cypress and hard pine in the south, as well as Douglas fir, red cedar, Port Orford cedar and Alaskan yellow cedar in the northwest. It may take some hunting to find what is required. Small country sawmills, what few there are left, are the most promising places to look. People have been known to take several years to accumulate enough lumber for their boat, and even to enjoy doing it, storing it away board by board and plank by plank.

White oak makes the best bending stock, by far, for the ribs and frames of small boats. For this purpose, always select heavy wood cut from the butt logs of young fast growing trees with large annual rings and a minimum of porous spring wood. For the rest of the boat high quality red oak that is dense and heavy with wide annual rings from young, fast growing trees, will do quite nicely when it can be found. Red oak in general has got an undeservedly bad name from some of its inferior species. Avoid oak that is light in weight, that is porous, and that smells bad, "piss oak" so-called. Good red oak is not rot prone when used for small boats, is strong, takes fastenings well and stays put, that is it does not warp and twist out of place, as white oak frequently does. An instance comes to mind in which a white oak keel in a small, lightly built launch changed shape setting up stresses that pulled the boat out of true.

Black locust, when it can be had, is a superior timber in all respects, dense, strong and lasting, and of handsome appearance when varnished. It is especially good for stems and keels. However, for lightweight construction its weight could be a drawback.

Tough and quite strong in spite of its ultra-light weight, northern white cedar (*arbor-vitae*) is the pick of planking wood for small craft. Although no longer obtainable in clear boards, those with a scattering of sound knots are quite acceptable except for work that is going to be finished bright, in fact the wood may be tougher for a few sound knots if they are not too large, or close together. Occasional loose knots can be reamed out with a tapered reamer and plugged. It is probably best to glue in the plugs. Southern cedar, known as Virginia cedar, and locally as juniper, has a pinkish cast, is slightly heavier than northern white and splits more readily. Because it still can be had knot-free, it is the logical choice for varnished work. It is available in wide boards.

When the United States Navy was building small wooden boats, cypress was the specified planking, a superior material for this purpose that does not rot, but is largely unobtainable now except in the south from a few local mills that saw it in limited quantities. Hard or long leaf pine is also a superior boat building timber. Impervious to rot, southern hard pine can substitute for oak for framing small boats, but is too heavy for planking them.

Eastern white pine, its sapwood excluded, makes prime planking stock. Somewhat heavier than cedar and stronger, it is adequately rot resistant. In early Colonial times, virgin stands of huge pine trees extended as far south as Long Island Sound. Now what is left is second and third growth, and is restricted to a relatively narrow strip extending from Maine and New Hampshire to Michigan's upper peninsula.

The demand for white pine for interior house finish has driven up the price almost out of reach and reason. Yet it is still occasionally possible to find a small country sawmill where pine boards can be picked up at a reasonable price, right off the saw, to put away to season. Pine boards not of planking quality with too much sap and too many knots, are just what are needed to cut up for molds, shores, braces, cleats and the like. The use of white pine for making battens has already been considered. As for sugar pine and Ponderosa, neither are suitable for planking, neither can stand exposure to the weather and both rot readily.

From what experience I have had with western red cedar, I have found it rather softer and weaker than I like for small-boat planking. Port Orford cedar, on the other hand, is the best, if any can be found these days. The old-growth quarter sawn Douglas fir we once were able to get was a beautiful wood to work. Clear, strong and rot resistant, it weighed somewhat heavy for small-boat planking, but made superior framing, floors, risers and even thwarts when a little extra weight was of no great consequence. You would be lucky, indeed, to find any Douglas fir of this quality today.

For spars, spruce, the strongest wood for its weight grown in North America, has no competitor. Formerly used in the construction of large vessels, spruce timbers acquired a bad reputation for not lasting. When spruce is used in the construction of small craft, this problem seems not to occur. An examination of scores of Adirondack guide-boats found no rot in either stems or frames. Both were made from the stump crooks of large red spruce trees, and some of these guide-boats had seen a lifetime of use. For laminating frames for small craft, I can recommend nothing better than thin strips of eastern red spruce. And this goes for laminated deck beams as well.

Two woods that bend as readily as white oak when steamed are white ash and rock elm, but white ash ribs have been known to rot and break, and Canadian rock elm, the first choice of British boat builders for bent frames, is little known or used in this country. Maple, beech and white birch rot readily when exposed to the weather, and are not to be considered as material for boats, while yellow birch, particularly the wood from large old-growth trees, lasts well and serves as a substitute for oak in eastern Maine and the Canadian Maritimes. Hard, close-grained and of a pale pinkish cast, yellow birch boards sometimes come wide enough to make entire transoms for 10 and 12 foot yacht tenders.

Hackmatack knees, so-called, the large root crooks of an American species of larch, were formerly an important commodity in the lumber trade, being supplied by the thousands to brace and support the interior construction of large wooden vessels. Frames for boats, particularly dories, were sawn from these knees, and they also provided stems for fancy yacht tenders. To get such a knee today, it would probably be necessary to go into the woods and dig it, which might just be a possibility for someone living in the part of the country where hackmatack is found.

Bob Cooper, a Nova Scotia farmer who built a replica of *Ocean Queen* in the off season, a large, double-ended sprit-rigged Bay of Fundy shad boat, and spent eight winters doing it, has said that one of the things he enjoyed the most was hunting out the numerous natural bends and crooks of timber required in the construction of this vessel. Maine salmon wherries were framed with white cedar root knees. Apple limbs make the best of boat knees, white oak crotch wood is ideal for breasthooks, and the list could go on. Such are the materials the old-timers put into their boats, but such materials are not plentiful today, and the time and labor expended in obtaining them, rules out their use by commercial boat shops which must show a profit.

Rot is not the menace for small boats that it is for larger wooden vessels, in considerable measure because of their more open construction which insures better ventilation, and opportunity for cleaning, allowing them to dry out and thus making it less easy for the invading rot fungi to gain a foothold. Rot generally starts in places like ill-fitting joints where moisture is trapped and collects in concentrations favorable to its growth. As the wood softens, more moisture accumulates and the infection spreads to surrounding construction, the softened parts acting like a sponge to hold more moisture and to spread the rot. The common expression "dry rot" is a misnomer. Neither dry wood nor wood that is completely soaked will rot.

The best insurance against rot is to build from rot-resistant species, to provide for adequate ventilation, to eliminate possible foci of infection where rot can start and flourish undetected and, of course, good care and maintenance. Most of the once widely used wood preservatives are now recognized as health hazards, which

In a fresh breeze Ned Costelloe and family enjoy some of the benefits of building and using a traditional small boat. Ned built this Dion dory from lines drawn up in 1952 and published in* The Dory Book *(photograph courtesty of Ned Costelloe).

places them out of bounds, but it is doubtful if many of them were ever very effective anyway. Indeed, unwarranted reliance on preservatives may well have contributed to neglect, producing the very results they were supposed to prevent.

There is one effective treatment for planking that is worth consideration. That is its saturation, after it is in place, both inside and out, with hot linseed oil. It is heated in a double boiler to keep its temperature below the flash point, but applied boiling hot nonetheless. This stabilizes and hardens the wood, especially cedar, and minimizes swelling and shrinking. On the best large double-planked yachts the inner layer of white cedar was soaked with hot oil and allowed to dry before interior construction was begun.

Summary

What started out to be a few directives and observations on the construction of wooden boats has gone on to become a lengthy chapter, and could go on almost indefinitely to fill many more chapters. So much not even touched upon remains to be considered, yet a start has been made.

Building and using small wooden boats is part of our heritage, something worth doing that our forebears did that we can still do, something to cherish and hold on to, something to enjoy while we still have it to enjoy. What will be left for the generations to come in the uncertain future that lies ahead? We can only hope that something will remain. But here and now, let there be no hesitation in making the most of the good that has been given us.

John Gardner in his 68th year demonstrating the shipwright's broad axe squaring off a white oak knee. He considered such tools to be "repositories of inherited technical knowledge" (Lester D. Olin photograph, M.S.M.).

INDEX